SELECTED TABLES IN MATHEMATICAL STATISTICS

Volume IV

This volume was prepared with the aid of:

C. Bingham, University of Minnesota

R. F. Gunst, Southern Methodist University

K. Hinkelmann, Virginia Polytechnic Institute
and State University

D. C. Hoaglin, Harvard University

W. Kennedy, Iowa State University

S. Pearson, Southern Methodist University

J. N. Srivastava, Colorado State University

N. S. Urquhart, New Mexico State University

R. H. Wampler, National Bureau of Standards

E. J. Wegman, University of North Carolina

with special assistance given by

J. S. Rao

SELECTED TABLES IN MATHEMATICAL STATISTICS

Volume IV

DIRICHLET DISTRIBUTION – TYPE 1

by

MILTON SOBEL, V. R. R. UPPULURI, and K. FRANKOWSKI

Edited by the Institute of Mathematical Statistics

Coeditors

D. B. Owen
Southern Methodist University

and

R. E. Odeh
University of Victoria

Managing Editor
J. M. Davenport
Texas Tech University

AMERICAN MATHEMATICAL SOCIETY
PROVIDENCE, RHODE ISLAND

AMS (MOS) subject classifications (1970).
Primary 62Q05; Secondary 62E15, 62H10.

International Standard Book Number 0-8218-1904-6
Library of Congress Card Number 74-6283

PREFACE

This volume of mathematical tables has been prepared under the aegis of the Institute of Mathematical Statistics. The Institute of Mathematical Statistics is a professional society for mathematically oriented statisticians. The purpose of the Institute is to encourage the development, dissemination, and application of mathematical statistics. The Committee on Mathematical Tables of the Institute of Mathematical Statistics is responsible for preparing and editing this series of tables. The Institute of Mathematical Statistics has entered into an agreement with the American Mathematical Society to jointly publish this series of volumes. At the time of this writing, submissions for future volumes are being solicited. No set number of volumes has been established for this series. The editors will consider publishing as many volumes as are necessary to disseminate meritorious material.

Potential authors should consider the following rules when submitting material.

1. The manuscript must be prepared by the author in a form acceptable for photo-offset. This includes both the tables and introductory material. The author should assume that nothing will be set in type although the editors reserve the right to make editorial changes.

2. While there are no fixed upper and lower limits on the length of tables, authors should be aware that the purpose of this series is to provide an outlet for tables of high quality and utility which are too long to be accepted by a technical journal but too short for separate publication in book form.

3. The author must, wherever applicable, include in his introduction the following:

(a) He should give the formula used in the calculation, and the computational procedure (or algorithm) used to generate his tables. Generally speaking, FORTRAN or ALGOL programs will not be included but the description of the algorithm used should be complete enough that such programs can be easily prepared.

(b) A recommendation for interpolation in the tables should be given. The author should give the number of figures of accuracy which can be obtained with linear (and higher degree) interpolation.

(c) Adequate references must be given.

(d) The author should give the accuracy of the table and his method of rounding.

(e) In considering possible formats for his tables, the author should attempt to give as much information as possible in as little space as possible. Generally speaking, critical values of a distribution convey more information than the distribution itself, but each case must be judged on its own merits. The text portion of the tables (including column headings, titles, etc.) must be proportional to the size 5–1/4″ by 8–1/4″. Tables may be printed proportional to the size 8–1/4″ by 5–1/4″ (i. e., turned sideways on the page) when absolutely necessary; but this should be avoided and every attempt made to orient the tables in a vertical manner.

(f) The table should adequately cover the entire function. Asymptotic results should be given and tabulated if informative.

(g) An example or examples of the use of the tables should be included.

4. The author should submit as accurate a tabulation as he can. The table will be checked before publication, and any excess of errors will be considered grounds for rejection. The manuscript introduction will be subjected to refereeing and an inadequate introduction may also lead to rejection.

5. Authors having tables they wish to submit should send two copies to:

>Dr. Robert E. Odeh, Coeditor
>Department of Mathematics
>University of Victoria
>Victoria, B. C., Canada   V8W 2Y2

At the same time, a third copy should be sent to:

>Dr. D. B. Owen, Coeditor
>Department of Statistics
>Southern Methodist University
>Dallas, Texas 75275

Additional copies may be required, as needed for the editorial process. After the editorial process is complete, a camera-ready copy must be prepared for the publisher.

Authors should check several current issues of *The Institute of Mathematical Statistics Bulletin* and *The AMSTAT News* for any up-to-date announcements about submissions to this series.

ACKNOWLEDGMENTS

The tables included in the present volume were checked at the University of Victoria. Dr. R. E. Odeh arranged for, and directed this checking with the assistance of Mrs. Amanda Nemec and Mr. Bruce Wilson. The editors and the Institute of Mathematical Statistics wish to express their great appreciation for this invaluable assistance. So many other people have contributed to the instigation and preparation of this volume that it would be impossible to record their names here. To all these people, who will remain anonymous, the editors and the Institute also wish to express their thanks.

To:

Judy Lynne Sobel

Shigeko Yoshino Uppuluri

Elaine Frankowski

Contents of VOLUMES I, II and III of this Series

TABLE OF CONTENTS

Table of Contents (*cont.*)

DIRICHLET DISTRIBUTION – TYPE 1

Milton Sobel

University of California at Santa Barbara

V.R.R. Uppuluri

Union Carbide Nuclear Division, Oak Ridge

and

K. Frankowski

University of Minnesota

ABSTRACT

The incomplete Type 1-Dirichlet integral is tabulated in a variety of useful forms. A rather long introductory section gives the essential properties that were used in the calculation and illustrates various ways of using tables. It also gives in Section 2 some mathematical results that are related to problems of interpolation in the table. In general, this Dirichlet integral can be used with most multinomial problems, especially those concerned with the maximum or minimum frequency in a homogeneous multinomial. Combinatorial aspects and relations to Stirling numbers of the second kind are also included; in fact these relations give rise to generalized Stirling numbers $(r > 1)$, which are listed in Table E along with the usual Stirling numbers $(r = 1)$. Table F gives exact and approximate values of n needed for a multinomial selection problem described in Section 4.1, the calculation of which depended on the type 1-Dirichlet integral.

Received by the editors August 1975 and in revised form February 1976 and July 1976. AMS(MOS) Subject Classifications (1970): Primary 62Q05; Secondary 62E15, 62H10.
This work was partially supported by the Energy Research and Development Administration under the auspices of the Union Carbide Nuclear Division.

§1. Introduction

    The incomplete Dirichlet integral of Type 1 is a direct generalization
of the incomplete beta distribution for the multinomial case. For the case of
two cells, the binomial case, it can be used to sum either tail of the binomial
or the negative binomial distribution. For more than two it gives rise as a
special case to the distribution of the minimum frequency in the multinomial.
Numerous applications of our Dirichlet tables for calculating multinomial
probabilities are given throughout the text.

    In Section 2 we define a generalized Dirichlet integral and develop its
basic properties and the recurrence relation it satisfies. Differential and
difference relations are developed in order to expand the integral in a Taylor
series.

    In Section 3 we show the relation between the Dirichlet integrals and
generalized Stirling numbers of the second kind (which are all integers). The
properties are developed and also the combinatorial interpretation as partitions
of n objects into b subsets.

    In Section 4 we discuss different applications of the tables:

    1). In 4.1 they are applied to a multinomial ranking problem.

    2). In 4.2 curtailment in the multinomial ranking problem is
         discussed.

    3). In 4.3 we consider the slippage problem in which one p-value
         is different from all the others.

    4). In 4.4 we give a number of different applications of the various
         tables to fundamental probability calculations dealing with the
         multinomial distribution.

§2. Incomplete Type 1-Dirichlet Integrals.

    In this section we define the integral (calling it an I function) and
present its relation to binomial and multinomial sums. We also give a probabilistic interpretation in terms of the minimum frequency in a multinomial
sample. Then we derive a basic recurrence relation on which our tables are

based. Other useful relations that are developed include a derivative of the
I function in terms of finite difference operators and the development of a
finite Taylor series expansion for the 1 function. Type 1 Dirichlet is
defined in the next section; corresponding definitions and results for Type 2
will be considered separately.

2.1.1. Definition and Preliminaries.

The incomplete Dirichlet integral of Type 1 is defined by the b-fold
integral

$$(2.1) \qquad I_p^{(b)}(r,n) = \frac{\Gamma(n+1)}{\Gamma^b(r)\Gamma(n+1-br)} \int_0^p \cdots \int_0^p (1 - \sum_{i=1}^b x_i)^{n-br} \prod_{i=1}^b x_i^{r-1} \, dx_i$$

where we assume that $0 \leq p \leq 1/b$, $n \geq rb$, b is an integer and n, b, r are
all positive; the only exception to the latter is the trivial case $b = 0$
(with no integrals), and in this case we take the value as 1 for any n, any
r and any p, including $p = 0$. In the applications and in our tables the
values of n and r will also be positive integers. Some of these assumptions
like $p \leq 1/b$ can be weakened (and will be in another paper) but such generalizations are not considered here.

This integral (2.1) represents a straightforward generalization of the
incomplete beta function and in fact, if we set $b = 1$, we obtain

$$(2.2) \qquad I_p^{(1)}(r,n) = I_p(r,n-r+1) = \frac{\Gamma(n+1)}{\Gamma(r)\Gamma(n+1-r)} \int_0^p x^{r-1}(1-x)^{n-r} \, dx,$$

where $I_p(a,b)$, without a superscript, denotes the usual incomplete beta
function in standard notation.

2.1.2 Relation with binomial and multinomial sums.

It is well known that the incomplete beta function (2.2) is used as a
standard tool and has an extremely large number of applications in probability
and statistics. In particular, it is used to sum the upper or lower tail of
the binomial (as well as the negative binomial) series. Thus, assuming r and
s are integers and $q = 1-p$, we have

$$(2.3) \qquad I_p^{(1)}(r,n) = \sum_{\alpha=r}^{n} \binom{n}{\alpha} p^\alpha q^{n-\alpha} = \sum_{\alpha=0}^{n-r} \binom{n}{\alpha} q^\alpha p^{n-\alpha} = 1 - I_q^{(1)}(n-r+1,n)$$

$$(2.4) \qquad I_p^{(1)}(r,n) = q^{n-r+1} \sum_{\alpha=r}^{\infty} \frac{\Gamma(n-r+1+\alpha)}{\Gamma(n-r+1)\alpha!} p^\alpha = p^r \sum_{\alpha=0}^{n-r} \frac{\Gamma(r+\alpha)}{\Gamma(r)\alpha!} q^\alpha$$

where the result (2.4) can be found in [12] or [16] and will not be proved
here. In a similar manner the incomplete Dirichlet integral (2.1) has
many applications, both in theoretical arguments as well as in applications,
whenever we are dealing with multinomial probabilities. In particular,
the expression (2.1) is the probability that the minimum frequency
$M = \underset{\sim}{M}_p(b,n)$ in a multinomial is at least r; here we assume there are
n observations and $b+1$ cells, b of which have a common cell
probability equal to p and the minimum frequency M is over these b cells.
By methods of inclusion-exclusion, we can easily relate the I-function (2.1)
to the cumulative distribution function (cdf) of the corresponding maximum
frequency; this will be done in a separate report and is considered here briefly
(cf. pp. 40, 41, 51).

In analogy with the identities in (2.3), the result $(cf. [9])$ stated above
relating $\underset{\sim}{M}_p(b,n)$ and (2.1) can be written (letting $S = \sum_{i=1}^{b} x_i$) as

$$(2.5) \qquad I_p^{(b)}(r,n) = \sum \frac{n!}{\prod_{i=1}^{b}(x_i!)(n - S)!} p^S (1-pb)^{n-S},$$

where the summation is over all b-tuples $(x_1, x_2, \ldots, x_b)$ with $x_i \geq r$
$(i = 1,2,\ldots,b)$ and $S \leq n$; for $n-S = 1-bp = 0$ take $0^0 = 1$.

It is pointed out in [12] and [3] that the generalization of (2.4)
leads to an incomplete Dirichlet integral of Type 2 and this will be considered in a separate report. The integral (2.1) also arose in a recent
paper [17] on the distribution of sparse and crowded cells in a multinomial
distribution, a sparse (resp., crowded) cell being one with frequency at
most u (resp., at least v). The probability density function (pdf),

cdf, and the factorial moments (both marginal and joint) are all naturally
expressible in terms of these I-functions; (see application 8 in section 4
below.) The I-function also arises in connection with the probability of a
correct selection when the goal is to select the cell with the lowest cell
probability (cf. Section 4.1).

As mentioned above the I-function is directly related to the minimum
frequency in a multinomial and by inclusion-exclusion to the corresponding
maximum frequency in a multinomial. The distribution of the maximum was
computed earlier by Steck (circa 1960) and this report [19] included means
and variances for the maximum; the report was not published. In addition,
Owen and Steck discussed the maximum of the multinomial in a paper [13]
where the moments of equi-correlated normal random variables are used
to approximate the moments of equi-correlated frequencies in a "homogeneous"
multinomial, i.e., one with equal cell probabilities. Some further references
dealing with the minimum and/or maximum of a multinomial are Greenwood and
Glasgow [5] Johnson and Young [8], and Kozelka [10].

It is easy to show that for the special case $k = 2$ and $p = 1/2$
we have

$$(2.6) \qquad I_{\frac{1}{2}}^{(2)}(r,n) = 2^{-n} \sum_{\alpha=r}^{n-r} \binom{n}{\alpha} = \begin{cases} 2I_{\frac{1}{2}}(r,n-r+1) - 1 & \text{for } n \geq 2r \\ 0 & \text{otherwise} \end{cases}$$

which is called the cumulative symmetric binomial distribution. This
was tabulated by Wijngaarden [20] and our numerical results are in perfect
agreement with his.

2.1.3. <u>A generalization and the corresponding probability interpretation.</u>

In order to derive the basic recurrence on which all of our calculations
depend, we first have to generalize (2.1) by introducing one more integer
parameter j ($j = 0,1,2,\ldots,b$). Although our primary interest and all our
table values are only for $j = 0$, the introduction of this parameter j
is fundamental for developing the simple recursion in (2.9) below. This

recursion has led to extremely efficient and highly accurate (14 correct decimals) values of the I-function; in fact we are forced to drop many decimals to put the tables in the form of a publishable report.

For p,n,r and b as above and for any integer j with $0 \leq j \leq b$, we define the (b-j)-fold integral

$$(2.7) \qquad I_p^{(b,j)}(r,n) = \frac{\Gamma(n+1)(p^r/r)^j}{\Gamma^b(r)\Gamma(n+1-br)} \int_0^p \cdots \int_0^p (1-jp- \sum_{\alpha=j+1}^b x_\alpha)^{n-br} \prod_{\alpha=j+1}^b x_\alpha^{r-1} dx_\alpha$$

and the value for j = b is taken to be the multinomial probability

$$(2.8) \qquad I_p^{(b,b)}(r,n) = \frac{n!}{(r!)^b (n-br)!} p^{br}(1-bp)^{n-br} \; .$$

In particular, for j = b = 0 the value is taken to be 1 for all r,n and p.

A simple probability interpretation exists that holds for both (2.7) and (2.1), as well as for (2.8). Consider a multinomial model with b cells (called blue cells) having equal probability $p \leq 1/b$ and one more cell to catch the remaining mass 1-bp, in case p < 1/b. Let n denote the number of observations to be taken. Then (2.7) represents the probability that a particular set of j of the blue cells have frequency equal to r and the remaining blue cells have frequency at least r. Then for j = 0 the value of (2.1) is the probability that the minimum of the blue cell frequencies is at least r. For j = b the value (2.8) then becomes the probability that every blue cell has frequency equal to r. For r = 0 the integrals (2.1) and (2.7) are not defined; using the above probability interpretation, we can then define the value of the I-function as being equal to $(1-jp)^n$ for all values of n,b,j and p. This is consistent with (2.8) for j = b.

2.2.1. Basic recurrence relation.

The basic recurrence formula for (2.7) is

$$(2.9) \qquad I_p^{(b,j)}(r,n) = \frac{n(1-jp)}{n-jr} I_p^{(b,j)}(r,n-1) + \frac{r(b-j)}{n-jr} I_p^{(b,j+1)}(r,n),$$

which holds for any j $(0 \le j \le b-1)$ and is used only for $n > br$. (Note that $n-jr > n-br \ge 0$ and hence $n-jr > 0$.)

We give a brief sketch of a derivation of (2.9). One factor $(1-jp- \sum x_\alpha)$ in the integrand in (2.7) can be separated and the result expanded as the difference of two integrals, one with coefficient $1-jp$ and the other containing $b-j$ equivalent integrals using a symmetry argument. After an integration-by-parts in the second term, we obtain the same integral as appears on the left side of (2.9) with the additional coefficient $-r(b-j)/(n-br)$; this then leads to the result given in (2.9) by a straightforward calculation.

There are two boundary conditions that are associated with the recurrence relation (2.9). One of these (for $j = b$) has already been given in (2.8). The other is the result for $n = br$, where it is easy to show by direct integration in (2.7) that for all j

$$(2.10) \qquad I_p^{(b,j)}(r,br) = \frac{(br)!}{(r!)^b} \ p^{br}.$$

Note that when $p < 1/b$ this condition for $j = b$ is consistent with (2.8) for $n = br$; when $p = 1/b$ and $n = br$ we take the indefinite form 0° in (2.8) to be 1 and the consistency still holds.

2.2.2. Reduction formula and other results.

It is of numerical as well as theoretical interest that the superscript j in (2.7) can be altered or brought to zero by a simple mathematical or a probabilistic argument (both of which are omitted), and we write for any i and j

$$(2.11) \qquad I_p^{(b,i+j)}(r,n) = \frac{n!}{(r!)^j(n-jr)!} \ p^{rj}(1-jp)^{n-jr} \ I_{p/(1-jp)}^{(b-j,i)}(r,n-jr).$$

We refer to (2.11) as the reduction formula and use it mostly with $i = 0$ to reduce the second superscript to zero.

In calculating $I_p^{(b,j)}(r,n)$ for any j, the value for $p = 1/b$ is of particular interest since there is a simple summation formula that expresses the value of I_p for every p in terms of $I_{1/b}$-values. In fact we can condition on the total number of observations in the blue cells and obtain for any p,b,j,r and n

$$(2.12) \quad I_p^{(b,j)}(r,n) = \sum_{\alpha=br}^{n} \binom{n}{\alpha}(bp)^{\alpha}(1-bp)^{n-\alpha} I_{1/b}^{(b,j)}(r,\alpha).$$

For $j = 0$ and $b = 2$ we can combine (2.6) and (2.12) to obtain

$$(2.13) \quad I_p^{(2)}(r,n) = \sum_{\alpha=2r}^{n} \binom{n}{\alpha} p^{\alpha}(1-2p)^{n-\alpha} \left[\sum_{\beta=r}^{\alpha-r} \binom{\alpha}{\beta} \right];$$

both (2.12) and (2.13) lead to polynomial expressions in p; this reduces to (2.6) for $p = 1/2$.

Another useful formula for $b > j$ that reduces b (even when $j = 0$) is

$$(2.14) \quad I_p^{(b,j)}(r,n) = \sum_{\alpha=(b-1)r}^{n-r} \binom{n}{\alpha}(1-p)^{\alpha} p^{n-\alpha} I_{p/(1-p)}^{(b-1,j)}(r,\alpha).$$

This is obtained by conditioning on the number of observations outside a particular blue cell; it reduces to (2.6) for $j = 0$, $p = 1/b$ and $b = 2$. Many such formulas for small values of b can be written but we shall not write any more since they are not used for calculations and in fact (2.12) is also not used in the calculations. However (2.12) is useful to obtain the I-function for general values of p as a function of the values tabled for the special case, $p = 1/b$.

2.3.1. <u>Derivative of the I-function in terms of finite difference operators</u>.

With the above goal in mind we also wish to give some general results on derivatives of the I-function which, in addition to being of theoretical interest, are useful to write a Taylor expansion in p and thus obtain an approximation for I_p for values of p not in our table.

For this purpose we consider only $j = 0$ and use the expression in
(2.1); the results for general j can easily be obtained in the same
manner but are not needed since our tables are only calculated for $j = 0$.
Let D_p^j denote the $j\underline{\text{th}}$ derivative operator with respect to p and let
Δ denote the usual finite difference operator, $\Delta f(x) = f(x+1)-f(x)$,
which operates on the argument n of the function $I_p^{(b)}(r,n)$. By straightforward differentiation in (2.1) we obtain for $j = 0$,

$$(2.15) \qquad D_p I_p^{(b)}(r,n) = \frac{n}{p} \Delta I_p^{(b)}(r,n-1),$$

where we assume that $p > 0$.

   To generalize (2.15) and obtain the $s\underline{\text{th}}$ derivative, we first use
induction to obtain a Leibnitz-type result in finite differences, namely
that for any s

$$(2.16) \qquad \Delta^s \{F(x)G(x)\} = \sum_{j=0}^{s} \binom{s}{j}[\Delta^j F(x)][\Delta^{s-j}G(x+j)].$$

This result (2.16) can then be used to prove the desired generalization
of (2.15), namely that for $j = 0$ and any s

$$(2.17) \qquad D_p^s I_p^{(b)}(r,n) = \frac{n^{[s]}}{p^s} \Delta^s I_p^{(b)}(r,n-s),$$

where $n^{[s]} = n(n-1)\ldots(n-s+1)$ and we assumed in (2.17) that $p > 0$
and $n \geq s + br$. A detailed proof of (2.16) as well as (2.17) is omitted.
   Thus (2.17) tells us that we can calculate any derivative with respect
to p by merely taking the appropriate differences in our basic table.
For the special case $s = 1$ with $j = 0$ we can combine (2.17) and (2.9)
to obtain for the first derivative

$$(2.18) \qquad D_p I_p^{(b)}(r,n) = \frac{n}{p} \Delta I_p^{(b)}(r,n-1) = \frac{rb}{p} I_p^{(b,1)}(r,n),$$

but this last result in (2.18) does not appear to generalize simply to higher derivatives. We note from (2.7) that for $p = 0$ the I-function (as well as its derivative) is zero unless $j = b$ and $r = 0$, in which case the I-value is 1.

Moreover, if we use the reduction formula (2.11) on the last member of (2.18) then we obtain for the first derivative,

$$(2.19) \qquad D_p I_p^{(b)}(r,n) = \frac{n!}{(r-1)!(n-r)!} \, p^{r-1} (1-p)^{n-r} b I_{p/(1-p)}^{(b-1)}(r,n-r)$$

$$= nb \, B_p(n-1,r-1) \, I_{p/(1-p)}^{(b-1)}(r,n-r),$$

where we assume that $b \geq 1$ and $r \geq 1$ and use $B_p(n,r)$ to denote the usual binomial probability of r successes in n trials when the common probability of success on a single trial is p. In particular, we note that all powers of p in the denominator in (2.17) and (2.18) must cancel since the I-function is a polynomial. Hence the Taylor expansion around any point p_0 must terminate at or before the power n of $p-p_0$.

2.3.2. Taylor series representation and numerical illustrations.

For any p and any p_0 (close to p) we can now write the Taylor expansion of (2.1) about p_0, using (2.17), as

$$(2.20) \qquad I_p^{(b)}(r,n) = \sum_{\alpha=0}^{n-br} \left(\frac{p-p_0}{p_0} \right)^{\alpha} \binom{n}{\alpha} \Delta^{\alpha} I_{p_0}^{(b)}(r,n-\alpha),$$

and we note that this is a finite exact expression if we use all $n-br+1$ terms.

To illustrate the typical speed of convergence, suppose we take $n = 10$ and consider the first three terms in the expansion (2.20).

Letting $b = 3$, $r = 2$, $p = .3$ and $p_0 = 1/3$, we obtain using Table A

$$(2.21) \quad I_{.3}^{(3)}(2,10) \sim .69349 - \frac{3\,(10)}{30}(.10882) + \frac{9\,(45)}{(30)^2}(-.02774)$$

$$= .69349 - .10882 - .01248 = .5722 .$$

The correct answer is .5730 so that the error is in the third decimal.
However if we change p to .33 then we can use the same table entries
and obtain

$$(2.22) \quad I_{.33}^{(3)}(2,10) \sim .69349 - \frac{3\,(10)}{300}(.10882) + \frac{9\,(45)}{(300)^2}(-.02774) = .68248 .$$

The correct answer is .68249 and the error is only 1 in the fifth
decimal. Thus in order to get excellent interpolation results
we will need p-values that differ by at most .01 in our table.

It is interesting to note that the exact expression (2.20) can also
be written as a finite linear combination of I-functions (without differences) with binomial weights. If we write $E - 1$ for Δ where
$E\,f(x) = f(x+1)$, etc., then the order of summation can be interchanged
and after a straightforward binomial summation we obtain the exact results

$$(2.23) \quad I_p^{(b)}(r,n) = \begin{cases} \sum\limits_{\beta=0}^{n-br} B_{p'}(n,\beta)\, I_{p_0}^{(b)}(r,n-\beta) & \text{for } p \le p_0 \\[2mm] \left(\dfrac{2p-p_0}{p_0}\right)^n \sum\limits_{\beta=0}^{n-br} (-1)^\beta\, B_{p''}(n,\beta)\, I_{p_0}^{(b)}(r,n-\beta) & \text{for } p \ge p_0 \end{cases}$$

where $B_p(n,r)$ is the usual binomial probability as in (2.19), $p' = (p_0-p)/p_0$
and $p'' = (p-p_0)/(2p-p_0)$.

Although (2.20) and (2.23) are both exact, the use of 3 terms in
each will clearly not give the same result for $n > 3$. If we look at
the same illustrations as above and use (2.23) we obtain

$$(2.24) \quad I_{.3}^{(3)}(2,10) \sim (.34868)(.69349) + (.38742)(.58467)$$

$$+ (.19371)(.44810) = .55512 \text{ (where } \Sigma \text{ coef.} = .92981),$$

(2.25) $I_{.33}^{(3)}(2,10) \sim (.90438)(.69349) + (.09135)(.58467)$

$$+ (.00415)(.44810) = .68245 \text{ (where } \Sigma \text{ coef.} = .99988).$$

These are clearly lower bounds to the correct answers and the sum of all
the coefficients would be unity in (2.23) if we summed up to n. Since
the I-function is monotonically increasing in the second argument, it
follows that by replacing the third binomial coefficient by 1 minus the
sum of the first two coefficients, we obtain an upper bound; this gives
.58658 in (2.24) and .68250 in (2.25). Averaging these upper and lower
bounds, we obtain .57085 for (2.24) and .68248 for (2.25). The first
one is 2 off in the third decimal and the second one is one off in
the fifth decimal.

Assuming that one more binomial coefficient (i.e., the fourth) is
available, we can improve the above estimates still further by using
the differences of I-values to <u>estimate</u> the fourth I-value in (2.23)
and then using the above technique of altering the fourth binomial
coefficient. We thus obtain two estimates (i.e., with and without the
alteration) which are then averaged. To estimate the fourth I-value we
assume that the second differences (based on our first three I-values)
are constant. Using the results in (2.24) and (2.25) this gives for p = .3

(2.26) $I_{.3}^{(3)}(2,10) \sim .55512 + (.05740)(.28378) = .57141$,

(2.27) $I_{.3}^{(3)}(2,10) \sim .55512 + (.07019)(.28378) = .57504$.

These average to .57322, which has 3 correct decimals and an error of 2
in the fourth decimal. We omit the application of this technique to
(2.25).

In summary, we can get a better approximation with (2.23) then
with (2.20), by making use of the knowledge that the coefficients sum
to unity. In addition, (2.23) depends on the tabular values and not on
differences and hence is easier to apply.

2.4 Normal Approximation to Dirichlet I.

Actually there are two different normal approximations to $I_p^{(b)}(r,n) =$ Prob $[\underline{M}_p(b,n) \geq r]$. One of these (Method I) uses the normal approximation to the multinomial; the other (Method II) uses a normal approximation to the Dirichlet distribution. Although more is known about the multinomial (i.e., its moments, correlations, etc.), we prefer to approximate the normal to the continuous Dirichlet density under the integral in (2.1) rather than to the discrete multinomial since Method II gives better results in general.

If we regard the Dirichlet as a density for b exchangeable random variables, the common smaller central moments of each variate are

$$(2.28) \qquad \mu_1 = \frac{r}{n+1}, \quad \mu_2 = \frac{r(n-r+1)}{(n+1)^2(n+2)}, \quad \mu_3 = \frac{2r(n-r+1)(n-2r+1)}{(n+1)^3(n+2)(n+3)},$$

and some joint central moments are (for all b for which the moment is defined)

$$(2.29) \quad \mu_{11} = \frac{-r^2}{(n+1)^2(n+2)}, \quad \mu_{21} = \frac{-2r^2(n-2r+1)}{(n+1)^3(n+2)(n+3)}, \quad \mu_{111} = \frac{4r^3}{(n+1)^3(n+2)(n+3)},$$

and it follows, in particular, that $\rho = -r/(n-r+1)$ is the correlation between any two of the b variates; some of these results are in Wilks [22]. Since the correlations are all negative, as in the multinomial case, this suggests that an arc sin (or angular) transformation would be in order. Thus we define

$$(2.30) \qquad U_i = 2 \text{ arc sin } \sqrt{X_i} \qquad (i=1,2,\ldots,b)$$

and a Taylor expansion of U_i about $X_i = r/(n+1)$ gives us

$$(2.31) \qquad U_i = 2 \text{ arc sin } \sqrt{\frac{r}{n+1}}$$

$$+ \frac{n+1}{\sqrt{r(n+1-r)}}\left(X_i - \frac{r}{n+1}\right) - \frac{(n+1-2r)(n+1)^2}{2[r(n+1-r)]^{3/2}}\frac{\left(X_i - \frac{r}{n+1}\right)^2}{2!} + \ldots$$

(2.32) $E\{U_i\} \approx 2 \arcsin \sqrt{\dfrac{r}{n+1}}$ $(i=1,2,\ldots,b)$

(2.33) $\sigma^2(U_i) \approx \dfrac{1}{n+2}$ $(i=1,2,\ldots,b)$

(2.34) $\rho(U_i,U_j) \approx -\dfrac{r}{n-r+1}$ $(i \neq j)$.

Letting U_i^* denote the standardized U_i-variables and letting

(2.35) $h = 2\left[\arcsin \sqrt{p} - \arcsin \sqrt{\dfrac{r}{n+1}}\;\right]\sqrt{n+2}$,

the normal approximation is obtained by finding

(2.36) $P\{U_i^* < h \quad (i=1,2,\ldots,b)\}$,

where the U_i^* are correlated as in (2.34).

For common <u>positive</u> ρ the normal approximation (cf.[13] and [4]) takes
the form

(2.37) $F(\rho) = \displaystyle\int_{-\infty}^{\infty} \Phi^b\left(\dfrac{y\sqrt{\rho}+h}{\sqrt{1-\rho}}\right)d\Phi(y)$,

where $\Phi(x)$ is the standard normal cdf; this is obtained by setting
$U_i^* = Y_i\sqrt{1-\rho} - Y\sqrt{\rho}$ in (2.36) with $Y, Y_1,\ldots,Y_b$ all independent standard
normal and finding the probability by first conditioning on Y. For negative
ρ it can be shown that the expression in (2.37) is still valid and that $F(\rho)$
for negative ρ is a real function, i.e., its imaginary part is equal to zero.
Moreover, $F(\rho)$ has (several) derivatives at zero so that a Taylor expansion
around $\rho = 0$ can be used to approximate the function for ρ less than and
close to 0. The latter condition is helped by the fact that our probability
interpretation only holds for $\rho > -1/(b-1)$ and hence a negative ρ is forced
to be close to zero. In fact, since $\Phi(z)$, as a function of the complex
variable z, is analytic over the whole complex plane, we can show that for
all $b > 0$ and h real, $F(\rho)$ is analytic for $|\rho| < 1/(b-1)$.
but this has not been shown.

The function $F(\rho)$ in (2.37) can be expanded around any value of ρ between $-1/(b-1)$ and 1 and we first write this expansion and the derivatives needed for it as general functions of ρ for expanding about $\rho = \rho_0$; later we set $\rho_0 = 0$. For convenience, we write $F(\rho)$ as $A_b(\rho,h)$ and let

$$(2.38) \qquad \rho_i = \frac{\rho}{1+i\rho}, \quad h_i = \frac{h\rho_i}{\rho}\sqrt{\frac{1-\rho}{1-\rho_i}} \qquad (i=1,2,\ldots)$$

and if we set $\rho = \rho_0$ then we use the notation ρ_{i0} and h_{i0}, respectively. The general Taylor expansion about $\rho = \rho_0$ takes the form

$$(2.39) \qquad A_b(\rho,h) = A_b(\rho_0,h) + (\rho-\rho_0)\left\{ \frac{\binom{b}{2}}{\sqrt{1-\rho_0^2}}\varphi(h)\varphi(h_{10})A_{b-2}(\rho_{20},h_{20}) \right\}$$

$$+ \frac{1}{2!}(\rho-\rho_0)^2 \left\{ \frac{\binom{b}{2}[h^2(1-\rho_0)+\rho_0(1+\rho_0)]}{(1+\rho_0)(1-\rho_0^2)^{3/2}}\varphi(h)\varphi(h_{10})A_{b-2}(\rho_{20},h_{20}) \right.$$

$$- \frac{3\binom{b}{3}(2+2\rho_0-\rho_0^2)}{(1-\rho_0)(1+\rho_0)^2(1+2\rho_0)^{3/2}}\varphi(h)\varphi(h_{10})\varphi(h_{20})A_{b-3}(\rho_{30},h_{30})$$

$$+ \left. \frac{6\binom{b}{4}}{(1+\rho_0)(1+2\rho_0)\sqrt{1-\rho_0}(1+3\rho_0)}\varphi(h)\varphi(h_{10})\varphi(h_{20})\varphi(h_{30})A_{b-4}(\rho_{40},h_{40}) \right\} + \ldots$$

where $\varphi(x)$ is the standard normal density. For $\rho_0 = 0$ this reduces to

$$(2.40) \qquad A_b(\rho,h) = \Phi^b(h) + \rho\binom{b}{2}\varphi^2(h)\Phi^{b-2}(h) + \frac{\rho^2}{2!}\left\{ \binom{b}{2}h^2\varphi^2(h)\Phi^{b-2}(h) \right.$$

$$\left. - 6\binom{b}{3}\varphi^3(h)\Phi^{b-3}(h) + 6\binom{b}{4}\varphi^4(h)\Phi^{b-4}(h) \right\} + \ldots \quad .$$

It is also possible to regard (2.37) as a first term of a normal approximation and develop correction terms but the reward in accuracy does not appear to warrant all the algebraic computation involved. The accuracy attained by using (2.40) can be assessed by simply considering one or two cases for which the exact answer is in our table; this we now do.

Illustrative Examples of the Use of (2.40)

For $n = 81$, $b = 3$, $r = 10$ and $p = 1/5$, from (2.35), we obtain $h = 1.948$, $\rho = -5/36 = -.138889$ and

(2.41) $A_3(-\frac{5}{36},1.948) \approx .92484 - .00145 + .00037 = .92375;$

the exact answer from Table B is .92372, an error of less than 1/2 of 10^{-4}.
We cannot expect to get such high accuracy throughout the table but the better
approximations are obtained when n is large, b is small and r is larger
than the entries in our table. Thus for n = 64, b = 5, r = 4 and p = 1/8
from (2.35) we obtain h = 1.7983, $\rho = -4/61 = -.065574$ and

(2.42) $A_5(-\frac{4}{61},1.7983) = .83221 - .00237 + .00021 = .8301,$

and the correct answer by Table B is .8382, so that our error is in the
second decimal. Taking another example, for n = 98, b = 9, r = 4 and
p = 1/10 we obtain h = 2.3873, $\rho = -4/95 = -.042105$ and

(2.43) $A_9(-\frac{4}{95},2.3873) = .92616 - .00076 + .00009 = .9255,$

and the correct answer by Table B is .9192, so that we are again close to
only 2-place accuracy.

In many cases the first two terms in (2.40) already give the accuracy
attainable since the second correction term is quite small and hence the ρ^2
term is usually not worth calculating. In a final example we take n = 105,
b = 2, r = 5 and p = 1/10 so that $\rho = -.05$ and using (2.35), h = 2.127.
Then we obtain

(2.44) $A_2(-\frac{1}{20},2.127) = .96686 - .00009 + .00001 = .9668;$

the correct answer is .9711, so that we again have only 2-place accuracy.
Since b = 2 we can check the result against bivariate normal table values
for $\rho = -.05$ and h = 2.127; using linear interpolation we get .9668
exactly as in (2.44). Thus the error we are obtaining by the Taylor expansion (2.40) about $\rho = 0$ is negligible and the entire error lies in the
fact that the Dirichlet distribution is not well fitted by a multivariate
normal distribution without the aid of complicated correction terms for the
deviation from normality. In contrast to this the use of the normal dis-

tribution to approximate the n-value needed to satisfy a P^*-requirement
led to very good results that are given in Table F in section 4.1.1.

§3. Generalized Stirling Numbers

In this section we study combinatorial aspects of the special
case $p = 1/b$. In this case our recursion formulas yield recurrence
relations for generalized Stirling Numbers (of the second kind).
We study some properties of these numbers and show that they have a simple
combinatorial interpretation. These numbers are related to the minimum
frequency of a homogeneous multinomial distribution. Analogous numbers
related to the maximum will appear with the appropriate tables.

3.1.1. Definition and Basic recurrence relation.

For any integers $b \geq 0$, j (with $0 \leq j \leq b$), $r \geq 1$ and $n \geq 0$
(with $n \geq br$, since $I = 0$ for $n < br$) such that b and n are
not both zero, we define

$$(3.1) \qquad S_{n,r}^{(b,j)} = \frac{b^n}{(b-j)!} I_{1/b}^{(b,j)}(r,n).$$

If $b = n = 0$ then $j = 0$ and we define the left side of (3.1) to be
1 for all r. If $b = 0 < n$ then (3.1) gives zero and if $n = 0 < b$
then (3.1) gives zero for $r > 0$ and δ_{0j} for $r = 0$.

Substituting (3.1) in (2.9) we obtain for all j,r,n, and b (with $j < b$ and $n > br$)

$$(3.2) \qquad S_{n,r}^{(b,j)} = \frac{n(b-j)}{n-jr} S_{n-1,r}^{(b,j)} + \frac{r}{n-jr} S_{n,r}^{(b,j+1)}.$$

The two boundary conditions for this recurrence are

$$(3.3) \qquad S_{br,r}^{(b,j)} = \frac{(br)!}{(b-j)!(r!)^b} \qquad\qquad (\, 0 \leq j \leq b) \,,$$

$$(3.4) \qquad S_{n,r}^{(b,b)} = \begin{cases} \dfrac{(br)!}{(r!)^b} & \text{for } n = br \\[2em] 0 & \text{otherwise .} \end{cases}$$

We note that (3.3) and (3.4) agree when both $j = b$ and $n = br$ hold.

3.1.2. Combinatorial interpretation.

For $r = 1$ and $j = 0$ in (3.1) we use the notation $S_{n,1}^{(b,0)} = S_n^{(b)}$ and these are the usual Stirling numbers of the second kind defined by the identity $x^n = \sum_{\alpha=1}^{n} S_n^{(\alpha)} x(x-1)\ldots(x-\alpha+1)$ for $n \geq 1$.

We claim that the number $S_{n,r}^{(b,j)}$ has a combinatorial interpretation for any j. When $r > 0$ it represents the number of "mixed" partitions of n distinct objects into exactly b subsets; the first j subsets are each to contain <u>exactly</u> r objects and the partition is ordered with respect to these subsets; the remaining $b-j$ subsets are each to contain <u>at least</u> r objects and the partition is unordered with respect to these subsets. There is, of course, no ordering <u>within</u> any of the b subsets. Thus for $j = 0$, $S_{n,r}^{(b)}$ represents the number of partitions in the usual unordered sense with all parts having size at least r. For $r = 0$ the definition (3.1) and this partition interpretation give different results; we consider both below.

It follows as a corollary of this combinatorial interpretation that the generalized Stirling numbers are all integers. In particular, the expression in (3.3) with $j = 0$ (and hence for any j) is an integer since it represents the number of partitions of br distinct objects into b subsets, each of size r. We defer the proof of this interpretation and apply it below to the essential case, $j = 0$.

For $p = 1/b$ the reduction formula (2.11) takes the slightly simpler form

$$(3.5) \qquad S_{n,r}^{(b,i+j)} = \frac{n!}{(r!)^j (n-jr)!} \, S_{n-jr,r}^{(b-j,i)},$$

which is again used principally with $i = 0$; note especially the simple
result for $j = 1$. Using (3.5) to remove the second superscript throughout
(3.2), we obtain for $n \geq br$ and $0 \leq j \leq b$

$$(3.6) \qquad S_{n-jr,r}^{(b-j)} = (b-j)\, S_{n-jr-1,r}^{(b-j)} + \binom{n-jr-1}{r-1}\, S_{n-jr-r,r}^{(b-j-1)}.$$

For $j = 0$ and $n \geq br > 0$ this reduces further to

$$(3.7) \qquad S_{n,r}^{(b)} = b\, S_{n-1,r}^{(b)} + \binom{n-1}{r-1}\, S_{n-r,r}^{(b-1)}.$$

Note that for $j = 0$ we can use (3.7) and (3.3) with $j = 0$ to
compute $S_{n,r}^{(b)}$ without the need of (3.4).

To prove the combinatorial interpretation we use (3.7), letting x
denote the one extra object in the set of size n on the left of (3.7)
that is not in the set of size $n-1$ on the right side of (3.7). We
can put x into any one of the b sets used in $S_{n-1,r}^{(b)}$ and this
explains the multiplication by b. In addition, we can combine x with
any subset of size $r-1$ and use the resulting set of size r as one
of the b subsets; this multiplies the number of possible partitions
of the remaining $n-r$ objects into $b-1$ subsets, each containing at least
r objects. Moreover every such partition is a new one and all possible
partitions are now included; this proves the combinatorial interpretation
since the boundary conditions are easily seen to represent the appropriate
number of partitions.

For $b = 2$ the last S in (3.7) equals 1 and we can get further
reduction and explicit expressions. In fact, by the combinatorial
interpretation the result for $b = 2$ (with $j = 0$) and $n \geq 2r$ must
be

$$(3.8) \qquad S_{n,r}^{(2)} = \frac{1}{2}\left[\binom{n}{r} + \binom{n}{r+1} + \cdots + \binom{n}{n-r}\right] = 2^{n-1} - \sum_{\alpha=0}^{r-1}\binom{n}{\alpha}.$$

However, the recurrence (3.7) through iteration gives another expression

for $b = 2$, namely

$$(3.9) \quad s_{n,r}^{(2)} = \binom{n-1}{r-1} + 2\binom{n-2}{r-1} + \cdots + 2^{n-2r}\binom{2r-1}{r-1} .$$

By an induction on n, it is easily shown that these two expressions

are equal for all r and $n \geq 2r$. In particular, the common value for

$n = 2r$ is $\binom{2r}{r}/2$.

3.1.3. Additional relations and a numerical illustration.

Some additional formulas for $j = 0$ are obtainable through the

probability interpretation of the I-function. Consider $r = 2$ first

with $n \geq 2b \geq 0$. Let $I_p^{(b,j)}(r,s,n)$ denote the probability that j

particular cells have frequency exactly r and the remaining $b-j$ cells

have frequency at least s; here we have n observations in all

and each blue cell has common probability p. Note that $s_{n,r}^{(1)} = 1$

for $n \geq r$ and 0 otherwise. Then by the probability interpretation

of I and (3.1), we have for $p = 1/b$ and $j = 0$

$$(3.10) \quad s_{n,2}^{(b)} = \frac{b^n}{b!} I_{1/b}^{(b)}(2,n) = \frac{b^n}{b!}\left[I_{1/b}^{(b)}(1,n) - \sum_{\alpha=1}^{b-1}\binom{b}{\alpha} I_{1/b}^{(b,\alpha)}(1,2,n)\right]$$

$$= s_n^{(b)} - \sum_{\alpha=1}^{b-1}\binom{n}{\alpha} \frac{(b-\alpha)^{n-\alpha}}{(b-\alpha)!} I_{1/(b-\alpha)}^{(b-\alpha)}(2,n-\alpha)$$

$$= s_n^{(b)} - \sum_{\alpha=1}^{b-1}\binom{n}{\alpha} s_{n-\alpha,2}^{(b-\alpha)} .$$

The generalization of (3.10) to $r > 2$ is not apparent; it can be

written as

$$(3.11) \quad s_{n,r}^{(b)} = s_n^{(b)} - \sum_{\gamma=1}^{r-1} \sum_{\alpha=1}^{[n/\gamma]} \binom{n}{\gamma\alpha} \frac{(\gamma\alpha)!}{\alpha!(\gamma!)^\alpha} s_{n-\gamma\alpha,\,\gamma+1}^{(b-\alpha)} ,$$

where, in the probability interpretation, the sum for any γ removes the probability that the minimum frequency is exactly γ. The terms in the double sum will be zero if $n - \gamma\alpha < (\gamma+1)(b-\alpha)$.

As a typical calculation using (3.2), (3.3) and (3.4), we consider $b = 3$, $j = 1$, $n = 7$ and $r = 2$. Then for $S_{7,2}^{(3,1)}$ we obtain

$$(3.12) \quad S_{7,2}^{(3,1)} = \frac{14}{5}S_{6,2}^{(3,1)} + \frac{2}{5}S_{7,2}^{(3,2)}; \quad S_{6,2}^{(3,1)} = 45;$$

$$S_{7,2}^{(3,2)} = \frac{7}{3}S_{6,2}^{(3,2)} = \frac{7}{3}(90) = 210, \qquad \text{since}$$

$$S_{6,2}^{(3,2)} = S_{6,2}^{(3,3)} = \binom{6}{2}\binom{4}{2} = 90;$$

Hence, combining these results,

$$(3.13) \quad S_{7,2}^{(3,1)} = \frac{14}{5}(45) + \frac{2}{5}(210) = 126 + 84 = 210$$

and a complete listing will confirm this result. It should be noted that if $a,b,c,d,e,f,g,$ denote the 7 objects in this example, then the partition $\{(a,b); (c,d), (e,f,g)\}$ is not the same as $\{(c,d); (a,b), (e,f,g)\}$ since the partition is ordered with respect to the 1st subset. However, the first of these partitions is the same as $\{(a,b); (e,f,g), (c,d)\}$ since the partition is unordered with respect to the last $b - j$ subsets. There is no ordering <u>within</u> any of the b subsets.

3.1.4. <u>Partitions into at most b parts.</u>

Another useful result that follows from (2.14) and (3.1) and holds for any j, n, r, b provided $b > j \geq 0$ and $r \geq 1$ is

$$(3.14) \quad S_{n,r}^{(b,j)} = \frac{1}{b-j} \sum_{\alpha (b-1)r}^{n-r} \binom{n}{\alpha} S_{\alpha,r}^{(b-1,j)}.$$

If we define $S_{n,0}^{*(b,j)} = S_{n,0}^{*(b-j)}$ as the number of partitions of n distinct objects into <u>at most</u> $b-j$ subsets, then $S_{n,0}^{*(b)}$ is not generally equal to $S_{n,0}^{(b)}$ defined by (3.1). From (3.1) we obtain

$S_{n,0}^{(b,j)} = S_{n,0}^{(b-j)} = b^n/(b-j)!$ and this generally disagrees with $S_{n,0}^{*(b,j)}$

for $b - j \geq 3$. The quantity S^* can be computed by many of our formulas

such as (3.14) but it is easier to apply the classical result (see

e.g. [14]) that the number of partitions of n into at most b is

equal to the number of partitions of $n + b$ into exactly b parts. Hence

$$(3.15) \qquad S_{n,0}^{*(b,j)} = S_{n,0}^{*(b-j)} = S_{n+b-j,1}^{(b-j)} = S_{n,1}^{(1)} + \ldots + S_{n,1}^{(b-j)}.$$

According to the generating function method (described below)

$S_{n,r}^{(b,j)}$ is the coefficient of $\theta^{b-j}\tau^n/n!$ in

$$(3.16) \qquad H(\tau,\theta) = \left(\frac{\tau^r}{r!}\right)^j \exp\left\{\theta(e^\tau - \sum_{\alpha=0}^{r-1} \frac{\tau^\alpha}{\alpha!})\right\}.$$

In particular, for $r = 1$ and $j = 0$ the sum $S_{n,1}^{(1)} + S_{n,1}^{(2)} + \ldots + S_{n,1}^{(b')}$

for $b' \geq n$ is the nth moment of a Poisson distribution with parameter

$\lambda = 1$. It follows from (3.15) and (3.16) that $S_{n,0}^{*(b,j)}$ is the coefficient

of $\theta^{b-j}\tau^{n+b-j}/(n+b-j)!$ in

$$(3.17) \qquad H^*(\tau,\theta) = \exp\{\theta(e^\tau - 1)\} .$$

In particular, if we fix $n' = n+b-j > 0$ and sum over b(or over $b-j$)

the sum $S_{n',1}^{*(0)} + S_{n',1}^{*(1)} + \ldots + S_{n',1}^{*(n')}$ is the n'^{th} moment of a Poisson

distribution with parameter $\lambda = 1$.

§4. Applications.

The tables associated with the incomplete Dirichlet integrals of

type 1 are found to be useful in several applications. In this section

we discuss applications to the following problems: (i) Multinomial

ranking problem , (ii) curtailment problem, (iii) two p-values problem,

and (iv) the computation of probabilities in various multinomial problems.

4.1. A Multinomial Selection Problem.

In a multinomial model with $k = b+1$ cells, suppose we wish to select the cell with the smallest cell probability and the procedure R (based on a fixed sample size n) is simply to select the cell with the smallest frequency using randomization among the cells tied for "first place" when there is more than one cell with the same lowest frequency. Here the number of blue cells (or cells of interest) is k.

Let the ordered cell probabilities be denoted by $p_{[1]} \leq p_{[2]} \leq \dots \leq p_{[k]}$; we also use $\underline{p}$ for $p_{[1]}$ in the discussion below. The problem is to find the smallest n-value such that the probability of a correct selection PCS using procedure R will be at least P^* (P^* is specified so that $(1/k) < P^* < 1$) whenever the difference $p_{[2]} - p_{[1]}$ is less than $\delta^* > 0$ (which is also specified and less than $1/b$). [It is shown in [1] that the corresponding requirement with a ratio of p's cannot be met for any fixed value of n, regardless of how large it might be]. In order to satisfy this requirement for all configurations we first have to put the p's into a so-called least favorable configuration i.e., one that minimizes the PCS subject to the δ^*-condition only. For this purpose we use the so-called slippage configuration and set $p_{[2]} = p_{[3]} = \dots = p_{[k]} = p$ (say). The condition that the p's sum to unity then gives

$$(4.1) \qquad \underline{p} = \frac{1 - b\delta^*}{b+1}, \qquad p = \frac{1 + \delta^*}{b+1}.$$

Actually we need not be concerned with the δ^*-condition except to note that $0 \leq \delta^* \leq 1/b$, but we do assume below the common p-value in our discussion and we prefer to use $\underline{p}$ and p rather than δ^*.

We wish to derive both an exact expression as well as upper and lower bounds for the PCS for the case of a common p-value, so that $bp + \underline{p} = 1$.

The upper (resp. lower) bound is obtained by deriving the PCS for procedure R under the assumption that randomization always (resp., never)

yields the correct answer. The exact value then uses the same format
with a deeper analysis. Let x denote the frequency in the best cell
i.e., the one with cell probability $\underline{p}$. Then, using our probability interpretation for the I-function, we have the bounds

$$(4.2) \quad \sum_{x=0}^{[\frac{n-b}{b+1}]} B_p(n,x) \, I_{1/b}^{(b)}(x+1, \, n-x) \leq PCS \leq \sum_{x=0}^{[\frac{n}{b+1}]} B_p(n,x) \, I_{1/b}^{(b)}(x, \, n-x)$$

where $B_p(n,r)$ is the usual binomial probability, and [x] is the
largest integer less than or equal to x (i.e., the integer part of x).
The upper limits on both sides of (4.2) could be omitted since they only
indicate that the I-function is zero above these values. When j cells
with probability p are tied with the best cell, then the randomization
probability is $1/(j+1)$ for getting a correct selection. Since this is
closer to zero than to 1 for each positive j (j = 1,2,...,b), it
follows that the left side of (4.2) will be closer to the exact PCS.

To obtain the exact PCS we first define the generalized I-function
$I_p^{(b,j)}(r, \, r+1, \, n)$ to be the probability that j specific blue cells
have frequency r and the remaining b-j blue cells have frequency at
least r+1 (i.e., greater than r). This can also be written as an
integral analogous to (2.7) but we omit this since it is not used. Then
the exact PCS for the generalized least favorable (GLF) configuration can
be written as

$$(4.3) \quad P\{CS|GLF\} = \sum_{x=0}^{[n/k]} B_p(n,x) \sum_{j=0}^{b} \frac{\binom{b}{j}}{j+1} \, I_{1/b}^{(b,j)}(x,x+1,n-x)$$

where the inside sum is positive only on the subset
$b(x+1) - n + x \leq j \leq [(n-x)/x]$ of the set [0,1,...,b]. Using a
reduction formula quite similar (in form and derivation) to (2.11)
and letting $\beta = j + 1$ we obtain

$$(4.4) \quad P\{CS|GLF\} = \sum_{x=0}^{[n/k]} \underline{p}^x \ p^{n-x} \ k^{n-1} \sum_{\beta=1}^{k} \binom{k}{\beta} \ I_{1/k}^{(k,\beta)}(x,x+1,n) \ .$$

The inside sum is easily seen to be equal to the probability that the minimum frequency is exactly x; hence for $0 \le \delta^* \le 1/b$

$$(4.5) \quad P\{CS \mid LF\} = k^{n-1} \sum_{x=0}^{[n/k]} p^{n-x} \ \underline{p}^x [I_{1/k}^{(k)}(x,n) - I_{1/k}^{(k)}(x+1,n)]$$

$$= (1 + \delta^*)^n [\frac{1}{k} - \frac{\delta^*}{1+\delta^*} \sum_{x=1}^{[n/k]} \left(\frac{1-b\delta^*}{1+\delta^*}\right)^{x-1} I_{1/k}^{(k)}(x,n)]$$

$$= (1+\delta^*)^n \left\{ \frac{1}{k} \left(\frac{1-b\delta^*}{1+\delta^*} \right)^{[n/k]} \right.$$

$$\left. + \frac{\delta^*}{1+\delta^*} \sum_{x=1}^{[n/k]} \left(\frac{1-b\delta^*}{1+\delta^*}\right)^{x-1} [1 - I_{1/k}^{(k)}(x,n)] \right\} .$$

Another expression for the $P\{CS|LF\}$ can be obtained from the first part of (4.5) by writing it as a sum on α where α is the number of cells whose frequency is equal to the minimum value x. This gives for $0 \le \delta* \le 1/b$

$$(4.6) \quad P\{CS|LF\} = \frac{(kp)^n}{k} \sum_{x=0}^{[n/k]} \left(\frac{\underline{p}}{p}\right)^x \sum_{\alpha=1}^{[n/x]} \binom{k}{\alpha} \frac{n! \left(\frac{1}{k}\right)^{\alpha x} \left(1 - \frac{\alpha}{k}\right)^{n-\alpha x}}{(x!)^\alpha \ (n-\alpha x)!} \ I_{1/(k-\alpha)}^{(k-\alpha)}(x+1,n-\alpha x).$$

Using the reduction formula (2.11) in reverse, we obtain

$$(4.7) \quad P\{CS|LF\} = n!(1+\delta*)^n \left\{ \sum_{\alpha=1}^{k-1} \binom{k}{\alpha} \frac{k^{\alpha-1}}{(n+\alpha)!} \sum_{x=0}^{[(n+\alpha-k)/k]} (x+1)^\alpha \left(\frac{1-b\delta*}{1+\delta*}\right)^x \ I_{1/k}^{(k,\alpha)}(x+1,n+\alpha) \right.$$

$$\left. + \left(\frac{1-b\delta*}{1+\delta*}\right)^{n/k} \frac{f(n,k)}{k^{n+1}\left[\left(\frac{n}{k}\right)!\right]^k} \right\} ,$$

where $f(n,k) = 1$ if n is a multiple of k and is zero otherwise.

For the special case $k = 2$ we can also obtain an even simpler result directly from (4.3) by noting that $I_1^{(1,j)}(x,x+1,n-x)$ in (4.3) equals 1 if $x + 1 - j \le n - x$ and is zero otherwise for $j = 0,1$. Hence for $k = 2$ we obtain for the $P(CS|GLF)$ for n odd and n even, respectively

$$(4.8a) \qquad \sum_{x=0}^{(n-1)/2} B_{\underline{p}}(n,x) = I_p^{(1)}(\frac{n+1}{2}, n) = I_p(\frac{n+1}{2}, \frac{n+1}{2})$$

$$(4.8b) \qquad \sum_{x=0}^{(n-2)/2} B_{\underline{p}}(n,x) + \frac{1}{2}\binom{n}{n/2}(\underline{p}\ p)^{n/2} = \frac{1}{2}[I_p^{(1)}(\frac{n}{2} + 1,n) + I_p^{(1)}(\frac{n}{2},n)]$$

$$= \frac{1}{2}\ [I_p(\frac{n}{2} + 1, \frac{n}{2}) + I_p(\frac{n}{2}, \frac{n}{2} + 1)]$$

where $p = 1 - \underline{p} = (1+\delta)/2$ for the GLF and $p = 1 - \underline{p} = (1+\delta)/2$ for the LF; here I without superscript is the usual incomplete beta function.

The equality of (4.8) and (4.5) with $k = 2$ is non-trivial; equations (4.8) appear in Alam and Thompson [1]. Using (4.8) the entries in Table F for $k = 2$ can be obtained from a good binomial table.

4.1.2. PCS in terms of generalized Stirling numbers.

We now use Stirling numbers of the second kind and generalized Stirling numbers of the second kind defined in section 3 to write (4.5) in a slightly different form. Recall the definition (3.1) for any j, r, n, b, with b and n not both zero

$$(4.9) \qquad S_{n,r}^{(b,j)} = \frac{b^n}{(b-j)!}\ I_{1/b}^{(b,j)}(r,n).$$

In terms of these S numbers we can use (4.5) and (3.1) to write

$$(4.10) \qquad P\{CS|GLF\} = b! \sum_{x=0}^{[n/k]} \underline{p}^x\ p^{n-x}(S_{n,x}^{(k)} - S_{n,x+1}^{(k)})$$

$$= p^n[k^{n-1} - b!\ (\frac{p-\underline{p}}{\underline{p}}) \sum_{x=1}^{[n/k]}(\frac{\underline{p}}{p})^x\ S_{n,x}^{(k)}].$$

For the LF configuration we use (4.1) and obtain

$$(4.11) \qquad P\{CS|LF\} = b_.!\left(\frac{1+\delta^*}{k}\right)^n \sum_{x=0}^{[n/k]} \left(\frac{1-b\delta^*}{1+\delta^*}\right)^x (s_{n,x}^{(k)} - s_{n,x+1}^{(k)})$$

$$= \left(\frac{1+\delta^*}{k}\right)^n [k^{n-1} - b_.!\left(\frac{k\delta^*}{1-b\delta^*}\right) \sum_{x=1}^{[n/k]} \left(\frac{1-b\delta^*}{1+\delta^*}\right)^x s_{n,x}^{(k)}].$$

The proof that the configuration (4.1) is LF is given by Alam and Thompson in [1]. It would be desirable to show for fixed $\underline{p}$ that (4.3) is strictly decreasing in p for $0 \le p \le 1/k$; this is clearly related to the proof that (4.1) is the LF configuration. The LF property tells us that (4.11) is a lower bound to the PCS over all configurations with $p_{[2]} = p_{[3]} = \cdots = p_{[k-1]} = p$ and $p - p_{[1]} \ge \delta^*$.

For $p = \underline{p}$ the common value is $1/k$ and by (4.10) the value of the PCS is $1/k$; this is regarded as the lim inf for $p \to 1/k$ through values $> 1/k$ since the concept of a correct selection is uniquely defined only when $p > p_{[1]}$.

For $\underline{p} = 0$ we obtain directly from (4.10) the result

$$(4.12) \qquad P\{CS|GLF, \underline{p} = 0\} = \frac{k^{n-1} - b_.!s_n^{(k)}}{b^n},$$

where $s_{n,1}^{(k)} = S_n^k$ are the usual Stirling numbers of the second kind.

Setting the final probability in (4.12) between 0 and 1 gives us the

Corollary: Stirling numbers of the second kind satisfy

$$(4.13) \qquad \frac{(b+1)^{n-1} - b^n}{b_.!} \le S_n^{(b+1)} \le \frac{(b+1)^{n-1}}{b_.!}.$$

The difference between the two bounds in (4.13) is $b^n/b_.!$; hence if n and b grow in such a way that this quantity tends to zero then the two bounds in (4.13) get asymptotically closer to the Stirling number between them and either bound in (4.13) can be used for an asymptotic evaluation of these numbers.

For the dual problem of selecting the cell with the largest cell probability, using the measure of distance $\delta = p_{[k]} - p_{[k-1]}$, we can obtain analogous results and they will be presented elsewhere.

Illustration 1: Suppose we take $n = 4$ observations on a nultinomial with $k = 2$ (i.e., a binomial) with cell probabilities $\underline{p} = 1/4$ and $p = 3/4$ (so that $\delta^* = p - \underline{p} = 1/2$ and $b = k - 1 = 1$). Using (4.5) and entries from Table A we obtain

$$(4.14) \quad P(CS|LF) = (\tfrac{3}{2})^4 \left[\tfrac{1}{2} - \tfrac{1}{3} (\tfrac{7}{8} + \tfrac{1}{3} \; \tfrac{3}{8}) \right] = (\tfrac{3}{2})^4 \; \tfrac{1}{6} = \tfrac{27}{32}.$$

The same result can be obtained from $(4.8b)$ and also from (4.11) together with Table E.

Illustration 2: Suppose we have a multinomial with $k = 5$ cells and we wish to find the cell with the smallest cell probability by taking $n = 30$ observations. What is the $P\{CS|LF\}$ when $\delta^* = .20$? Using (4.5) and Table A we obtain

$$(4.15) \quad P\{CS|LF\} = (1.2)^{30} \left\{ \tfrac{1}{5} - \tfrac{.9938125}{6} - \tfrac{.9476476}{6^2} - \tfrac{.786046}{6^3} \right.$$
$$\left. - \tfrac{.46075}{6^4} - \tfrac{.11827}{6^5} - \tfrac{.00147}{6^6} \right\} = .956906 \, ,$$

which agrees with the last entry $.957$ in Table 2 of Alam and Thompson [1].

Illustration 3: For $k > 2$ the equation (4.5) with Table A can be used up to $N = 30$ provided we use a good extrapolation formula to obtain those entries greater than $.999$, that are not in our table. Good results are obtained by assuming that the table values (say T_L) is of the form

$$(4.16) \quad T_L = 1 - e^{-(AL + C)},$$

where A and C are positive constants. Thus for $k = 4$ and $r = 1$
(treating $N = 27$ as $L = 0$ and $N = 28$ as $L = 1$), we obtain for
$N = 29$ (or $L = 2$) the T_2 value

$$(4.17) \quad T_2 = 1 - e^{-(2A + C)} = .9990476$$

where $C = - \ln(1 - T_0) = 6.381135$ and $A = -C - \ln(1 - T_1) = .287682$.
The answer in (4.17) agrees with our table value to all the seven
digits shown.

By this method we obtain for $N = 30$ the I-value .9992857 and,
using this in (4.5) for $k = 4$, $\delta^* = .1$ and $N = 30$, we obtain

$$(4.18) \quad P\{CS|LF\} = (1.1)^{30}[\frac{1}{4} - \frac{1}{11} \sum_{x=1}^{7} (\frac{7}{11})^{x-1} I_{1/4}^{(4)}(x,30)] = .592657 ,$$

which agrees with .593 in Table 2 of [1] for $c = .1$ and $k = 4$.

Remark: In the usual ranking and selection problem k denotes the
number of independent populations and, for given δ^* and P^*, the required
sample size n is strictly increasing with k. In the present multinomial
problem a glance at Table F shows that this is not so. In general, for
fixed δ^* and P^*, the sample size increases to a maximum and then decreases.
In most applications the value of k is not at the disposal of the statistician
so that he cannot vary it to get the smallest value of n. In addition
we cannot set any $p_{[i]}$'s equal to zero for $i \geq 2$ since each one has to be
at least $p_{[1]}+\delta^*$; moreover the LF property for the configuration (4.1) was
shown in [1]. Even if the experimenter could add cells, it would have to
be assured that they had cell probabilities equal to at least $p_{[1]} + \delta^*$;
this cannot be done without affecting the basic model of the problem. Hence
this strange phenomenon that n appears to reach a maximum and then diminish

as k increases is not a contradiction, but only an unusual property. Moreover

for fixed δ^* and varying k the sample size never gets very small, since k can

only go up to $(1+\delta^*)/\delta^*$; the latter follows from the fact that $\underline{p} = (1-b\delta^*)/k$

is assumed to be non-negative in our model. If we insist on having $\rho > 0$

in (4.20) below then k only goes up to $1/\delta^*$ as is indicated in (4.21) below.

4.1.3 <u>Normal Approximation</u>

A useful normal approximation is obtained by considering the LF-
configuration and writing the inequalities that lead to a correct selection
in the form

$$(4.19) \qquad \frac{(F_{(1)} - n\underline{p}) - (F_{(i)} - n\underline{p})}{\sqrt{nA}} \; < \; \frac{(p-\underline{p})\sqrt{n}}{\sqrt{A}} \; = \; \frac{\delta^*\sqrt{n}}{\sqrt{A}} \, ,$$

where $A = \underline{p}q + 2\underline{p}p + pq$, $F_{(i)}$ is the random frequency associated with the
ordered cell probability $p_{[i]}$ and nA is easily shown to be the variance of
$F_{(1)} - F_{(i)}$ in the LF-configuration. The common correlation between these

differences, given by

$$(4.20) \qquad\qquad \rho = 1 - \frac{p}{\underline{p}q + 2\underline{p}p + pq} = 1 - \frac{p}{A} = \frac{1 - \delta^* k}{2 - \delta^* k}$$

does not depend on i or n. We assume the condition that $\rho > 0$ in the LF
configuration and this reduces to

$$(4.21) \qquad \delta^* < 1/k$$

To get the normal approximation we define k independent N(0,1)
random variables V_i (i=1,2,...,k) and write the asymptotic equivalent of
(4.19)

$$(4.22) \qquad V_i\sqrt{1-\rho} - V_1\sqrt{\rho} < h \quad (i = 2,3,...,k)$$

where $h = \delta^* \sqrt{n}/\sqrt{A}$ from (4.19) and ρ is given in (4.20). Letting $\Phi(x)$ denote the standard normal cdf, it follows that our PCS is asymptotically given by

$$(4.23) \qquad P\{CS|LF\} \approx \int_{-\infty}^{\infty} \bar{\Phi}^{k-1}\left(\frac{x\sqrt{\rho} + h}{\sqrt{1 - \rho}}\right) \, d\bar{\Phi}(x),$$

which we now set equal to P^* and solve for $h = h(k,\rho,P^*)$ by using available tables. Hence, using (4.1) to evaluate A, we have from the above

$$(4.24) \qquad n = \left(\frac{h}{\delta^*}\right)^2 A = \left(\frac{h}{\delta^*}\right)^2 (1+\delta^*)\left(\frac{2}{k} - \delta^*\right)$$

and, of course if n is not an integer and we are not using a randomized strategy, then the next larger integer is used. This result (4.24) is equivalent to the normal approximation in [1].

For selected values of k, δ^* and P^* the above normal approximation (4.24) is given in Table F along with the exact value based on (4.5); both are given to 2 decimals and the user can attain P^* exactly by randomizing appropriately between the two integers that straddle the tabled value of n.

For P^* close to one (and b not too large) we can replace (4.23) by the simpler equation

$$(4.25) \qquad \bar{\Phi}(-h') = \frac{1-P^*}{b},$$

which does not depend on ρ and requires only a table of the standard normal cdf. Then we use (4.24) to compute n with h' replacing h.

This generally gives conservative results, i.e., an upper bound for n. For example with k = 3 (so that b=2), $P^* = .99$ and $\delta^* = .10$ we obtain 413.83 (or 414) by using (4.25), 409.79 (or 410) by using (4.24) and the exact result is 409.77 (or 410) using (4.5). Problem 32 near the end of section 4 is an application of this analysis and gives a good idea of the accuracy of (4.25) relative to (4.24).

4.1.4 Sample Sizes needed for the Multinomial Ranking Problem

In order to implement the procedure described in section 4.1 we need
a table of sample sizes n required for specified values of δ^* and P^*;
we now give such a table for k = 2(1)10. Although this is now computable
by using (4.5) together with Table B, it cannot be done without considerable
computation and it is desirable to include this table to complete the discussion. Such a table for k = 2(1)5 was already given in [1], but because
(1) the manner of computation was not given, (2) the information as to whether
the results are exact or approximate was not stated, (3) we use different
values of δ^* and P^*, and (4) we have some overlapping entries that differ from the results in [1], we feel justified to include our table F here.
In each case we report n-values to two decimals so that the user who does
not wish to randomize between two successive integers merely takes the next
larger integer. In most cells we include 2 entries, the top entry is the
result obtained by the normal approximation (4.24) and the lower entry
is the exact result obtained by using the last expression in (4.5).

4.2 Curtailment in the Multinomial Selection Problem

Any treatment of the fixed sample size multinomial problem for
finding the cell with the smallest (resp., largest) probability is not
complete without a discussion of the expected number of observations
actually used to terminate the procedure R. The value of n is the
maximum number needed (some tables of n are given in [1] and [2]
respectively) and this is an upper bound on the expected number since
we can often terminate (as soon as the final decision is determined)
before taking all n observations. We refer to this as "curtailment"
and let N denote the random number of observations actually needed
when n is prescribed. The distribution of N depends on the true
p-values; we consider only the case in which $p_{[2]} = p_{[3]} = \cdots = p_{[k]} = p$
(say) and $p_{[1]} = \underline{p} \leq p$ ($\underline{p}$ and p arbitrary otherwise). The evalu-

ation of $E\{N\}$ was completely overlooked in [1] which deals with selecting the cell with the smallest probability and also in [2] which deals with a different formulation for selecting the cell with the largest

probability. An exact expression for $E\{N\}$ appears to be almost impossible without the use of the I-functions and their generalizations; we consider here only the first of these two problems; corresponding results have also been obtained for the second problem. We give our results without detailed derivations.

Let $\{x\}$ denote the smallest integer equal to or greater than x. Let $p_1 = \underline{p}/(1-p)$ and $p_2 = p/(1-p)$; we write $\underline{p_2}$ to indicate that p_2 is repeated b times. Let $P_m(CS)$ (resp., $P_m(IS)$) denote the probability of terminating procedure R after exactly m observations without ties for first place and with a correct (resp., incorrect) selection. Then $P_m(S) = P_m(CS) + P_m(IS)$ is the probability of terminating after exactly m observations without these ties; m starts at $\{\frac{b(n+1)}{b+1}\}$ and runs to n. Exact expressions for $P_m(CS)$ and $P_m(IS)$ are given in terms of I-functions by

$$(4.26) \quad P_m(CS) = (1-\underline{p}) \sum_{y=\{\frac{bn-1}{b+1}\}}^{m-1} B_{1-\underline{p}}(m-1,y) \ [\ I_{1/b}^{(b)}(n-y,y+1)$$

$$- I_{1/b}^{(b)}(n-y+1,y)],$$

$$(4.27) \quad P_m(IS) = b(1-p) \sum_{y=\{\frac{bn-1}{b+1}\}}^{m-1} B_{1-p}(m-1,y) \ [I_{p_1,\underline{p_2}}^{(b)}(n-y,y+1)$$

$$- I_{p_1,\underline{p_2}}^{(b)}(n-y+1,y)].$$

It should be noted that for $b = 1$ the value of p_1 is 1 and $\underline{p_2}$ is not present, so that the value of $I_1^{(1)}(r,s)$ is 1 if

$s \geq r$ and 0 otherwise. For $b \geq 2$

$$(4.28) \quad I_{p_1,p_2}^{(b)}(r,s) = \sum_{i=(b-1)r}^{s-r} B_{1-p_1}(s,i)\, I_{1/(b-1)}^{(b-1)}(r,i).$$

For $m < n$ we can only terminate without ties for first place, but for $m = n$ we can terminate with or without such ties. Hence the expected number of observations needed to terminate procedure R for any such configuration is

$$(4.29) \quad E\{N\} = \sum_{m=\{\frac{b(n+1)}{b+1}\}}^{n} m P_m(S) + n\left[1 - \sum_{m=\{\frac{b(n+1)}{b+1}\}}^{n} P_m(S)\right].$$

Here the last factor in square brackets is the probability of termination at $m = n$ with ties for first place. For the special case $b = 1$ with n odd, the probability of ending with a tie is zero.

4.2.1. Special configurations for the Curtailment Problem.

In addition to LF, there are two interesting cases for computing $E\{N\}$:

Case i) $\underline{p} = p = 1/k$ (Equal p's with 1 cell tagged as best)

Case ii) $\underline{p} = 0$, $p = 1/b$ (Equal p's with 1 cell vanishing).

It is important to note that in Case i) we tag one cell as best even though all the p's are equal; actually, we are interested in the limit of $E\{N\}$ as $\underline{p} \to 1/k$ from below, keeping a common value p for the b remaining cells. These two cases are extremes in the sense that Case i) gives us the maximum $E\{N\}$ over all possible configurations of p-values and Case ii) gives us the minimum $E\{N\}$ over all configurations with a common p-value in b cells. In general, both results are less than n and Case ii) is not LF.

In Case i) we note that

$$(4.30) \quad P_m(IS) = b\, P_m(CS), \quad P_m(S) = k\, P_m(CS),$$

$$(4.31) \quad P_m(CS) = \frac{b}{k} \sum_{y=\{\frac{bn-1}{b+1}\}}^{m-1} B_{(b/k)}(m-1,y) \left[I_{1/b}^{(b)}(n-y,y+1) - I_{1/b}^{(b)}(n-y+1,y) \right]$$

$$(4.32) \quad E\{N|\underline{p} = \frac{1}{k}\} = k \sum_{m=\{\frac{b(n+1)}{b+1}\}}^{n} mP_m(CS) + n \left[1 - k \sum_{m=\{\frac{b(n+1)}{b+1}\}}^{n} P_m(CS) \right]$$

and hence we need not use (4.27) or (4.28) to obtain $E\{N\}$ in this

case. For $b = 2$ and $n = 5$ we obtain from (4.31) and (4.32)

$$(4.33) \quad P_4(CS) = \frac{6}{81}, \quad P_5(CS) = \frac{14}{81}, \quad E\{N|\underline{p} = 1/3\} = \frac{4(18) + 5(42) + 5(21)}{81} = \frac{43}{9}.$$

This is a % saving of only 4.4% but this is a 'worst case' for % saved

since $P_{[1]} = P_{[2]}$ and we will see below (in (4.40)) that the limit

$(n \rightarrow \infty)$ of the % saved is zero in this case.

In Case ii) we note that $P_m(IS) = 0$ since we cannot get an

incorrect selection except through ties for first place. Hence, setting

$y = m-1$ in (4.26) since the binomial factor is zero otherwise, we have

$$(4.34) \quad P_m(S) = P_m(CS) = I_{1/b}^{(b)}(n-m+1,m) - I_{1/b}^{(b)}(n-m+2,m-1)$$

and again we don't have to use (4.27) or (4.28) to obtain $E\{N\}$. In

Case i) we find using (4.34) that the probability of a correct

selection without ties for first place is

$$(4.35) \quad \sum_{m=\{\frac{b(n+1)}{b+1}\}}^{n} P_m(CS) = I_{1/b}^{(b)}(1,n)$$

and this is given as a multiple of a Stirling number of the second kind

in (3.1). The sum of terms of the form $mP_m(S) = mP_m(CS)$ in (4.29)

also simplifies and we easily obtain for Case ii

$$(4.36) \quad E\{N|\underline{p} = 0\} = n - \sum_{m=\{\frac{b(n+1)}{b+1}\}}^{n-1} I_{1/b}^{(b)}(n-m+1, m)$$

$$= n - \sum_{j=2}^{[\frac{n+1}{b+1}]} I_{1/b}^{(b)}(j, n+1-j),$$

which shows clearly that there is always a positive saving.

The sum in (4.36) vanishes for $n < 2b + 1$ as we need at least 2 in every cell except one and then we get a saving of 1 observation if $n = 2b + 2$ and $N = 2b + 1$. Thus if $b = 2$ and $n = 5$ then $E\{N\}$ has the value $5 - \binom{4}{2}(\frac{1}{2})^4 = 5 - 3/8 = 4.625$, which is a saving of 7.5%. We see below (in (4.40)) that the limiting $(n \to \infty)$ percent saved is 33 1/3% for this configuration. For $b = 1$ the result in (4.36) yields $n - [\frac{n-1}{2}] = \{\frac{n+1}{2}\}$, which is clearly correct.

In general, the sum in (4.36) is easily calculated with the help of a table of I-values.

Suppose for example that $n = 30$ and we have $k = 4$ cells (so that $b = 3$). We want $E\{N|\underline{p} = 0\}$, i.e., for Case ii). Using (4.36) and Table A we obtain

$$(4.37) \quad E\{N|\underline{p} = 0\} = n - \sum_{j=2}^{7} I_{1/3}^{(3)}(j, 31 - j)$$

$$= 30 - (.99964 + .99615 + .97530 + .89291 + .66981$$

$$+ .28376) = 30 - 4.81757 = 25.18243,$$

which amounts to a 16% saving in sample size due to curtailment when $\underline{p} = 0$ and 3 cells each have probability 1/3. If we take $\delta^* = 1/3$ then the result in (4.37) is also $E\{N\}$ for the LF configuration and, using (4.5) and the extrapolated answer in Illustration 3 above, the PCS for the LF configuration is

(4.38) $P\{CS|LF\} = (\frac{4}{3})^{30} [\frac{1}{4} - \frac{1}{4} (.9992857)] = .99996.$

Of course in case i) we expect to get a very small saving; using
(4.31) and (4.32) we obtain for this illustration (k = 4, b = 3, n = 30)

(4.39) $P_{24}(CS) = .000285,$ $P_{25}(CS) = .001999,$ $P_{26}(CS) = .007280,$

$P_{27}(CS) = .018321,$ $P_{28}(CS) = .035830,$ $P_{29}(CS) = .057879,$

$P_{30}(CS) = .080559,$ $\sum_{m=24}^{30} P_m(CS) = .20215,$ $E\{N\} = 29.0986.$

Here a correct selection (CS) means selecting the tagged cell and,
of course, $P(CS) = .25$ if we combine the results with and without
ties at termination.

4.2.2. <u>Asymptotic Per-cent Saving due to Curtailment.</u>

We refer to $n - E\{N\}$ as the saving and to $(n-E\{N\})/n$ as the
fraction saved; the percent saved is of course the latter multiplied
by 100. Asymptotically (i.e., as $n \to \infty$) the minimum number m of
observations actually needed also approaches infinity. For large
values of n and m we have approximately (using the symbol $\approx$ for
this) $x + n - m \approx (m-x)/b$ with probability $\to 1$; here x denotes
the frequency of the best cell (i.e. the one with probability $\underline{p}$) and
$(m-x)/b$ denotes the common limiting value of the b other cell
frequencies. Putting $x = m\,\underline{p}$ we obtain

(4.40) $\text{Lim } \dfrac{n - E\{N\}}{n} = \text{Lim } \dfrac{n-m}{n} = 1 - \dfrac{b}{(b+1)(1-\underline{p})} = 1 - \dfrac{1}{k\underline{p}} = \dfrac{\delta}{1+\delta}\,;$

this is the asymptotic fraction saved for configurations with a common
$p \geq \underline{p}$ and $\delta = p - \underline{p}$; for the LF configuration we set $\delta = \delta^{*}$. It
is a maximum at $\underline{p} = 0$ (or $p = 1/b$) and $b = 1$, in which case
there is a 50% saving asymptotically. Although the amount of this

saving is generally not large, it should be noted that the fraction
saved does not vanish as n grows larger and, except for very special
situations, there is always a positive saving in E{N}. Hence the
effect of curtailment should not be overlooked in treating the fixed
sample size multinomial selection problem.

§ 4.3. The Problem of Two or More Different p-Values.

Let p_j denote the cell probability for the jth blue cell
$(j = 1,2,\ldots,b)$ and assume that $p_1 + p_2 + \ldots + p_b \leq 1$. For a total
of n observations on this multinomial the probability that the frequency
N_j in the jth cell is at least r is

$$(4.41) \quad I_{\underset{\sim}{p}}^{(b)}(r,n) = \frac{\Gamma(n+1)}{\Gamma^b(r)\Gamma(n+1-br)} \int_0^{P_1} \ldots \int_0^{P_b} (1 - \sum_1^b x_i)^{n-br} \prod_1^b x_i^{r-1} dx_i$$

where $\underset{\sim}{p}$ denotes the vector $(p_1,p_2,\ldots,p_b)$ and we assume that $n \geq br$
and $r \geq 1$. We avoid the more general form with different r-values for
each cell and with a second superscript j although they lead to straightforward generalizations.

The case of a common p in (4.41) has already been treated; we now
consider two different p-values, namely $p_1 = p_2 = \ldots = p_{b-1} = p$ (say) and
$p_b = p'$. For $s \geq 1$ the sth derivative of (4.41) with respect to p'
can be shown to be of the form

$$(4.42) \quad D_p^s \cdot \{I_p^{(b)}(r,n)\} = n^{[s]} \sum_{j=max(0,s-r)}^{Min(s-1,n-br)} (-1)^j \binom{s-1}{j} B_{p'}(n-s,r-s+j) I_{p/(1-p')}^{(b-1)}(r,n-r-j)$$

where $n^{[s]}$ is defined after (2.17) and we use this for small values of
s to expand (4.41) in a Taylor series about p' = p. The result is

$$(4.43) \quad I_{R}^{(b)}(r,n) = I_{p}^{(b)}(r,n) + \sum_{s=1}^{n-(b-1)r} \binom{n}{s}(p'-p)^s \left[\sum_{j}(-1)^j \binom{s-1}{j} B_p(n-s,r-s+j) \right.$$

$$\left. \cdot I_{p/(1-p)}^{(b-1)}(r,n-r-j) \right]$$

where the limits on j for the inside summation are given in (4.42).

In fact, the finite sum indicated in (4.43) already gives us the exact

result for this I-function. The first two or three terms in the summation

usually give a satisfactory approximation based on I-values taken from

Table A or B .

Illustration 1: Consider a multinomial with 4 blue cells having cell

probabilities .25, .25, .25 = p and .20 = p'. Based on $n = 5$ observations

what is the probability that all 4 blue cells are non-empty. Using (4.43)

we obtain

$$(4.44) \quad I_{(.25,.25,.25,.20)}^{(4)}(1,5) = I_{1/4}^{(4)}(1,5) - \frac{1}{4} B_{1/4}(4,0) I_{1/3}^{(3)}(1,4) - .025 B_{1/4}^{(3,0)} I_{1/3}^{(3)}(1,3)$$

$$= \frac{63}{320} = .196875,$$

where the numerical entries for the I-values are obtained from Table A

or E and the binomial $B_p(n,r)$ is zero for $r < 0$ and for $r > n$.

Since $n - (b-1)r = 2$ the result in (4.34) is exact.

Illustration 2: Consider the case $n = 15$, $b = 3$, and $r = 1$ with $p = .25$

and $p' = .20$. Using (4.43)

$$(4.45) \quad I_{(\frac{1}{4},\frac{1}{4},\frac{1}{5})}^{(3)}(1,15) = I_{1/4}^{(3)}(1,15) - \frac{15}{20} B_{1/4}(14,0) I_{1/3}^{(2)}(1,14)$$

$$- \frac{\binom{15}{2}}{400} B_{1/4}(13,0) I_{1/3}^{(2)}(1,13) - \frac{\binom{15}{3}}{8000} B_{1/4}(12,0) I_{1/3}^{(2)}(1,12)$$

$$+ \ldots \approx .9388,$$

where all the I-entries are obtained from Table B with the appropriate

arguments. The exact answer can be obtained from (4.41) by some "persistent"

integration and the value is

$$(4.46) \quad 1 - 2\left(\tfrac{3}{4}\right)^{15} + \left(\tfrac{1}{2}\right)^{15} - \left(\tfrac{4}{5}\right)^{15} + 2\left(\tfrac{11}{20}\right)^{15} - \left(\tfrac{3}{10}\right)^{15} = .938374 \, .$$

Here we would have to sum up to s = 13 (i.e., use 14 terms) to get an

exact result; the result in (4.45) is based on the 4 terms shown.

Another exact expression for $I_{\underset{\sim}{p}}^{(b)}(r,n)$ is given in (4.49) below.

It should also be noted that the general Taylor series method developed

in Section 2.3.2 can also be used if we expand both p and p' around

an average $\bar{p}$ preferably between these two and close to both of them. For the

present problem that expansion is more involved and the development in this

section is easier to apply.

§ 4.4. Use of the Tables to Compute Related Probabilities.

The tables of the I-function can be used to compute many other

multinomial-related probabilities. A complete discussion of this is not possible

without introducing other functions, e.g., we introduce $J_p^{(b,j)}(r,n)$ which

is the probability that j specific blue cells have frequency exactly r   and

the remaining b-j blue cells have frequency less than r when n balls are

independently distributed by a multinomial law with b blue cells having

common cell probability p ≤ 1/b .

1. We first note that the tables give $I_p^{(b,j)}(r,n)$ only for j = 0

and hence our first usage of the tables is the one already mentioned,

namely that we can use the reduction formula (2.11) to write $I_p^{(b,j)}(r,n)$

in terms of I-functions with j = 0.

2. Let $N_1, N_2, \ldots, N_j$ denote the (random) frequencies in j specific blue cells with common cell probability p and let

$$(4.47) \quad I_p^{(b,j)}(r,s,n) = P\{N_1 = r, N_2 = r, \ldots, N_j = r, \text{Min}(N_{j+1}, \ldots, N_b) \geq s \mid n, p\},$$

based on a total of n observations. By conditioning on the result for these j cells, we easily obtain

$$(4.48) \quad I_p^{(b,j)}(r,s,n) = [_{r,r,\ldots,r,n-jr}^{\qquad n}] \, p^{jr}(1-jp)^{n-jr} I_{\left(\frac{p}{1-jp}\right)}^{(b-j,0)}(s,n-jr)$$

and a similar result holds if the r-values are not all equal. For $j = 0$, we obtain a trivial identity in (4.48).

3. In another usage suppose that $b - 1$ of the b blue cells have cell probability p and one (say, the bth cell) has cell probability p'. Then the tables are not directly applicable but we can apply them with the

help of the following formula:

$$(4.49) \quad P\{\text{Min}(N_1, N_2, \ldots, N_{b-1}, N_b) \geq r \mid n, p, p'\}$$

$$= \sum_{\alpha=r}^{n-(b-1)r} P\{\text{Min } N_1, N_2, \ldots, N_{b-1}) \geq r, \, N_b = \alpha \mid n, p, p'\}$$

$$= \sum_{\alpha=r}^{n-(b-1)r} B_{p'}(n,\alpha) I_{\left(\frac{p}{1-p'}\right)}^{(b-1)}(r, n-\alpha) \, ,$$

where $B_p(n,x)$ denotes the usual binomial, $\binom{n}{x} p^x (1-p)^{n-x}$. In general, in (4.49) the ratio $p/(1-p')$ will not be the reciprocal of an integer, even though p and p' are, and hence interpolation in the I-table will usually be required.

4. In using the I-table to compute the J-function defined above we restrict our attention to the case $p \leq 1/(b-j)$ and employ the formula

$$(4.50) \quad J_p^{(b,j)}(r,n) = \sum_{\beta=j}^{b} (-1)^{\beta-j} \binom{b-j}{\beta-j} I_p^{(\beta,j)}(r,n) \, .$$

This is obtained by the principle of inclusion-exclusion; we omit the proof. The superscript j on the I-function can again be reduced to zero by a straightforward application of the reduction formula (2.11).

For $j = 0$ and $p = 1/b$ the left side of (4.50) denotes the probability that the maximum frequency in a homogeneous multinomial (based on n observations) is less than r, which is obviously equivalent to the cdf of the maximum frequency (with argument $r-1$). This cdf was tabulated by Steck [19] in an unpublished table. All of our techniques can also be applied to the study and computation of the J-function; however, the basic recurrence formula for the J-function is not as stable as the one above in (2.9) for the I-function.

5.   Another usage of our I-function tables is to compute means and variances of the minimum frequency $Min(n,b)$ in a multinomial based on n observations and b equi-probable cells, each with common probability $p \leq 1/b$. Since $Min(n,b)$ is a nonnegative random variable, we have for the mean and variance, respectively

$$(4.51) \qquad E\{Min(n,b)|p\} = \sum_{\alpha=1}^{[n/b]} I_p^{(b)}(\alpha,n) \ ,$$

$$(4.52) \qquad Var\{Min(n,b)|p\} = \sum_{\alpha=1}^{[n/b]} (2\alpha-1)I_p^{(b)}(\alpha,n) - (E\{Min(n,b)|p\})^2 \ .$$

Numerical values for both of these are given in Table D for $p = \frac{1}{b},\ldots,\frac{1}{b+9}$ with $b = 2(1)10$ and for $n = b(1)b + 54$. In addition to being of interest per se, these will also be useful in a study of the asymptotic behavior of $Min(n,b)$. The analogous quantities for the maximum frequency were computed by Steck in [19]. Another paper [6] contains some means and variances of the minimum frequency for $b = 2(1)10$, $n = 2(1)15$ and $p = 1/b$ to only 4 decimals; these overlap only slightly and are in agreement with our values.

Illustration 1:   In a homogeneous multinomial with $b = 10$ cells

how large an n-value is needed to make the expected minimum frequency at

least 1 (resp., at least 2). Using Table D the two answers are easily

seen to be $n = 36$ and $n = 52$, respectively. The corresponding

variances for the minimum frequency are .4653 and .7717 respectively;

these can be used to obtain a prediction interval for the minimum frequency.

6.   It has already been noted (see e.g. Jordan [9]) that for a homo-

geneous multinomial with b cells and n observations the probability

of getting no empty cells is

$$(4.53)\quad I_{1/b}^{(b)}(1,n) = \sum_{\alpha=0}^{b}(-1)^{\alpha}\binom{b}{\alpha}\left(\frac{b-\alpha}{b}\right)^{n} = \Delta^{b}\left(\frac{x}{b}\right)^{n}\bigg]_{x=0} = \frac{b!}{b^{n}}S_{n}^{(b)}\ ,$$

where Δ operates on x and $S_{n}^{(b)}$ are the Stirling numbers of the

second kind.

7.   Let the random variable $\nu_{b,r}$ denote the number of observations

required to get at least r observations in each blue cell (for the first

time). It can be shown that

$$(4.54)\qquad P\{\nu_{b,r} \leq n\} = I_{p}^{(b)}(r,n);$$

the probability "density" of $\nu_{b,r}$ is then

$$(4.55)\qquad P\{\nu_{b,r} = n\} = I_{p}^{(b)}(r,n) - I_{p}^{(b)}(r,n-1) = \Delta\,I_{p}^{(b)}(r,n-1)$$

and this is simply related to $D_{p}I_{p}^{(b)}(r,n)$ by (2.15). Thus our tables

can be used to find distribution values and properties of $\nu_{b,r}$ for the

sequential waiting problem (or coupon collecting problem).

Illustration 1:   Mice in a certain experiment can die of 5 different

causes with probability p_i for the ith cause of death. If the p_i

are all equal to $p = 1/5$ how many mice do we need to have probability

at least $P^{*} = .95$ that there will be at least r observations (i.e.,

deaths) in each of the five categories. (This information is useful to help the experimenter to set up a reasonable design before the experiment

is carried out.) From Table C for $r = 5$ and 10, respectively, and

for $b = 5$, and $p = 1/5$ we find that, $n = 55$ mice and 89 mice, res-

pectively, if randomization is not used.

This is equivalent to asking how large an n-value is needed so that

the probability of getting at least r in each cell at or before the nth

observation should be at least P^*. This application is given in (4.54)

and is usually called the sequential waiting problem.

8.   In a recent paper [17] on "Sparse and Crowded Cells in a Homogeneous

Multinomial", a sparse cell is defined as one with frequency at most u

while a crowded cell is one with frequency at least v. The I-function

is used to evaluate the probability distribution, the cdf., the moments

(and also the corresponding joint functions) for both the number S of

sparse blue cells and the number C of crowded blue cells based on n

observations. For instance, it is shown that for common cell probability

$p \leq 1/b$ in all the b blue cells

$$(4.56) \quad P\{S = s\} = \binom{b}{s} \sum_{i=0}^{s} (-1)^i \binom{s}{i} I_p^{(b-s+i)}(u+1,n) \ ,$$

$$(4.57) \quad P\{C = c\} = \binom{b}{c} \sum_{j=0}^{b-c} (-1)^j \binom{b-c}{j} I_p^{(c+j)}(v,n) \ .$$

Descending factorial moments of C and S are given by

$$(4.58) \quad E\{C^{[m]}\} = b^{[m]} I_p^{(m)}(v,n) \ ,$$

$$(4.59) \quad E\{S^{[m]}\} = b^{[m]} \sum_{\alpha=0}^{m} (-1)^\alpha \binom{m}{\alpha} I_p^{(\alpha)}(u+1,n) = b^{[m]} J_p^{(m)}(u+1,n)$$

and our tables are useful to find these values.

Illustration 1: In a homogeneous multinomial with $b = 10$ blue cells

and common cell probability $p \leq 1/b$, how many observations are needed

to raise the expected number of crowded cells to $C^* = 9$ (say), if a cell is defined as crowded when it has at least $v = 3$ observations in it (assume $p = .05$ or $.10$)? Using (4.58) with $m = 1$ we have to find the smallest n-value such that $I_p^{(1)}(3,n) \geq .9$ for $p = .05$ and for $p = .10$. By Table C the answers are $n = 105$ and 52, respectively.

9.   Let V denote the number of empty (i.e., vacant) cells in a multi-nomial with b blue cells having common probability $p \leq 1/b$ for each of these b cells. Let S (resp., C) denote the number of sparse (resp., crowded) cells defined by having a frequency at most u (resp., at least v) when the total number of observations is n. Find the first two moments (as well as the variance) of $V, S,$ and C.

We express V as $V_1 + V_2 + \ldots + V_b$ where V_i is 1 if the ith blue cell is vacant and zero otherwise. Then we easily obtain

(4.60) $EV = bEV_1 = b(1-p)^n,$

(4.61) $EV^2 = bEV_1^2 + b(b-1)EV_1V_2 = b(1-p)^n + b(b-1)(1-2p)^n,$

(4.62) $\sigma^2(V) = b(1-p)^n\left[1 - b(1-p)^n\right] + b(b-1)(1-2p)^n.$

Similarly for $C = C_1 + C_2 + \cdots + C_b$ we obtain

(4.63) $E(C) = bEC_1 = bI_p^{(1)}(v,n),$

(4.64) $E(C^2) = bE(C_1^2) + b(b-1)E(C_1C_2) = bI_p^{(1)}(v,n) + b(b-1)I_p^{(2)}(v,n),$

(4.65) $\sigma^2(C) = bI_p^{(1)}(v,n)\left[1 - bI_p^{(1)}(v,n)\right] + b(b-1)I_p^{(2)}(v,n).$

For $S = S_1 + \cdots + S_b$ where the S_i are similarly defined we use the symbol $J_p^{(b)}(r,n)$ defined in (4.50) and obtain

(4.66) $E(S) = bE(S_1) = bJ_p^{(1)}(u+1,n) = b\left[1 - I_p^{(1)}(u+1,n)\right] = bI_{1-p}^{(1)}(n-u,n),$

(4.67) $E(S^2) = bE(S_1^2) + b(b-1)E(S_1S_2) = bJ_p^{(1)}(u+1,n) + b(b-1)J_p^{(2)}(u+1,n)$

$= bI_{1-p}^{(1)}(n-u,n) + b(b-1)\left[1 - 2I_p^{(1)}(u+1,n) + I_p^{(2)}(u+1,n)\right],$

(4.68) $\sigma^2(S) = bI_{1-p}^{(1)}(n-u,n)\left[1 - bI_{1-p}^{(1)}(n-u,n)\right]$

$+ b(b-1)\left[1 - 2I_p^{(1)}(u+1,n) + I_p^{(2)}(u+1,n)\right].$

If we take $u = 0$ in the definition of S, then the results agree with those for V.

10. Continuing with 22, find the first two moments of $C + S$.

$$(4.69) \qquad E(C + S) = E(C) + E(S) = b\left[I_p^{(1)}(v,n) + I_{1-p}^{(1)}(n-u,n)\right]$$

$$(4.70) \qquad E\left\{(C+S)^2\right\} = b\left[E(C_1^2) + E(S_1^2)\right] + b(b-1)\left[E(C_1C_2) + E(S_1S_2)\right]$$

$$+ bE(C_1S_1) + 2b(b-1)E(C_1S_2).$$

The first four were obtained above, $E(C_1S_1) = 0$ and we need only consider the last expectation. We have

$$(4.71) \qquad E(S_1C_2) = P\{f_1 \leq u, \ f_2 \geq v\} = P\{f_2 \geq v\} - P\{f_1 \geq u + 1, \ f_2 \geq v\}$$

$$= I_p^{(1)}(v,n) - I_p^{(2)}(\{u+1,v\},n).$$

Using (4.55) to increase the first component of $\underline{r}$, we have, assuming $u + 1 \leq v$,

$$(4.72) \qquad I_p^{(2)}(\{u+1,v\},n) = I_p^{(2)}(v,n) + \sum_{\alpha=1}^{v-u-1} B_p(n,v-\alpha)I_{p/(1-p)}^{(1)}(v,n-v+\alpha).$$

Hence the final result for $E\{(C + S)^2\}$ in (4.70)

$$(4.73) \qquad E\{(C + S)^2\} = b\left[I_p^{(1)}(v,n) + I_{1-p}^{(1)}(n-u,n)\right] + b(b-1)\left[I_p^{(2)}(v,n) +\right.$$

$$1 - 2I_p^{(1)}(u+1,n) + I_p^{(2)}(u+1,n)\right] + 2b(b-1)\left[I_p^{(1)}(v,n) - I_p^{(2)}(v,n)\right.$$

$$\left.- \sum_{\alpha=1}^{v-u-1} B_p(n,v-\alpha)I_{p/(1-p)}^{(1)}(v,n-v+\alpha)\right].$$

The variance of $C + S$ is now obtainable from (4.69) and (4.73). Without the use of the I-functions this would be difficult. Statistics like $C + S$ could be used with great ease to test homogeneity in a multinomial. Equation (4.69) shows that the I-function table is useful to calculate the properties of such statistics. These results should be compared with related results in reference [17].

11. Prove the identity for all p with $p \le 1/b$ and all integers b, n

$$(4.74) \quad \sum_{j=0}^{b} \left\{ \sum_{\alpha=0}^{b-j} (-1)^{\alpha} \binom{b-j}{\alpha} [1 - (j + \alpha)p]^n \right\} = 1.$$

The probability (based on n observations) that j specific blue cells each have <u>exactly</u> r observations and the remaining b-j blue cells each have <u>at least</u> s observations is

$$(4.75) \quad I_p^{(b,j)}(r,s,n) = \frac{n! \, (p^r/r)^j}{\Gamma^j(r) \, \Gamma^{b-j}(s) \, [n-jr-(b-j)s]!} \cdot$$

$$\int_0^p \cdots \int_0^p (1-jp- \sum_{j+1}^{b} x_\alpha)^{n-jr-(b-j)s} \prod_{j+1}^{b} x_\alpha^{s-1} dx_\alpha \cdot$$

where for $r = 0$ and any j we replace $r\Gamma(r)$ by $\Gamma(r+1) = 1$ and the result then also holds for $r = 0$. For the special case $s = 1$ and $r = 0$ the integration of (4.75) is straightforward and leads to the sum inside the braces in (4.74). Thus the inside sum is the probability of <u>exactly</u> j empty cells (among the b blue cells) when n observations are taken and the b blue cells have common probability $p \le 1/b$. Since these events for $j = 0,1,\ldots,b$ are mutually exclusive and exhaustive they must sum to unity.

12. As a generalization of our basic multinomial model, we assume that $b_i \ge 1$ blue cells each have cell probability $p_i > 0$ ($i = 1,2,\ldots,t$), so that

$$(4.76) \quad \sum_{i=1}^{t} b_i = b \quad \text{and} \quad Q = 1 - \sum_{i=1}^{t} b_i p_i \ge 0$$

and if $Q > 0$ then we have one non-blue cell with probability Q. We use $\underset{\sim}{b}$ for $(b_1, b_2, \ldots, b_t)$, $\underset{\sim}{p}$ for $(p_1, p_2, \ldots, p_t)$ and define the t-vectors $\underset{\sim}{j} = (j_1, j_2, \ldots, j_t)$, $\underset{\sim}{r} = (r_1, r_2, \ldots, r_t)$ and $\underset{\sim}{s} = (s_1, s_2, \ldots, s_t)$ with integer components such that $0 \le j_i \le b_i$, $0 \le R = \sum_{i=1}^{t} j_i r_i$, $0 \le S = \sum_{i=1}^{t} (b_i - j_i)s_i$ and $R + S \le n$. For convenience we also set

$$(4.77) \qquad P = \sum_{i=1}^{t} j_i p_i, \quad X = \sum_{i=1}^{t} \sum_{\alpha=j_i+1}^{b_i} x_{i\alpha} \quad \text{and} \quad \oint_0^{p_i} \text{ denotes a } (b_i - j_i)\text{-fold}$$

integral.

The probability (based on n observations) that among the cells of type i there are j_i specified blue cells with frequency <u>exactly</u> r_i and that the remaining $b_i - j_i$ blue cells have frequency <u>at least</u> s_i $(i = 1,2,\ldots,t)$ is given by

$$(4.78) \qquad I_{\underset{\sim}{p}}^{(b,j)} (\underset{\sim}{r}, \underset{\sim}{s}, \underset{\sim}{n}) = \frac{n! \prod_{i=1}^{t} (p_i^{r_i}/r_i)^{j_i}}{(n-R-S)! \prod_{i=1}^{t} \{\Gamma^{j_i}(r_i)\Gamma^{b_i-j_i}(s_i)\}} \cdot$$

$$\oint_0^{p_1} \oint_0^{p_2} \cdots \oint_0^{p_t} (1-P-X)^{n-R-S} \prod_{i=1}^{t} \left\{ \prod_{\alpha=j_i+1}^{b_i} x_{i\alpha}^{s_i-1} \, dx_{i\alpha} \right\}.$$

This is a direct generalization of the integral in (4.75) and as in that case the more interesting application is when $r_i \leq s_i$ $(i = 1,2,\ldots,t)$. For the special case $r_i = 0$, $s_i = 1$ $(i = 1,2,\ldots,t)$ the integral in (4.78) represents the probability that j_i specified cells of the b_i blue cells of type i are vacant $(i = 1,2,\ldots,t)$ and when multiplied by $\prod_{i=1}^{t} \binom{b_i}{j_i}$ it will give the probability of having exactly j_i vacant cells of type i $(i = 1,2,\ldots,t)$. As in (4.75), for $r_i = 0$ we set $r_i \Gamma(r_i) = \Gamma(r_i + 1) = 1$ in (4.78); then $R = 0$ and $S = \sum_{i=1}^{t} (b_i - j_i)$ is the dimension of the integral. A straightforward integration in (4.78) yields for this case

$$(4.79) \qquad I_{\underset{\sim}{p}}^{(b,j)} (\underset{\sim}{0}, \underset{\sim}{1}, n) = \Sigma(-1)^{\gamma} \left[\prod_{i=1}^{t} \binom{b_i-j_i}{\gamma_i} \right] (1 - P - \sum_{i=1}^{t} \gamma_i p_i)^n,$$

where the outside summation is over vectors $(\gamma_1, \gamma_2, \ldots, \gamma_t)$ such that $\gamma_i \geq 0$ and $0 \leq \gamma \equiv \sum_{i=1}^{t} \gamma_i \leq b - J$, and $J = \sum_{i=1}^{t} j_i$; note that $\gamma_i + j_i \leq b_i$ since the combinatorial coefficient is zero otherwise. Hence the desired probability Z of exactly j_i vacant cells of type i $(i = 1,2,\ldots,t)$ is

$$(4.80) \qquad Z = \Sigma \, (-1)^{\gamma} \left\{ \prod_{i=1}^{t} \left[\gamma_i, j_i^{\, b_i}, b_i - \gamma_i - j_i \right] \right\} \, (1 - \sum_{i=1}^{t} (j_i + \gamma_i) p_i)^n$$

where the square bracket denotes the usual multinomial coefficient $b_i! / [\gamma_i! j_i! (b_i - \gamma_i - j_i)!]$ and the outside sum is as in (4.79).

A result equivalent to (4.80) was recently obtained by Johnson and Kotz [7] by a different method; their emphasis is on the occupied cells rather than the vacant cells, but the results are immediately translatable from either form to the other.

Now suppose a cell in the ith category is crowded if its frequency is at least v_i ($i = 1, 2, \ldots, t$). Let $C_i = C_{i,1} + C_{i,2} + \ldots + C_{i,b_i}$ denote the number of crowded cells of type i as in problem 9 above. Then we obtain for $i = 1, 2, \ldots, t$

$$(4.81) \quad E\{C_i\} = b_i E\{C_{i,1}\} = b_i \, I_{p_i}^{(1)} (v_i, n) \, ,$$

$$(4.82) \quad E\{C_i^2\} = b_i E\{C_{i,1}^2\} + b_i (b_i - 1) E\{C_{i,1} C_{i,2}\}$$

$$= b_i I_{p_i}^{(1)} (v_i, n) + b_i (b_i - 1) I_{p_i}^{(2)} (v_i, n) \, .$$

$$(4.83) \quad E\{C_i C_j\} = \sum_{\beta=1}^{b_j} \sum_{\alpha=1}^{b_i} E\{C_{i,\alpha} C_{j,\beta}\} = b_i b_j I_{p_i, p_j}^{(2)} (\{v_i, v_j\}, n) \, ,$$

where

$$(4.84) \quad I_{p_i, p_j}^{(2)} (\{v_i, v_j\}, n) = \frac{n!}{\Gamma(v_i) \Gamma(v_j) (n - v_i - v_j)!} \int_0^{p_i} \int_0^{p_j} (1-x-y)^{n-v_i-v_j}$$

$$x^{v_i-1} y^{v_j-1} dxdy$$

(For the special case $v_i = v_j = r$ (say) the Taylor expansion results in Section 4.3 can be used to evaluate (4.84). It follows from above that

(4.85) $\sigma^2(C_i) = b_i I_{p_i}^{(1)}(v_i,n) [1 - b_i I_{p_i}^{(1)}(v_i,n)] + b_i(b_i - 1) I_{p_i}^{(2)}(v_i,n)$,

(4.86) $\sigma(C_i,C_j) = b_i b_j [I_{p_i,p_j}^{(2)}(\{v_i,v_j\},n) - I_{p_i}^{(1)}(v_i,n) I_{p_j}^{(1)}(v_j,n)]$.

A similar analysis holds for the sparse cells of type i and as a special case for the vacant cells of type i. If we regard the t vectors

(4.87) $\underset{\sim}{C}_i = (C_{i,1}, C_{i,2}, \ldots, C_{i,b_i})$ $(i = 1,2,\ldots,t)$

as a collection of vector random variables then the above results give us the mean vectors and the variance-covariance matrix. In this sense we can regard this model as a multivariate form of the multinomial.

13. In a multinomial b blue cells having common cell probability $p \leq 1/b$ a crowded cell is defined in [17] as one with frequency at least v. The probability (based on n observations) that the number C of crowded cells is exactly c was given below in (4.57).

The corresponding result for the model in problem 12 above with b_i cells of type i (i.e., having cell probability p_i) and using v_i to define crowdedness in the ith cell $(i = 1,2,\ldots,t)$ will now be obtained. Let $\underset{\sim}{\alpha} = (\alpha_1,\ldots,\alpha_t)$ be a vector of t integers $(\alpha_i \geq 0)$ whose sum is α and similarly let $\underset{\sim}{c} = (c_1,\ldots,c_t)$ be a vector of t integers $(c_i \geq 0)$ whose sum is c. Let $p(\underset{\sim}{\alpha}+\underset{\sim}{c})$ (resp, $v(\underset{\sim}{\alpha}+\underset{\sim}{c})$) denote a vector with $\alpha+c$ components in which the ith component of $\underset{\sim}{\alpha}+\underset{\sim}{c}$ indicates the number of times p_i (resp, v_i) is repeated; note that we can assume that $0 \leq \alpha_i + c_i \leq b_i$ $(i=1,2,\ldots,t)$. Let $\Sigma_{\underset{\sim}{\alpha}|\alpha}$ denote a sum over all ordered t-tuples $\underset{\sim}{\alpha}$ with $\alpha_i \geq 0$ and fixed sum α, i.e., over compositions of α. Let $\underset{\sim}{C} = (C_1,C_2,\ldots,C_t)$ denote the random vector in which C_i is the number of crowded cells of type i. Then

$$(4.88) \quad P\{\underline{C} = \underline{c}\} = \{ \prod_{i=1}^{t} \binom{b_i}{c_i} \} \sum_{\alpha=0}^{b-c} (-1)^{\alpha} \sum_{\underline{\alpha}|\alpha} \{ \prod_{i=1}^{t} \binom{b_i - c_i}{\alpha_i} \} I(v(\underline{\alpha}+\underline{c}),n) \frac{(\alpha+c)}{p(\underline{\alpha}+\underline{c})}$$

$$= \sum_{\alpha=0}^{b-c} (-1)^{\alpha} \sum_{\underline{\alpha}|\alpha} \{ \prod_{i=1}^{t} [\begin{smallmatrix} b_i \\ \alpha_i, c_i, b_i - \alpha_i - c_i \end{smallmatrix}] \} I(v(\underline{\alpha}+\underline{c}),n) \frac{(\alpha+c)}{p(\underline{\alpha}+\underline{c})} .$$

Similar generalizations can be obtained for the results in [17] on $P\{C \leq c\}$, $P\{S = s\}$, $P\{S \leq s\}$, $P\{V = v\}$ and $P\{V \leq v\}$ where S (resp., V) is the number of sparse (resp., vacant) cells. In each case the I-function is found to be useful.

14. The distribution of the maximum frequency in a homogeneous multinomial with b blue cells having common cell probability $p \leq 1/k$ (i.e. the J-function) can be obtained by inclusion-exclusion methods from our table of the I-function; this was recently used in a paper [18] on Bonferroni bounds of the same degree. (The same idea can also be used in reverse depending on which table is available, i.e. we could use tables of the J-function, if available, to compute I-values or bounds on I-values, but the computation becomes tedious for large values of b.)

Letting Max (b,n) denote the maximum frequency among the b blue cells based on n observations, we have

$$(4.89) \qquad P\{Max(b,n) \geq r\} = \sum_{\alpha=1}^{b} (-1)^{\alpha-1} \binom{b}{\alpha} I_p^{(\alpha)}(r,n) = 1 - J_p^{(b)}(r,n)$$

where in (4.89) we need only sum up to the integer part of n/r since $I = 0$ otherwise.

Illustration 1: Suppose $b = 3$, $p = 1/5$, $r = 2$ and $n = 15$. Then by (4.89) we obtain from Table B

$$(4.90) \qquad P\{Max(3,15) \geq 2\} = 3(.832874 - .681891) + .546491 = .99944.$$

15. An urn contains bM objects of b distinct types and there are exactly M
of each type. We now draw a random sample of size n without replacement and
ask for the probability that our sample contains at least r balls of each type,
where $n \geq br$. This is the hypergeometric analogue of our multinomial problem and
it often arises in practical applications. Thus a farmer may have six varieties
of seeds all in the same seed box. Without sorting the seeds, how many does
he have to plant to have a preassigned probability P^* that he has planted at
least r of each, if the six varieties appear in the seed box with given frequencies
(or proportions).

Our tables do not answer this problem, but it is of interest to see that
as $M \to \infty$, the answers approach the answers given by our table, so that we can
give an approximate answer to such questions. In fact, our answers give conservative bounds in the sense that the tabled I value is a lower bound and hence
the n value (called N in the table) is an upper bound on the sample size needed
to reach a specified P^*.

We can also relate the sequential random variable for this hypergeometric
model with that of our multinomial model. We assert that the percentile P^* of
the distribution of the number of observations needed to get at least r of
each variety (for the first time) approaches the value of n in Table C corresponding
to the same P^*; here we assume a common large M and the same b and r.

Note that M gets used only indirectly by assuming equal M values and hence
using $p = 1/b$ and taking the limit as $M \to \infty$.

To see this convergence numerically, we give a few values for $b = 3$, $r = 1$,
$n = 10$, and $M = 50, 100, 500, 1000, \infty$.

Convergence of Probabilities as $M \to \infty$

M	P{No Empty Cells|b=3, r=1, n=10}	Sample Size n Needed to Reach P^*= .95
50	.9556	10
100	.9519	10
500	.9488	11
1000	.9484	11
∞	.9480	11

Percentiles for
$\nu_{b,r}$ (based on M in each category)

M	90th Percentile	95th Percentile
5	6.37	7.17
10	7.28	8.50
∞	8.43	10.11

From Table C for
b=3, r=1, p=1/3, P^*= .90, .95.

16. Another related multinomial problem is to find the smallest fixed sample size n such that for all configurations of p-values with range (i.e., $p_{[k]} - p_{[1]}) \le \delta^*$ the probability that each of the k cells contains r or more observations is at least P^*. Here we are treating all the cells as blue cells.

In the multinomial ranking problem of Section 4.1 the condition was $p_{[2]} - p_{[1]} \ge \delta^*$, so that none of our tables were constructed to solve the present problem, which we will call the 'range problem.' Nevertheless, we conjecture that the configuration in (4.1) with b = k-1 is also least favorable for the range problem. Hence we can approximate the result by using Table C appropriately. As in Section 4.1 we assume that $\delta* < 1/b$.

Illustration 1: How many observations n are needed to get 5 or more observations in each of k = 4 cells with probability at least $P^* = .90$, if the range of the p_i-values is at most $\delta^* = .2$. The least favorable configuration, using (4.1), is conjectured to be $p_{[1]} = .1 = \underline{p}$ (say) and $p_{[2]} = p_{[3]} = p_{[4]}$ $= .3 = p$ (say). If this is least favorable, then we can approximate the result by combining the three cells with probability .3 into a single cell with probability .9 and disregard the condition on these three cells. Then Table C with $b = 1$, $p = 1/10$, $r = 5$ and $P^* = .90$ shows that 77.9 (or the next integer $n = 78$) is a first approximation to the required number n.

A more general Taylor series needed to approximate the probability attained in the above configuration will be developed in a separate paper; since $p > 1/b = .25$, we cannot utilize (4.43) in its present form for this purpose. However, for the least favorable configuration we can give a simple argument to check the answer $n = 78$. Letting F_i denote the frequency in the cell with probability $p_{[i]}$ $(i = 1,2,3,4)$, we have

$$(4.91) \quad 1 - P\left\{\text{Min}(F_1,F_2,F_3,F_4) \geq 5\right\} = P\left\{\text{Min}(F_1,F_2,F_3,F_4) < 5\right\}$$

$$= P\left\{F_1 < 5\right\} + P\left\{\text{Min}(F_2,F_3,F_4) < 5 \leq F_1\right\}$$

$$\leq P\left\{F_1 < 5\right\} + \text{Min}\left[P\left\{\text{Min}(F_2,F_3,F_4) < 5\right\}, P\left\{\text{Min}(F_2,F_3,F_4) \leq F_1\right\}\right].$$

Hence we can use either Table C (with $b = 3$, $r = 5$, p between 1/3 and 1/4) or Table F (with $k = 4$, $\delta^* = .20$ and $P^* > .99$) to infer that the second term above is zero to at least 2 decimal places; by Table C it is less than .0001. From Table B (with $b = 1$, $p = .1$, $r = 5$) we have for $n \geq 78$ (and not for any smaller n)

$$(4.92) \quad P\left\{F_1 \geq 5\right\} = .9006 \Rightarrow P\left\{F_1 < 5\right\} < .0994.$$

It follows from (4.91) that for $n \geq 78$ (and not for any smaller n).

$$(4.93) \quad P\left\{\text{Min}(F_1,F_2,F_3,F_4) < 5\right\} < .0994 + .0001 < .10 .$$

Hence for $n \geq 78$ (and not for any smaller n)

$$(4.94) \quad P\left\{\text{Min}(F_1, F_2, F_3, F_4) \geq 5\right\} \geq .90 .$$

Thus $n = 78$ is the required value. Note that this result cannot be directly obtained from Table F, which was set up for a different goal. If we used only Table F above then the answer would have been $n = 78$ or 79.

17. Some fairly involved multinomial probabilities become quite easy when computed with the help of I-functions. To illustrate this we extend the I-function to include vector arguments $\underset{\sim}{r}$ for the argument r. Here $\underset{\sim}{r} = (r_1, r_2, \ldots, r_b)$ and for simplicity we keep a common cell probability $p \leq 1/b$. The probability interpretation for $I_p^{(b,j)}(\underset{\sim}{r}, n)$ is that the αth cell should have frequency <u>exactly</u> $r_\alpha (\alpha = 1, 2, \ldots j)$ and the βth cell should have frequency <u>at least</u> $r_\beta (\beta = j+1, j+2, \ldots, b)$. The form of the resulting integral is given in (4.78) by setting $t = b$, $b_i = 1$ and $p_i = p$ for all i $(1 \leq i \leq b)$, $j_i = 1$ for $1 \leq i \leq j$ and 0 for $j + 1 \leq i \leq b$; we also have to equate r_i with r_α for $1 \leq i \leq j$ and s_i with r_β for $j + 1 \leq i \leq b$. If the r-values are not widely dispersed (see illustration below) then the following recurrence relation will be useful, and for a large class of n-values our 'forward' tables A and B can be used.

For $r_{j+1} > 1$ we integrate by parts with respect to x_{j+1} in (2.7) and obtain for any p $(0 < p \leq 1/b)$, any j $(0 \leq j < b)$ and any $b \geq 1$

$$(4.95) \qquad I_p^{(b,j)}(\underset{\sim}{r}, n) = I_p^{(b,j)}(\underset{\sim}{r'}, n) - \frac{r_{j+1}}{(n+1)p} I_p^{(b,j+1)}(\underset{\sim}{r}, n+1)$$

$$= I_p^{(b,j)}(\underset{\sim}{r'}, n) - B_p(n, r_{j+1} - 1) I_{p/(1-p)}^{(b-1,j)}(\underset{\sim}{r''}, n+1-r_{j+1})$$

where $\underset{\sim}{r'}$ reduces the component r_{j+1} of $\underset{\sim}{r}$ by one and $\underset{\sim}{r''}$ eliminates r_{j+1} from $\underset{\sim}{r}$. If $\underset{\sim}{r}$ has one component larger than all the others by exactly one, then we can use (4.95) as is, but if the difference is more than one or if there are further differences the result (4.85) is easy to extend. An example will illustrate the method.

Illustration 1: With 30 tosses of a 4-sided fair die (marked 1, 2, 3 and 4), what is the probability of observing each even number at least five times and observing the face marked 3 at least 7 times? Iterating (4.95) twice with $j = 0$, $p = 1/4$, $n = 30$, $b = 3$ and $\underset{\sim}{r} = (r_2, r_3, r_4) = \{5, 7, 5\}$, we obtain

$$(4.96) \quad I_{1/4}^{(3)}(\{5, 7, 5\}, 30) =$$

$$= I_{1/4}^{(3)}(5, 30) - B_{1/4}(30, 5)I_{1/3}^{(2)}(5, 25) - B_{1/4}(30, 6)I_{1/3}^{(2)}(5, 24)$$

$$= .7157269 - (.1047285)(.9077400) - (.1454562)(.8815946)$$

$$= .492427.$$

If we make use of the fact that for $r_{j+1} = 1$

$$(4.97) \quad I_p^{(b-1, j)}(\underset{\sim}{r}'', n) = I_p^{(b, j)}(\underset{\sim}{r}', n)$$

then (4.85) also holds for $r_{j+1} = 1$, as can be seen by direct integration in (2.7).

If we integrate by parts in (2.7) in the other direction the result is

$$(4.98) \quad I_p^{(b, j)}(\underset{\sim}{r}, n) = I_p^{(b, j)}(\underset{\sim}{r}''', n) + I_p^{(b, j+1)}(\underset{\sim}{r}, n)$$

$$= I_p^{(b, j)}(\underset{\sim}{r}''', n) + B_p(n, r_{j+1})I_{p/(1-p)}^{(b-1, j)}(\underset{\sim}{r}'', n-r_{j+1})$$

where $\underset{\sim}{r}'''$ adds one to the component r_{j+1} of $\underset{\sim}{r}$ and $\underset{\sim}{r}''$ has already been defined above. This is useful when we wish to increase one or two components of $\underset{\sim}{r}$.

Illustration 2: With 55 tosses of a 6-sided fair die, what is the probability of observing the even numbers at least 5 times and each of the faces marked three and five at least 4 times? Using an iteration of (4.98) which is slightly different from that used above where the iteration applied to the same cell, we obtain (setting $\underset{\sim}{r} = \{r_2, r_3, r_4, r_5, r_6\} = \{5,4,5,4,5\}$)

(4.99) $I_{1/6}^{(5)}(\underset{\sim}{r}, 55) =$

$= I_{1/6}^{(5)}(\{5,5,5,4,5\}, 55) + B_{1/6}(55,4)I_{1/5}^{(4)}(\{5,5,4,5\}, 51)$

$= I_{1/6}^{(5)}(5,55) + B_{1/6}(55,4)\{2I_{1/5}^{(4)}(5,51) + B_{1/5}(51,4)I_{1/4}^{(3)}(5,47)\}$

$= .8228552 + (.0240977)\{2(.9366391) + (.0111459)(.9881654)\}$

$= .868262.$

18.   If five ordinary fair dice are used in each of t tosses, what is the probability of seeing each of the six faces (not necessarily on the same toss) at least three times? Here the answer is simply $I_{1/6}^{(6)}(3,5t)$, which can be found in Table A.

In the above problem how large should t be in order to have probability at least $P^* = .90$ of observing this event? Here we use Table C with $b = 6$, $p = 1/6$, $r = 3$ and $P^* = .90$, take the smallest multiple of 5 that is greater than the tabulated value and divide it by 5. Since the table gives 43.46, the answer is $t = 9$. If we then use $n = 45$ and look for the forward probability in Table A we find that it is .9195 for $n = 45$ and .8403 for $n = 40$; this checks the result $t = 9$.

19.   In throwing 13 ordinary fair dice simultaneously, a success on a single trial requires that we obtain each even face at least twice. What is the probability P of at least 5 successes in 40 such trials? To solve this problem we first find the probability θ of success on a single trial; this is $\theta = I_{1/6}^{(3)}(2,13)$.

Since this does not appear in Table B we will compute it by using (2.12) with five terms, each requiring Table A. We have from Table A

$$(4.100) \qquad \theta = I_{1/6}^{(3)}(2,13) = \left(\tfrac{1}{2}\right)^{13} \sum_{\alpha=6}^{13} \binom{13}{\alpha} I_{1/3}^{(3)}(2,\alpha) = .2406787.$$

Hence the required probability P is now given by $I_{\theta}^{(1)}(5,40)$. Since θ is not a caption in Table B we use the Taylor series in (2.20) with $p = \theta$ and $p_0 = \tfrac{1}{4}$. Four terms suffice to get six decimal accuracy and we then obtain

$$(4.101) \quad I_{\theta}^{(1)}(5,40) = I_{1/4}^{(1)}(5,40) + (4\theta-1)\binom{40}{1}\Delta I_{1/4}^{(1)}(5,39) + (4\theta-1)^2\binom{40}{2}\Delta^2 I_{1/4}^{(1)}(5,38)$$

$$+ (4\theta-1)^3\binom{40}{3}\Delta^3 I_{1/4}^{(1)}(5,37)$$

$$= .9839578 - .0050768 - .0007257 - .0000598$$

$$= .978096.$$

20. In a toss of 5 ordinary fair dice we use poker terminology and poker ranking (two pair < three of a kind < straight < full house < four of a kind < five of a kind). What is the probability P of getting 2 pair or better in a single toss?

The solution in terms of I-functions is given by

$$(4.102) \quad P = \binom{6}{2}I_{1/6}^{(2)}(2,5) + 2I_{1/6}^{(5)}(1,5) + \binom{6}{1}I_{1/6}^{(1)}(3,5) - \binom{6}{1}\binom{5}{1}I_{1/6}^{(2)}(\{3,2\},5).$$

Applying (4.88) to the last term, this becomes

$$(4.103) \quad P = 15I_{1/6}^{(2)}(2,5) + 2I_{1/6}^{(5)}(1,5) + 6I_{1/6}^{(1)}(3,5) - 30B_{1/6}(5,2)I_{1/5}^{(1)}(3;3).$$

Using (2.9), (2.10) and (2.11) (or using (2.12)), we now obtain

$$(4.104) \quad P = \left(\tfrac{1}{6}\right)^5 \{2100 + 240 + 1656 - 300\} = \frac{3696}{6^5} = \frac{77}{162}$$

and this is .4753 to four decimal places. Other elementary methods can be applied here and, of course, they do yield the same result.

21. How many natural numbers less than 10^9 have the odd digits (i.e., the digits 1, 3, 5, 7, 9) each appearing at least once? Using (2.12) with $p = .1$, $b = 5$, $r = 1$, $n = 9$ and Table A we obtain

$$(4.105) \quad I_{1/10}^{(5)}(1,9) = \frac{1}{2^9} \sum_{\alpha=5}^{9} \binom{9}{\alpha} I_{1/5}^{(5)}(1,\alpha) = .049974120.$$

This is the <u>probability</u> that each odd digit will be present at least once if the nine digits represent nine observations in a multinomial with ten cells, each having probability 1/10. Hence, multiplying by 10^9 to get the total number of "successful cases", we obtain $\theta = 49,974,120$. Another method is to let A_j denote the event that $2j-1$ does <u>not</u> appear in the 9 digit number, and compute the probability of the complement of $A_1 \cup A_2 \cup A_3 \cup A_4 \cup A_5$. Using inclusion-exclusion we obtain

$$(4.106) \qquad \theta = \sum_{\alpha=0}^{5} (-1)^\alpha \binom{5}{\alpha} (10 - \alpha)^9 = 49,974,120,$$

as was obtained from (4.105).

22. How many natural numbers less than 10^9 and consisting of odd digits only have each odd digit appearing at least once? How many natural numbers less than 10^{13} have each of the ten possible digits $(0,1,\ldots,9)$ appearing at least once? How many natural numbers less than 10^{25} have each of the 10 digits appearing at least twice?

Using (3.1) with $j = 0$ in the form

$$(4.107) \qquad\qquad b^n I_{1/b}^{(b)}(r,n) = b! S_{n,r}^{(b)}$$

and Table E we obtain respectively for the three above answers

$5!(6,951) = 834,120$ by taking $r = 1$, $b = 5$, $n = 9$,

$10!(39,325) = 142,702,560,000$ by taking $r = 1$, $b = 10$, $n = 13$,

$10!(22,298,773,748,501,250)$ by taking $r = 2$, $b = 10$, $n = 25$.

We note that Table E is useful in many interesting combinatorial problems.

23. In a 20-digit natural number (≥ 0) formed by selecting each digit at random from the set $\{0,1,\ldots,9\}$, let F (resp., f) denote the maximum

(resp., minimum) of the frequencies of the ten possible digits. What are
the values of $E\{f\}$, $E\{F\}$ and $E\{F-f\}$?

By Table D we obtain directly for $b = 10$, $p = 1/10$ and $n = 20$

$$(4.108) \qquad\qquad E\{f\} = .2147611,$$

the calculation of which is based on (4.51) Using the definition of
$J_p^{(b)}(r,n)$ with $j = 0$ in (4.50) and at the beginning of Section 4.4, we
have, as in the derivation of (4.51),

$$(4.109) \quad E\{F \mid b,n,p\} = \sum_{\alpha=1}^{n} P\{F \geq \alpha\}$$

$$= n - \sum_{\alpha=1}^{n} J_p^{(b)}(\alpha,n) = \sum_{\beta=1}^{b} (-1)^{\beta-1} \binom{b}{\beta} \sum_{\alpha=1}^{[n/\beta]} I_p^{(\beta)}(\alpha,n)$$

$$= \sum_{\beta=1}^{b} (-1)^{\beta-1} \binom{b}{\beta} E\{\text{Min freq} \mid \beta,n,p\}.$$

For $b = 10$, $n = 20$ and $p = 1/10$ this gives

$$(4.110) \quad E\{F\} = \binom{10}{1}(2) - \binom{10}{2}(1.2228634) + \binom{10}{3}(.8841388) - \binom{10}{4}(.6837416)$$
$$+ \binom{10}{5}(.5479205) - \binom{10}{6}(.4484905) + \binom{10}{7}(.3718237) - \binom{10}{8}(.3101420)$$
$$+ \binom{10}{9}(.2587159) - \binom{10}{10}(.2147611) = 4.4098555$$

Hence from (4.108) and (4.110)

$$(4.111) \qquad E\{F-f\} = E\{F\} - E\{f\} = 4.1950944.$$

24. In the fixed sample size aspect of Problem 15 suppose we add the
complication that 90% of our seeds germinate regardless of the variety.
For the multinomial case (i.e., $M = \infty$) with $b = 3$ varieties and requiring
at least $r = 1$ of each, what is the sample size n needed to reach
$p^* = .95$ of satisfying the requirement?

Since our previous answer without the condition was $n = 11$ we now
expect the answer to be about $11/.9$ or about 12.2, so that we can try
$n = 12$ and 13. For $n = 12$ we obtain

(4.112) $P\{No\ Empty\ Cells\} = \sum\limits_{\alpha=3}^{12} B_{.9}(12,\alpha)I_{1/3}^{(3)}(1,\alpha)$

$$= (.2824295)(.9768836) + \cdots + (.0000002)(.2222222)$$

$$= .958526.$$

Since $n = 11$ gives something less than .95 by the above method, the required answer is $n = 12$. We could also have used Table C and calculated $10.11/.9 = 11.23$, and the correct answer will usually be the smallest integer greater than this quantity.

25. An icosahedral (or 20-sided) die is marked j on each of two sides for $j = 0,1,\ldots,9$. If five such balanced dice are tossed, we denote the outcomes:

(a)   exactly 1 pair   by $(2^1 1^3)$,

(b)   exactly 2 pair   by $(2^2 1^1)$,

(c)   exactly three of a kind   by $(3^1 1^2)$,

(d)   full house   by $(3^1 2^1)$,

(e)   straight   by (S),

(f)   nothing   by (N), etc.

Arrange these events in order of increasing probability.

In terms of I-functions and using (4.91) and (4.94) we can write

(4.113) $P(2^1 1^3) = \binom{10}{1}\binom{9}{3}I_{1/10}^{(4)}(\{2, 1, 1, 1\}, 5)$

$$= 840\left[I_{1/10}^{(4)}(1,5) - B_{1/10}(5,1)I_{1/9}^{(3)}(1,4)\right] = .50400,$$

(4.114) $P(2^2 1^1) = \binom{10}{2}\binom{8}{1}I_{1/10}^{(3)}(\{2, 2, 1\}, 5)$

$$= 360\left[I_{1/10}^{(3)}(2,5) + B_{1/10}(5,1)I_{1/9}^{(2)}(2,4)\right] = .10800,$$

(4.115) $P(3^1 1^2) = \binom{10}{1}\binom{9}{2}I_{1/10}^{(3)}(\{3,1,1\},5)$

$$= 360\left[I_{1/10}^{(3)}(1,5) - B_{1/10}(5,1)I_{1/9}^{(2)}(1,4) - B_{1/10}(5,2)I_{1/9}^{(2)}(1,3)\right]$$

$$= .07200,$$

(4.116) $P(3^1 2^1) = \binom{10}{1}\binom{9}{1}I_{1/10}^{(2)}(\{3,2\},5)$

$$= 90\left[I_{1/10}^{(2)}(2,5) - B_{1/10}(5,2)I_{1/9}^{(1)}(2,3)\right] = .00900,$$

(4.117) $P(4^1 1^1) = \binom{10}{1}\binom{9}{1}I_{1/10}^{(2)}(\{4,1\},5) = .00450,$

(4.118) $P(5^1) = \binom{10}{1}I_{1/10}^{(5)}(5,5) = .00010$

(4.119) $P(S) = 6I_{1/10}^{(5)}(1,5) = .00720$

(4.120) $P(N) = \left[\binom{10}{5} - 6\right]I_{1/10}^{(5)}(1,5) = 246(.00120) = .29520$

As a check we note that 8 probabilities above add to one. Hence, the required order is

(4.121) $(5^1) < (4^1 1^1) < (S) < (3^1 2^1) < (2^2 1^1) < (N) < (2^1 1^3),$

where both S and N appear in strange places.

26.   Observations are taken sequentially on a multinomial with b blue cells, each having common probability $p \leq 1/b$. What is the expected number of observations required to obtain for the first time a frequency of at least r in each of these b blue cells?

Putting j=0 in (2.9) we have for the probability that the desired event occurs exactly on the nth observation

(4.122) $I_p^{(b)}(r,n) - I_p^{(b)}(r,n-1) = \dfrac{rb}{n}I_p^{(b,1)}(r,n) = \dfrac{rb}{n}B_p(n,r)I_{p/q}^{(b-1)}(r,n-r)$

Hence, multiplying by n and summing over n, the desired expectation E is

$$(4.123) \quad E = b \sum_{n=br}^{\infty} \frac{n! \, p^r}{\Gamma^b(r)\,(n-br)!} \int_0^p \cdots \int_0^p (1-p-\Sigma x_\alpha)^{n-br} \prod_2^b x_\alpha^{r-1} \, dx_\alpha \ .$$

We can now proceed either by summing under the integral sign or by using the reduction formula and leaving the sum outside the integral sign. The former yields a Dirichlet integral of Type 2, namely

$$(4.124) \quad E = \frac{b}{p} \left\{ \frac{(br)!}{\Gamma^b(r)} \int_0^1 \cdots \int_0^1 \frac{\prod_2^b x_\alpha^{r-1}\,dx_\alpha}{(1+\Sigma_2^b x_\alpha)^{br+1}} \right\} = \frac{br}{p} C_1^{(b-1)}(r,r+1) \ (\text{say}).$$

(Note that p enters in a very simple manner!),
but the tables of the C-integral are not yet available. The second method yields

$$(4.125) \quad E = br \sum_{n=br}^{\infty} B_p(n,r) I_{p/q}^{(b-1)}(r,n-r).$$

Useful bounds for E can be obtained by replacing all the I-values for $n \geq c$ by either 1 (for the upper bound) or by $I_{p/q}^{(b-1)}(r,c-r)$ (for the lower bound) in (4.125) here of course $c \geq br$ and $q = 1-p$. In both cases the resulting series is summable by taking complements of the extreme members of (2.4) with n replaced by c and r replaced by r+1. We obtain for any $c \geq br$

$$(4.126) \quad \frac{br}{p} I_{p/q}^{(b-1)}(r,c-r)[1 - I_p^{(1)}(r+1,c)] < E - Q < \frac{br}{p}[1 - I_p^{(1)}(r+1,c)],$$

where

$$(4.127) \quad Q = br \sum_{n=br}^{c-1} B_p(n,r) I_{p/q}^{(b-1)}(r,n-r).$$

Thus the I-function is seen to be a basic tool for evaluating and/or bounding other Dirichlet integrals.

Another method of finding the exact value of E for $r \geq 1$ is by means of an algorithm that works well on any computer. For example with $r = 2$, we let $H(x,y) = H_2(x,y|b,p)$ denote the additional expected number of observations required if x of the b cells still have to be seen once and y of the b cells still have to be seen twice. Then for $x \geq 0$, $y \geq 0$ and $0 < x+y \leq b$ it is easy to show that

$$(4.128) \quad H(x,y) = \frac{1/p}{x+y} + \frac{x}{x+y} H(x-1,y) + \frac{y}{x+y} H(x+1,y-1).$$

The boundary conditions are

$$(4.129) \quad H(x,0) = H(x) \quad \text{and} \quad H(0,0) = H(0) = 0,$$

where $H(x) = H_1(x)$ indicates the same problem with $r = 1$. The required expection E is $H(0,b)$. This method can be used for any r but we then require r arguments for the function $H_r(x_1, x_2, \ldots, x_r)$.

For the case $r = 1$, $b = 6$ and $p = 1/6$ this method gives

$$(4.130) \quad \begin{array}{l} H(6) = 1 + H(5); \; H(5) = 1.2 + H(4); \; H(4) = 1.5 + H(3); \\ H(3) = 2 + H(2); \; H(2) = 3 + H(1); \; H(1) = 6; \; H(0) = 0. \end{array}$$

This leads to $H(6) = 14.7$ also obtained below. For $r = 2$, $b = 6$ and $p = 1/6$ we obtain in the order indicated below

$$H(j,0) = H(j), \; 1 \leq j \leq 6; \; H(5,1) = 17.1500, \; H(4,1)$$
$$= 16.4400, \; H(3,1) = 15.6250,$$

$$H(2,1) = 14.6667, \; H(1,1) = 13.5000, \; H(0,1) = 12.0000;$$

$$(4.131) \quad H(4,2) = 17.9034, \; H(3,2) = 16.7802, \; H(2,2) = 15.0069, \; H(1,2)$$
$$= 11.3889, \; H(0,2) = 16.5000;$$

$$H(3,3) = 18.8063, \; H(2,3) = 17.7091, \; H(1,3) = 16.1024, \; H(0,3) = 13.3889;$$

$$H(2,4) = 19.8334, \; H(1,4) = 18.8877, \; H(0,4) = 17.6024;$$

$$H(1,5) = 20.8758, \; H(0,5) = 20.0877;$$

$$H(0,6) = 21.8758.$$

For the special case $r = 1$ the exact result for E is more easily obtained as the sum of expectations of independent (but not identically distributed) geometric chance variables, namely for $p \le 1/b$ and $r = 1$

$$(4.132) \quad E = \frac{1}{p} \sum_{j=1}^{b} \frac{1}{j} \quad .$$

For example with a fair die $b=6$, $p=1/6$ and for $r=1$ we obtain

$6(1 + 1/2 + 1/3 + 1/4 + 1/5 + 1/6) = 14.7$ (which happens to be the air pressure at ground level in $lbs/inch^2$). The difference of the upper and lower bounds above for (say) $c = 38$ in this case is

$$(4.133) \quad \frac{br}{p} [1 - I_p^{(1)} (r+1, c)][1 - I_{p/2}^{(b-1)} (r, c-r)] = 36(.00843)(.00104)$$
$$= .00032,$$

so that the 2 bounds will agree to 3 decimal places. For the corresponding problem with $r = 2$, $b = 6$, $p = 1/6$ and taking $c = 38$ again, the exact answer 21.8758 was obtained in (4.131) and (using Tables A and B) the difference of the bounds in (4.133) is .04199, so that the bounds will agree to only 1 decimal place. Thus for larger r we have to use a larger value of c to get more accuracy. In general for $r > 1$ the exact answers are not available and the bounds in (4.126) are quite useful.

One further expression for $r = 1$ and $p \le 1/b$ is of interest, although it does not involve I-functions. Viewed as a coupon-collecting problem $p < 1/b$ allows the coupons to be missing with probability $1 - pb$ and the probability p_n of getting a full set of b coupons in exactly n purchases, using inclusion-exclusion, is

$$(4.134) \quad p_n = \sum_{\alpha=0}^{b} (-1)^\alpha \binom{b}{\alpha} (1 - \alpha p)^n$$

and for the special case $p = 1/b$ this reduces to

$$(4.135) \quad p_n = b^{-n} [\Delta^n x^n]_{x=0} = \frac{b!}{b^n} S_n^{(b)},$$

where $S_n^{(b)}$ is the usual Stirling number of the second kind. Using (4.134) the value of E for the sequential problem with $s = 1$ and $p \leq 1/b$ leads to

$$(4.136) \qquad E = \sum_{n=b}^{\infty} nbp(1-p)^{n-1} \sum_{\alpha=0}^{b-1} (-1)^{\alpha} \binom{b-1}{\alpha} (1 - \frac{\alpha p}{1-\alpha})^{n-1}$$

$$= b - 1 + \frac{1}{p} \sum_{\beta=1}^{b} \frac{(-1)^{\beta-1}}{\beta} \binom{b}{\beta} (1 - \beta p)^{b-1} =$$

$$= \frac{1}{p} \sum_{\beta=1}^{b} \frac{(-1)^{\beta}}{\beta} \binom{b}{\beta}.$$

This yields the same result as in (4.132) although the resulting identity is not immediately obvious.

27. The digits of a 9-digit number are independent discrete uniform from 0 to 9. Let f_i denote the frequency of the digit i $(i = 1,2,3)$ and let $f_{[i]}$ $(i = 1,2,3)$ denote the corresponding ordered values $(f_{[1]} \leq f_{[2]} \leq f_{[3]})$ of these three frequencies. What is the expectation of $f_{[i]}$ $(i = 1,2,3)$?

By Table D with $b = 3$, $p = 1/10$ and $n = 9$ we obtain the exact result

$$(4.137) \quad Ef_{[1]} = .203950918.$$

By summing the complement of (4.40) with $j = 0$ from $r = 1$ to $r = n$ we obtain

$$(4.138) \quad E\{\text{Max freq}|b,n,p\} = \sum_{\alpha=1}^{b} (-1)^{\alpha-1} \binom{b}{\alpha} E\{\text{Min freq}|\alpha,n,p\}$$

and hence for $b = 3$, $p = 1/10$ and $n = 9$, we obtain for our example

$$(4.139) \quad E\{f_{[3]}\} = 3(.900000000) - 3(.397937610) + .203940918 = 1.710128088.$$

Since each f_i has expection $.9$ the sum of three such f's has expectation 2.7 and hence

$$(4.140) \quad E(f_{[2]}) = 2.7 - Ef_{[1]} - Ef_{[3]} = .785930994.$$

An independent derivation of (4.107) or (4.108) is not conceptually difficult but it does require a bit of work. We note that $Ef_{[2]}$ is not equal to the common value of $Ef_i = .9$. (It would be interesting to know under what conditions we obtain equality and when we obtain the same direction inequality as obtained above.)

28. The souvenir of Harrah's Club in Lake Tahoe is a 6-sided die with a hole in it, such that the faces marked 1 and 2 are symmetrical with respect to the hole and should have a common probability p, not necessarily equal to 1/6. Let $f_i (i=1,2)$ denote these two frequencies if the die is tossed $n = 10$ times, and let $f_{[i]}(i=1,2)$ denote the corresponding ordered values. What is $E\{f_{[i]}\}(i=1,2)$ if $p = 1/5, 1/4, 1/3$?

From Table D with $b = 2$, $n = 10$ and the above values of p we obtain

(4.141) $E(f_{[1]}) = 1.217266688$

(4.142) $E(f_{[2]}) = 4 - E(f_{[1]}) = 2.782733312$ $\Big\}$ for $p = 1/5$,

(4.143) $E(f_{[1]}) = 1.619014740$

$E(f_{[2]}) = 5 - Ef_{[1]} = 3.380985260$ $\Big\}$ for $p = 1/4$,

(4.144) $E(f_{[1]}) = 2.3094379244$

$E(f_{[2]}) = 20/3 - Ef_{[1]} = 4.3572287423$ $\Big\}$ for $p = 1/3$.

29. How many nonnegative numbers $< 10^9$ have the digits 1, 2, and 3 each appearing at least (resp., exactly) r times $(r=0,1,\ldots)$?

Using the notation of illustration 25 the required number is 10^9 times $P\{f_{[1]} \geq r\}$ (resp. $P\{f_{[1]} = r\}$), if each of the $n = 9$ digits is regarded as a discrete uniform chance variable with values $0,1,\ldots,9$. From Table B with $b = 3$ and $p = 1/10$ we obtain

r	$10^9 I_{1/10}^{(3)}(r,9)$	$-10^9 \Delta I_{1/10}^{(3)}(r,9)$
0	1,000,000,000	799,961,890
1	200,038,110	196,136,982
2	3,901,128	3,899,448
3	1,680	1,680

30 . How many nonnegative numbers $< 10^9$ have the digits 1, 2 and 3 appearing at most r times (resp., have the maximum of these three frequencies equal to r)?

Using (4.50) and the fact that the required number is 10^9 times $P\{f_{[3]} < r + 1\} = 10^9 J_{1/10}^{(3)}(r+1,9)$ (resp., $P\{f_{[3]} = r\} = 10^9 \Delta J_{1/10}^{(3)}(r,9)$), we obtain

r	$10^9 J_{1/10}^{(3)}(r+1,9)$	$10^9 \Delta J_{1/10}^{(3)}(r,9)$
0	40,353,607	40,353,607
1	433,183,618	392,830,011
2	844,186,798	411,003,180
3	975,022,594	130,835,796
4	997,327,240	22,304,646
5	999,807,298	2,480,058
6	999,991,006	183,708
7	999,999,754	8,748
8	999,999,997	243
9	1,000,000,000	3

31. In a single toss of d ordinary fair dice we want the minimum frequency to be at least 1 (or the number of vacant cells to be zero). What value of d is required to make the probability of this event at least P^* (take $P^* = .75$, .90, .95 and .99).

This is clearly the same as finding the smallest number of tosses of a single die needed to satisfy the same requirement. From Table C for $b = 6$,

$p = 1/6$ and $r = 1$ we obtain 18, 23, 27 and 36, respectively, for these values of P^*. An example like this with $r > 1$ was considered in Problem 18 above.

32. In the American version of roulette there are 38 "cells" (in the European version there are only 37 cells). Assuming the cell probabilities are not necessarily equal, we want to determine the minimum number of observations required to have a PCS of at least $P^* = 3/4$ to find the cell with the smallest cell probability, when the two smallest p's differ by at least $\delta^* = 1/76 = .5(1/38) \cong .013$.

With P^* only $3/4$ and 38 too high to use our exact tables, it is necessary to give an approximation of (4.23) by expanding $\{1 - [1 - \Phi(y)]\}^b$ to 4 terms, so that the result will again be conservative. Let $h'' = h''(b,\rho,P^*)$ be the solution of

$$(4.145) \qquad \frac{1-P^*}{b} = \Phi(-h'') - \binom{b-1}{2} F(-h''|3,1/3) + \frac{1}{3}\binom{b-1}{2} F(-h''|4,1/3),$$

where $F(h|k,\rho)$ is the right side of (4.23) and we have also used (4.20) and the given value of $\delta^* = 1/76$ to obtain $\rho = 1/3$ exactly; here $\Phi(x)$ is the standard normal cdf. The one-term approximation (4.23) set equal to P^*, gives us $h' = 2.47$ for this example and, using this, the more accurate (4.145) gives $h'' = 2.459$ (by trial and error). Finally using (4.24) with h'' for h, we have

$$(4.146) \qquad n = (2.459)^2 (76)^2 \left(\frac{77}{76}\right)\left(\frac{3}{76}\right) = 1396.8,$$

and the one-term approximation gives the upper bound 1684. Hence approximately 1400 observations are required on the very same roulette table (without any interim adjustments by the management) in order to satisfy our requirement. We note that higher values of P^* and smaller values of δ^* would lead to even larger values of n and would be quite impractical.

It may also be of interest to fix n at 1000 in (4.24) and solve for δ^*.
This leads to a quadratic in δ^* that is easily solvable and we denote it by
δ since it is not specified. For the above example with k = 38 and $P^* = 3/4$
we obtain

$$(4.147) \quad \delta = \frac{2.459}{76} \left[\frac{\sqrt{1600(2.459)^2 + 30,400 - 36(2.459)}}{1000 + (2.459)^2} \right] = .015 \quad .$$

Hence n = 1000 observations will be sufficient if we specified $\delta^* = .015$
in the requirement for the PCS in the multinomial ranking problem with
$P^* = 3/4$ and k = 38.

33. In a recent note of Robbins [15] some interest was generated in the
number of singletons in a multinomial, i.e., cells containing exactly 1
observation out of a sample of fixed size n. Let W denote the number
of singleton blue cells and, as in 22, let V denote the number of empty
blue cells in our model with b blue cells having common probability
$p \leq 1/b$. The joint probability that V = v and W = w is given by

$$(4.148) \quad P\{V=v, \; W=w\} = \begin{bmatrix} b \\ v,w \end{bmatrix} \frac{n!}{(n-w)!} p^w [1-(v+w)p]^{n-w} \; I_{\frac{p}{1-(v+w)p}}^{(b-v-w)} (2,n-w)$$

where $\begin{bmatrix} b \\ v,w \end{bmatrix}$ denotes the multinomial coefficient $b!/[v!w!(b-v-w)!]$;
this is positive for $\max\left[0, b - \left(\frac{n+w}{2}\right)\right] \leq v \leq b - w$ and $0 \leq w \leq \min(b,n)$.
The marginal probabilities for V and W can be obtained by an appropriate
sum in (4.148), but we also have

$$(4.149) \quad\quad\quad P\{V=v\} = \binom{b}{v} (1-vp)^n \; I_{\frac{p}{1-vp}}^{(b-v)} (1,n).$$

Hence this latter result is equal to the result from (4.148), namely

$$(4.150) \quad P\{V=v\} = \binom{b}{v} \sum_{w=0}^{b-v} \binom{b-v}{w} \frac{n!}{(n-w)!} p^w [1-(v+w)p]^{n-w} \; I_{\frac{p}{1-(v+w)p}}^{(b-v-w)} (2,n-w).$$

The marginal for W is given by

$$(4.151) \quad P\{W = w\} = \frac{n!}{(n-w)!} \binom{b}{w} p^w \sum_{v=0}^{b-w} \binom{b-w}{v} [1-(v+w)p]^{n-w} \, I_{\frac{p}{1-(v+w)p}}^{(b-v-w)}(2, n-w)$$

for $0 \le w \le \min(b,n)$. The fact that the sum of the probabilities add to 1 in (4.150) and (4.151) gives rise to new identities.

The expected number of singletons can be obtained from (4.151) but, more simply, we let $X_i = 1$ if the ith blue cell is a singleton and $X_i = 0$ otherwise. Then $W = X_1 + \cdots + X_b$ and

$$(4.152) \quad EW = nbpq^{n-1} = \sum_{w=0}^{\min(b,n)} \frac{n!}{(n-w)!} \, w \binom{b}{w} p^w A,$$

where $q = 1-p$ and A is the summation in (4.151); this is another identity for I-functions. From (4.149) and (4.60) we obtain

$$(4.153) \quad EV = bq^n = \sum_{v=1}^{\min(b,q/p)} v \binom{b}{v} (1-vp)^n \, I_{\frac{p}{1-vp}}^{(b-v)}(1,n),$$

which is another identity for I-functions.

In particular, the probabilities that $W = 0$ and $V = 0$ are

$$(4.154) \quad P\{W = 0\} = \sum_{v=0}^{b} \binom{b}{v} (1-vp)^n \, I_{\frac{p}{1-vp}}^{(b-v)}(2,n),$$

$$(4.155) \quad P\{V = 0\} = I_p^{(b)}(1,n).$$

The zero-one approach for W above also yields the second moment and variance of W in the form

$$(4.156) \quad EW^2 = nbpq^{n-1} + b(b-1)n(n-1)p^2(1-2p)^{n-2},$$

$$(4.157) \quad \sigma^2(W) = nbpq^{n-1}(1-nbpq^{n-1}) + b(b-1)n(n-1)p^2(1-2p)^{n-2};$$

the corresponding results for V are in (4.61) and (4.62). For the special case $bp = 1$ there is some simplification in (4.157) and in particular for $b = n = 2$ we set 0^0 equal to 1.

Similar results are also obtainable for occupied blue cells, i.e., having at least 1 observation, and for non-singleton blue cells; we omit these.

It may be of some interest to ask for the value of n that maximizes EW. For $p = 1/b$ it is easy to show that the maximum for EW occurs when $n = b$ and $b - 1$; the common value for EW is then $(b-1)^{b-1}/b^{b-2}$ For $p \to 0$ and fixed b the maximum occurs near $n = 1/p$ and the value is b/e. If $p \to 0$ and $bp \to 1$ then the maximum again occurs at $n = b$ (or $b-1$) and $E(W/b) \to 1/e$, i.e., the _proportion_ of singletons tends to the constant $1/e$.

34. Let C_v denote the number of blue cells containing at least v observations (call these crowded); let S_u denote the number of blue cells containing at most u observations (call these sparse). What is the probability for some fixed $v \geq 1$ that $C_v = c$ _and_ $S_o = b-c$ for $c \geq 1$ (i) in a fixed number of observations $n \geq 1$ and (ii) in a sampling experiment that never stops. For (i) with n observations this says that exactly c cells are crowded and the rest are all empty. For (ii) it says that at the time when we obtain $c \geq 1$ crowded blue cells the remaining blue cells are all empty. In (i) if we set $v = 2$ and sum over C we obtain the probability of no blue singletons after taking n observations. In (ii) if we set $c = b$ the answer is clearly 1.

For (i) the result, letting $\theta = 1 - (b-c)p$, is

$$(4.158) \qquad P\{C_v = c, \quad S_o = b-c \mid n\} = \binom{b}{c} \theta^n I_{p/\theta}^{(c)}(v,n).$$

Summing this over c $(1 \le c \le [n/v])$ would give the probability that after n observations the smallest <u>positive</u> frequency is at least v.

For (ii) we use (4.78) with (4.158) and sum on n to obtain

$$(4.159) \quad P\{S_o = b - c \text{ when } C_v = c\} = \begin{bmatrix} b \\ c-1,1 \end{bmatrix} p \sum_{n=cv}^{\infty} \theta^{n-1} I_{p/\theta}^{(c,1)}(v-1,v,n-1)$$

$$= \begin{bmatrix} b \\ c-1,1 \end{bmatrix} (p\theta^{c-1})^v \frac{\Gamma(cv)}{\Gamma^c(v)} \int_0^{p/\theta} \cdots \int_0^{p/\theta} \sum_{n=cv}^{\infty} \binom{n-1}{n-cv} \left[\theta(1-\frac{p}{\theta}-\sum_1^{c-1} x_i)\right]^{n-cv} \prod_1^{c-1} x_i^{v-1} dx_i$$

$$= \begin{bmatrix} b \\ c-1,1 \end{bmatrix} (p\theta^{c-1})^v \frac{\Gamma(cv)}{\Gamma^c(v)} \int_0^{p/\theta} \cdots \int_0^{p/\theta} \frac{\prod_1^{c-1} x_i^{v-1} dx_i}{(1-\theta+p+\theta\sum_1^{c-1} x_i)^{cv}}$$

$$= \left(\frac{1}{b-c+1}\right)^v \begin{bmatrix} b \\ c-1,1 \end{bmatrix} \frac{\Gamma(cv)}{\Gamma^c(v)} \int_0^{(b-c+1)^{-1}} \cdots \int_0^{(b-c+1)^{-1}} \frac{\prod_1^{c-1} y_i^{v-1} dy_i}{(1+\sum_1^{c-1} y_i)^{cv}}$$

$$= \left(\frac{1}{b-c+1}\right)^v \begin{bmatrix} b \\ c-1,1 \end{bmatrix} C_{1/(b-c+1)}^{(c-1)}(v,v)$$

this type of Dirichlet integral also appeared in (4.125).

For the special case c = b, since the answer is clearly 1, we obtain from (4.159) as a corollary

$$(4.160) \qquad\qquad C_1^{(b-1)}(v,v) = \frac{1}{b} \quad \text{for all } v \ge 1.$$

For c = 0 the answer in (4.159) is clearly zero. Suppose we set v = 2 and take the complement of (4.159). It will represent the probability that there are singletons present at the time when we first observe c <u>crowded</u> cells, i.e., cells with at least 2 observations in them.

35. Let C_v be defined as in 34 and let S_u' denote the number of cells with frequency less than u. Using the function $J_p^b(r,n)$ defined in (4.50) we can write the probability distribution of C_v and that of S_u' in terms

of I and J functions. Let α denote the total number of observations in the blue cells with frequency less than v. Then, letting $\theta = 1 - (b - c)p$ as before,

$$(4.161) \qquad P\{C_v = c\} = \binom{b}{c} \sum_{\alpha=0}^{\min(A, B)} \binom{n}{\alpha} (1 - \theta)^{\alpha} \theta^{n-\alpha} I_{p/\theta}^{(c)}(v, n-\alpha) J_{p/(1-\theta)}^{(b-c)}(v, \alpha),$$

where $A = (b-c)(v-1)$ and $B = n - cv$.

Using the notation $\theta = 1 - (b - s)p$ we have the corresponding result

$$(4.162) \qquad P\{S_u' = s\} = \binom{b}{s} \sum_{\alpha=0}^{\min(A', B')} \binom{n}{\alpha} \theta^{\alpha}(1 - \theta)^{n-\alpha} J_{p/\theta}^{(s)}(u, \alpha) I_{p/(1-\theta)}^{(b-s)}(u, n-\alpha)$$

where $A' = s(u - 1)$ and $B' = n - (b - s)u$. These two equations are identical if we replace v, c and θ by u, b-s and 1-θ, respectively. Joint probabilities for C_v and S_u' can also be written in terms of I and J functions; we omit these.

36. The referee of this write-up has requested that we try to apply the I and J functions to a Disjoint Test and a Scan Test (cf.[11] and [21]).

The Disjoint Test breaks up the unit interval into b disjoint (and exhaustive) cells and requires the probability P. that some cell contains at least r observations. This is simply the complement of our J-function i.e.,

$$(4.163) \quad P_0 = 1 - P\{ \text{Max freq} < r \mid b, n \} = 1 - J_{1/b}^{(b)}(r, n)$$

where the J-function is defined in #4, Section 4.4 and its relation to I is given in (4.50).

The application to the Scan Test is much more involved. Here we use a movable 'window' of fixed length d (with $0 \leq d \leq 1$) and the random variable S of interest is the maximum number of observed points that fit inside the window, i.e., that fit inside <u>any</u> interval of length d. We wish to compute $P_1 = P\{ S \geq s \}$ as a function of d and the total number of points n. The notation in [21] is N, n, p for our n, s, d, respectively. We claim that the Dirichlet integrals are useful in computing

approximations as well as exact values of these probabilities. In particular,

a J-function table (we have only tabulated the I-function) would provide

an approximation to the above probability. To show this let $0 \le X_1 \le X_2$

$\le \cdots \le X_n \le 1$ denote the ordered observations on the uniform $\cup (0,1)$

distribution. Let $Y_i = X_i - X_{i-1}$ and for $1 \le s \le n$ let $Z_i^{(s)} = Y_i + Y_{i-1}$

$+ \cdots + Y_{i-s+2} = X_i - X_{i-s+1}$, so that each Z_i spans exactly s points.

Then P_1 is given by

$$(4.164) \quad P_1 = P\{ S \ge s \mid d,n \} = P\{ Z_n^{(s)} < d \ \text{ or } \ Z_{n-1}^{(s)} < d \ \text{ or } \cdots \text{ or } \ Z_s^{(s)} < d \}$$

$$= 1 - P\{ Z_n^{(s)} > d, \ Z_{n-1}^{(s)} > d,\ldots, Z_s^{(s)} > d \}$$

$$= 1 - P\left\{ \text{Min}\left(Z_n^{(s)}, Z_{n-1}^{(s)},\ldots, Z_s^{(s)} \right) > d \right\} \ .$$

If these Z-variables had no overlapping Y-values in common with each

other then we could write the probability in the last line above exactly

as a J-function . Our approximation is obtained by keeping the same

n-value and disregarding the overlap. This gives the approximation $(\approx)$

$$(4.165) \quad P_1 \approx 1 - J_d^{(n-s+1)} \, (s-1,n) = \sum_{\alpha =1}^{n-s+1} (-1)^{\alpha-1} \binom{n-s+1}{\alpha} I_d^{(\alpha)} (s-1,n) \ .$$

(Note that r corresponds to s-1 in the Dirichlet integrals.)

Thus for $n = 6$, $s = 5$ and $d = .2$ this gives

$$(4.166) \quad P_1 \approx 2 \, I_{.2}^{(1)} \, (4, 6) - I_{.2}^{(2)} \, (4,6) \ = 2(.017) - 0 = .034,$$

and the exact answer appears to be .031 (cf. reference [8] at the end

of [21]). The closeness of this approximation has not been extensively

studied, but it does appear to be better for small d . Thus for the

example below with $n = 6$, $s = 3$ and variable d $(0 \le d \le .2)$ we obtain

from (4.165) the polynomial

$$(4.167) \quad P_1 \approx 4d^2 (15 - 38d - 90d^2 + 336d^3 - 160d^4)$$

which yields .0059 (exact value is .0057) for d = .01 and .997 (exact
value is .808) for d = .2 . It is not known whether this approximation
always leads to an upper bound as it has in the above calculations.

It is also possible to obtain bounds for the Scan probability
$P\{S \geq s\}$ by considering some new event that implies the desired event
(cf. (4.164)). Thus if we let E denote the event:

(4.168) $\left\{\text{Either } \left(Y_6 < d/2 \text{ and } Y_5 < d/2\right)\right.$

$\left.\text{or } \left(Y_5 < d/2 \text{ and } Y_4 < d/2\right) \text{ or } \cdots \text{ or } \left(Y_3 < d/2 \text{ and } Y_2 < d/2\right)\right\}$

then it follows easily by inclusion-exclusion that

(4.169) $P_1 \geq P(E) = 4I_{d/2}^{(2)}(1,6) - 3I_{d/2}^{(3)}(1,6) - I_{d/2}^{(4)}(1,6) + I_{d/2}^{(5)}(1,6)$

$= 1 - (1-d)^6 - 3(1 - \frac{3d}{2})^6 + 4(1 - 2d)^6 - (1 - \frac{5}{2}d)^6$

$= d^2(30 - 105d + \frac{525}{4}d^2 - \frac{315}{8}d^4 - \frac{373}{16}d^5).$

However the numerical answers are not at all close, since we
obtain .0029 (instead of .0057) for d = .01 and .556 (instead of
.808) for d = .2. Although these results are not very close, (4.169)
does illustrate nicely the use of I-functions.

Similarly, for an upper bound we use the event E' that $Y_6 < d/2$
or $\cdots$ or $Y_2 < d/2$ and obtain

(4.170) $P_1 \leq P(E') = 1 - J_{d/2}^{(5)}(1,6) = \sum_{\alpha=1}^{5}(-1)^{\alpha-1}\binom{5}{\alpha}I_{d/2}^{(\alpha)}(1,6)$

$= 1 - \left(1 - \frac{5d}{2}\right)^6$

which gives .141 for d = .01 and .984 for d = .2 .

For the exact computation of P_1 we use the first line of (4.164)
and again apply the principle of inclusion-exclusion. We consider only
the case of d close to zero; more precisely, assume that $0 \leq d \leq 1/(n-1)$.

To illustrate the method consider Example 3 of [21] in which $n = 6$, $s = 3$ and $d = .01$ [The authors in [21] devote 2 large pages to this example but fail to give the numerical result ___ an unbelievable oversight!] From (4.164) for $d \le 1/(n-1)$

$$(4.171) \quad P_1 = \sum_{i=s}^{n} P\left\{ Z_i^{(s)} < d \right\} - \sum_{i<j} P\left\{ Z_i^{(s)} < d, \; Z_j^{(s)} < d \right\} + \cdots$$

Two such probabilities are equal by exchangeability if the corresponding events concern the same number of Z's with the same overlapping structure. Thus for $n = 6$, $s = 3$ and any d we have, after combining similar terms,

$$(4.172) \quad P_1 = 4P\left\{ Y_6 + Y_5 < d \right\} - 3P\left\{ Y_6 + Y_5 < d, \; Y_5 + Y_4 < d \right\}$$

$$- 3P\left\{ Y_6 + Y_5 < d, \; Y_4 + Y_3 < d \right\} + 2P\left\{ Y_6 + Y_5 < d, \; Y_5 + Y_4 < d, \; Y_4 + Y_3 < d \right\}$$

$$+ 2P\left\{ Y_6 + Y_5 < d, \; Y_5 + Y_4 < d, \; Y_3 + Y_2 < d \right\}$$

$$- P\left\{ Y_6 + Y_5 < d, \; Y_5 + Y_4 < d, \; Y_4 + Y_3 < d, \; Y_3 + Y_2 < d \right\};$$

call these terms T_1, T_{21}, T_{22}, T_{31}, T_{32} and T_4, respectively. When there is no overlap we have simply

$$(4.173) \quad T_{22} = -3 \, I_d^{(2)}(2,6) \quad \text{and} \quad T_1 = 4 \, I_d^{(1)}(2,6).$$

In the remaining cases we have to introduce appropriate inequalities into the evaluation of each integral. Thus T_{21} is given by the 3-fold integral

$$(4.174) \quad T_{21} = -3\left(\frac{6!}{3!}\right) \int_0^d \int_0^{d-y} \int_0^{d-y} (1 - x - y - z)^3 \, dx \, dz \, dy$$

$$= -3 + 3(1-2d)^6 + 36 \, d(1-d)^5 \, ;$$

here we assumed that $d < 1/3$ since the limits of integration must be worked out more carefully for $d > 1/3$. If we assume that $d \le .2$ and add the corresponding six results in (4.172) we obtain a fairly simple polynomial $P_1 = P_1(d)$ given for $0 \le d \le \frac{1}{3}$ by

(4.175) $P_1(d) = 1 - 5(1 - 2d)^6 + 4(1 - 3d)^5$ $(0 \le d \le \frac{1}{3})$.

For $d = .2$ the value of $P_1(.2)$ is .808, in agreement with a tabled value in [21]; for $d = .01$ the value is .0057. Note that $P_1(0) = P_1'(0) = 0$ and hence $P_1(d)$ behaves near $d = 0$ like a parabola that goes through the origin, has derivative zero, and opens upward. It is monotonically increasing in d, reaching 1 at $d = .5$, and has a point of inflexion between 0 and 1/5. These polynomials (4.175) may be of interest in addition to the tabled values for specific values of d. In general the polynomial would change for $d \ge \{\frac{n}{2}\}^{-1}$ whenever 1/d takes on an integer value until d reaches the value $\{(n-s+1)/(s-1)\}^{-1}$ when P_1 reaches the value 1; here $\{x\}$ is the smallest integer not less than x.

The above discussion is based on the following useful result. Let $X_1 \le X_2 \le \cdots \le X_n$ be the ordered version of independent observations on the uniform $U(0,1)$ distribution and let

(4.176) $Y_j = X_{jr+1} - X_{(j-1)r+1}$ $\left(j = 1, 2, \ldots, \left[\frac{n-1}{r}\right]\right)$

denote sample spacings with X-subscripts in (4.176) differing by exactly r. Denote the integer part of $(n-1)/r$ by b. Then the joint density of these b sample spacings is exactly the Dirichlet (type 1) density given in (2.1) (without the integral signs). The same result holds if b is any integer less than $(n-1)/r$ where b is still the number of Y-values as in (4.176). Then we obtain the marginal density of these b chance variables, which is again Dirichlet, type 1.

Moreover, if we define X_0 to be identically zero and treat $X_1 - X_0 = X_1$ as an additional unit spacing, then the number of unit intervals among the X's is increased from n-1 to n. Then b can take on any integer value up to the integer part of n/r and the Dirichlet density in (2.1) is still valid. However the model described in (4.176) is more appropriate for the Scan Test.

Thus, although our tables may not be directly applicable to the exact evaluation of the overall Scan probabilities, the Dirichlet (type 1) integrals are useful and for some parts of the computation, as shown in (4.173), our tables are applicable.

References

[1] Alam, K. and Thompson, J. R. (1972). On selecting the least probable
     multinomial event. Ann. Math. Statist. 43 1981-1990.

[2] Bechhofer, R. E., Elmaghraby, S. A. and Morse, N. (1959). A single-
     sample multiple-decision procedure for selecting the multinomial
     event which has the largest probability. Ann. Math. Statist. 30
     102-119.

[3] Cacoullos, T. and Sobel, M. (1966). An inverse-sampling procedure
     for selecting the most probable event in a multinomial distribution.
     In Multivariate Analysis I, (P. R. Krishnaiah ed). pp. 423-455,
     Academic Press, New York.

[4] Dunnett, C. W. and Sobel, M. (1955). Approximations to the
     probability integral and certain percentage points of a
     multivariate analogue of student's t distribution. Biometrika
     42 258-260.

[5] Greenwood, R. E. and Glasgow, M. O. (1950). Distribution of Maximum and
     Minimum frequencies in a sample drawn from a Multinomial Distribution.
     Ann. Math. Statist. 21 416-424.

[6] Gupta, S. S. and Nagel, K. (1967). "On selection and ranking procedures
     and order statistics from the multinomial distribution" Sankhya 29B
     1-34.

[7] Johnson, N. L. and Kotz, S. (1975). "On a multivariate generalized
     occupancy model", private communication.

[8] Johnson, N. L. and Young, D. H. (1960). Some applications of
     two approximations to the multinomial distribution. Biometrika
     47 463-469.

[9] Jordan, C. (1964). Calculus of Finite Differences. 2nd ed. Dover
     Publications, New York.

[10] Kozelka, R. M. (1956). Approximate upper percentage points for
      extreme values in multinomial sampling. _Ann. Math. Statist._
      _27_ 507-512.

[11] Naus, J. I. (1966). A power comparison of two tests of non-random
      clustering _Technometrics_ _8_ 493-517.

[12] Olkin, I. and Sobel, M. (1965). Integral expressions for tail
      probabilities of the multinomial and negative multinomial
      distributions. _Biometrika_ _52_ 167-179.

[13] Owen, D. B. and Steck, G. P. (1962). Moments of order statistics
      from the equicorrelated multivariate normal distribution.
      Ann. Math. Statist. _33_, 1286-1291.

[14] Riordan, J. (1958). _An Introduction to Combinatorial Analysis_.
      John Wiley & Sons, Inc., New York.

[15] Robbins, H. E. (1968). Estimating the total probability of the
      unobserved outcomes of an experiment. _Ann. of Math. Statist._
      _39_ 256-257.

[16] Sobel, M. and Weiss, G. (1972). Play-the-winner rule and inverse
      sampling in selecting the better of two binomial populations.
      JASA _66_, 545-551.

[17] Sobel, M. and Uppuluri, V. R. R. (1974). Sparse and crowded cells
      and Dirichlet distributions. _Ann. Statist._ _2_ 977-987.

[18] Sobel, M. and Uppuluri, V. R. R. (1972). On Bonferroni-type
      inequalities of the same degree for the probability of unions
      and intersections. _Ann. Math. Statist._ _43_ 1549-1558.

[19] Steck, G. P. (1960). (Unpublished) Tables of the distribution
      and moments of the maximum of homogeneous multinomial
      distributions. (b = 2(1)50, n = 1(1)50).

[20] van Wijngaarden, A. (1950). Table of the cumulative symmetric
      binomial distribution. _Indagationes Mathematicae_ _53_ 1-12.

[21] Wallenstein, S. R. and Naus, J. I. (1974). Probabilities

        for the size of largest clusters and smallest intervals.

        JASA 69 690-697.

[22] Wilks, S. S. (1962). Mathematical Statistics. John Wiley

        & Sons, Inc., New York. 179.

TABLE A

$(p = 1/b)$

This table gives the value of the incomplete Dirichlet integral of type 1

$$I_p^{(b)}(r,n) = \frac{n!}{\Gamma^b(r)(n-br)!} \int_0^p \cdots \int_0^p (1 - \sum_1^b x_i)^{n-br} \prod_1^b x_i^{r-1} dx_i$$

as defined in (2.1) for $p = 1/b$, $b = 2(1)10$, $r = 1(1)10$, $n \geq br$.

Stopping Rule: Increase n from br by ones until $I \geq .999$.

Note 1: In this table entries less than 10^{-5} are given in exponential notation to avoid too many zeros.

Note 2: The symbols n, b, r and p in the text appear as N, B, R and P, respectively in all the tables.

A1

DIRICHLET 1   PROBABILITY (P=1/B)   B= 2

N	0	1	2	3	4
			R= 1		
0+	0	0	.5000000000	.7500000000	.8750000000
5+	.9375000000	.9687500000	.9843750000	.9921875000	.9960937500
10+	.9980468750	.9990234375	.9995117188	.9997558594	.9998779297
			R= 2		
0+	0	0	0	0	.3750000000
5+	.6250000000	.7812500000	.8750000000	.9296875000	.9609375000
10+	.9785156250	.9882812500	.9936523438	.9965820313	.9981689453
15+	.9990234375	.9994812012	.9997253418	.9998550415	.9999237061
			R= 3		
5+	0	.3125000000	.5468750000	.7109375000	.8203125000
10+	.8906250000	.9345703125	.9614257813	.9775390625	.9870605469
15+	.9926147461	.9958190918	.9976501465	.9986877441	.9992713928
			R= 4		
5+	0	0	0	.2734375000	.4921875000
10+	.6562500000	.7734375000	.8540039063	.9077148438	.9426269531
15+	.9648437500	.9787292480	.9872741699	.9924621582	.9955749512
20+	.9974231720	.9985103607	.9991445541	.9995117188	.9997228384
			R= 5		
10+	.2460937500	.4511718750	.6123046875	.7331542969	.8204345703
15+	.8815307617	.9231872559	.9509582520	.9691162109	.9807891846
20+	.9881820679	.9928026199	.9956564903	.9974005222	.9984561205
25+	.9990894794	.9994664788	.9996892512	.9998200089	.9998962842
			R= 6		
10+	0	0	.2255859375	.4189453125	.5760498047
15+	.6982421875	.7898864746	.8565368652	.9037475586	.9364318848
20+	.9586105347	.9733963013	.9830994606	.9893779755	.9933892488
25+	.9959226847	.9975060821	.9984862804	.9990877658	.9994538873
			R= 7		
10+	0	0	0	0	.2094726563
15+	.3927612305	.5455017090	.6676940918	.7621154785	.8329315186
20+	.8846817017	.9216461182	.9475212097	.9653103352	.9773441553
25+	.9853667021	.9906446934	.9940753877	.9962808341	.9976842999
30+	.9985690936	.9991220897	.9994649473	.9996759365	.9998048744
			R= 8		
15+	0	.1963806152	.3709411621	.5193176270	.6407165527
20+	.7368240356	.8107528687	.8661994934	.9068603516	.9360853434
25+	.9567147493	.9710407257	.9808427095	.9874590486	.9918699414
30+	.9947771206	.9966731071	.9978975984	.9986812728	.9991786047
			R= 9		
15+	0	0	0	.1854705811	.3523941040
20+	.4965553284	.6166896820	.7137212753	.7899603844	.8484103680
25+	.8922478557	.9244813025	.9477610141	.9643018618	.9758804552
30+	.9838751983	.9893261595	.9929996333	.9954486159	.9970649444
35+	.9981217745	.9988067570	.9992471029	.9995280133	.9997059231
			R=10		
20+	.1761970520	.3363761902	.4765329361	.5951271057	.6925437450
25+	.7704770565	.8313624561	.8779218793	.9128414467	.9385716543
30+	.9572260547	.9705506265	.9799383930	.9864690132	.9909588145
35+	.9940118794	.9960668270	.9974367920	.9983419475	.9989349804
40+	.9993204517	.9995691430	.9997284608	.9998298448	.9998939553

DIRICHLET 1   PROBABILITY (P=1/B)   B= 3

N	0	1	2	3	4
			R= 1		
0+	0	0	0	.2222222222	.4444444444
5+	.6172839506	.7407407407	.8257887517	.8834019204	.9221155312
10+	.9480262155	.9653338753	.9768836051	.9845871884	.9897241651
15+	.9931492343	.9954327532	.9969551455	.9979700893	.9986467236
20+	.9990978149	.9993985430	.9995990285	.9997326857	.9998217904
			R= 2		
5+	0	.1234567901	.2880658436	.4481024234	.5846669715
10+	.6934918458	.7768124778	.8390319904	.8847341473	.9179206472
15+	.9418171015	.9589131771	.9710807938	.9797034715	.9857915543
20+	.9900763165	.9930833736	.9951883899	.9966585881	.9976832819
25+	.9983961161	.9988911412	.9992343592	.9994719717	.9996362472
			R= 3		
5+	0	0	0	0	.0853528426
10+	.2133821064	.3520804755	.4824806517	.5957425189	.6893822644
15+	.7643060200	.8229056053	.8679877847	.9022415436	.9280147958
20+	.9472537060	.9615195723	.9720374065	.9797530023	.9853876332
25+	.9894859570	.9924559205	.9946009588	.9961454118	.9972542597
30+	.9980482535	.9986153933	.9990195625	.9993069720	.9995109400
			R= 4		
10+	0	0	.0652000881	.1695202290	.2900679473
15+	.4106156657	.5215363094	.6182721327	.6995940062	.7661848468
20+	.8196612802	.8619697636	.8950492814	.9206644369	.9403391799
25+	.9553457997	.9667216718	.9752979432	.9817315054	.9865358115
30+	.9901084891	.9927550196	.9947084536	.9961454802	.9971993068
35+	.9979698496	.9985317049	.9989403274	.9992367792	.9994513539
			R= 5		
15+	.0527396268	.1406390048	.2466763496	.3573644552	.4635346790
20+	.5598659600	.6439354511	.7152518135	.7744710095	.8228375798
25+	.8618224938	.8929075732	.9174692495	.9367251013	.9517177256
30+	.9633193764	.9722470907	.9790821730	.9842905359	.9882420091
35+	.9912276988	.9934750495	.9951605872	.9964204946	.9973592536
40+	.9980566193	.9985731873	.9989547964	.9992359822	.9994426658
			R= 6		
15+	0	0	0	.0442752422	.1201756575
20+	.2145993884	.3163353948	.4170381448	.5112499257	.5959278810
25+	.6698145010	.7328449729	.7856725688	.8293260052	.8649811459
30+	.8938197921	.9169495467	.9353639664	.9499279287	.9613779351
35+	.9703306451	.9772954050	.9826881930	.9868454628	.9900370519
40+	.9924777462	.9943373530	.9957492900	.9968177888	.9976238518
45+	.9982301225	.9986848260	.9990249297	.9992786579	.9994674789
			R= 7		
20+	0	.0381510024	.1049152566	.1899160431	.2837569115
25+	.3789669508	.4702450120	.5542676377	.6292961979	.6947440572
30+	.7507910564	.7980791366	.8374914993	.8700030645	.8965857098
35+	.9181527367	.9355299619	.9494440708	.9605216435	.9692943980
40+	.9762077243	.9816306403	.9858660111	.9891603440	.9917127824
45+	.9936831191	.9951987779	.9963607899	.9972488347	.9979254416
50+	.9984394537	.9988288570	.9991230730	.9993448010	.9995114902
			R= 8		
20+	0	0	0	0	.0335145958
25+	.0930960996	.1703313822	.2572628691	.3472369581	.4352383710
30+	.5178617847	.5930808242	.6599401521	.7182507832	.7683293683
35+	.8107950273	.8464217888	.8760377086	.9004599487	.9204559470
40+	.9367226560	.9498777943	.9604587603	.9689261859	.9756700858
45+	.9810172507	.9852390130	.9885588417	.9911594470	.9931892222
50+	.9947679523	.9959917816	.9969374713	.9976660043	.9982256028
55+	.9986542317	.9989816568	.9992311236	.9994207183	.9995644634

A 3

DIRICHLET 1   PROBABILITY (P=1/B)   B= 3

N	0	1	2	3	4
			R= 9		
25+	0	0	.0298826986	.0836715562	.1544120536
30+	.2352945579	.3203934420	.4050271041	.4858233655	.5605978596
35+	.6281408699	.6879812222	.7401681243	.7850899708	.8233347427
40+	.8555887150	.8825668273	.9049674440	.9234449643	.9385949745
45+	.9509478989	.9609681992	.9690570307	.9755569068	.9807573897
50+	.9849011538	.9881899984	.9907905499	.9928395031	.9944483269
55+	.9957074142	.9966896845	.9974536743	.9980461569	.9985043438
60+	.9988577169	.9991295424	.9993381130	.9994977596	.9996196716
			R=10		
30+	.0269608348	.0759805343	.1412163466	.2167836224	.2973924650
35+	.3787022540	.4574389105	.5313466485	.5990455813	.6598518348
40+	.7135979918	.7604748112	.8009027255	.8354338007	.8646807478
45+	.8892679935	.9097997231	.9268404177	.9409042679	.9524506920
50+	.9618839103	.9695551021	.9757661043	.9807739317	.9847956268
55+	.9880131125	.9905778396	.9926151015	.9942279495	.9955006806
60+	.9965018995	.9972871709	.9979012942	.9983802336	.9987527422
65+	.9990417186	.9992653313	.9994379447	.9995708769	.9996730168

DIRICHLET 1   PROBABILITY (P=1/B)   B= 4

N	0	1	2	3	4
			R= 1		
0+	0	0	0	0	.0937500000
5+	.2343750000	.3808593750	.5126953125	.6229248047	.7113647461
10+	.7806015015	.8339881897	.8747591972	.9057033062	.9290944040
15+	.9467292577	.9600011688	.9699779889	.9774720477	.9830983137
20+	.9873208743	.9904892252	.9928662036	.9946492951	.9959867925
25+	.9969900050	.9977424590	.9983068219	.9987301053	.9990475734
			R= 2		
5+	0	0	0	.0384521484	.1153564453
10+	.2162933350	.3260421753	.4334378242	.5318641663	.6182409525
15+	.6918069720	.7531399559	.8034837004	.8443295967	.8771776288
20+	.9034132286	.9242542549	.9407378996	.9537286707	.9639362072
25+	.9719365154	.9781931823	.9830768693	.9868823961	.9898432673
30+	.9921437766	.9939289479	.9953126127	.9963839208	.9972125554
35+	.9978528876	.9983472744	.9987286652	.9990226574	.9992491128
			R= 3		
10+	0	0	.0220298767	.0715970993	.1440891623
15+	.2310125157	.3238929715	.4160183016	.5028353781	.5817035334
20+	.6514353884	.7118384817	.7633397847	.8067092011	.8428705980
25+	.8727808362	.8973577266	.9174412917	.9337766383	.9470101739
30+	.9576935628	.9662917597	.9731928150	.9787180549	.9831318418
35+	.9866505017	.9894502410	.9916740172	.9934374089	.9948335700
40+	.9959373715	.9968088381	.9974959805	.9980371188	.9984627778
45+	.9987972282	.9990597352	.9992655658	.9994268004	.9995529844
			R= 4		
15+	0	.0146829989	.0499221962	.1048366120	.1747990041
20+	.2539412858	.3367252133	.4186887120	.4966426949	.5685707430
25+	.6334056806	.6907833105	.7408217549	.7839439390	.8207446409
30+	.8518964213	.8780865408	.8999771488	.9181821703	.9332557039
35+	.9456880436	.9559065145	.9642791599	.9711199454	.9766946035
40+	.9812265632	.9849026289	.9878782227	.9902821031	.9922205321
45+	.9937809054	.9950348776	.9960410296	.9968471279	.9974920256
50+	.9980072537	.9984183461	.9987459389	.9990066780	.9992139667
			R= 5		
20+	.0106708694	.0373480430	.0806984501	.1382674202	.2060619785

A 4

DIRICHLET 1   PROBABILITY (P=1/B)   B= 4

N	0	1	2	3	4
			R= 5		
25+	.2797571744	.3554368787	.4299457509	.5009773152	.5670114337
30+	.6271830038	.6811327239	.7288673362	.7706414505	.8068641202
35+	.8380287175	.8646626620	.8872930378	.9064243598	.9225252884
40+	.9360217173	.9472942410	.9566785167	.9644674359	.9709143411
45+	.9762367567	.9806202811	.9842224094	.9871761480	.9895933431
50+	.9915676852	.9931773817	.9944875068	.9955520463	.9964156639
55+	.9971152159	.9976810428	.9981380643	.9985067039	.9988036661
60+	.9990425872	.9992345782	.9993886767	.9995122215	.9996111621
			R= 6		
20+	0	0	0	0	.0082023046
25+	.0292939450	.0646036108	.1129348835	.1715836078	.2372294928
30+	.3065746238	.3767232748	.4453552505	.5107591218	.5717836346
35+	.6277500975	.6783534441	.7235678453	.7635646214	.7986451338
40+	.8291884632	.8556122577	.8783445856	.8978045761	.9143898312
45+	.9284688954	.9403773911	.9504167300	.9588545684	.9659263855
50+	.9718377363	.9767668556	.9808673925	.9842711243	.9870905557
55+	.9894213424	.9913445111	.9929284616	.9942307517	.9952966729
60+	.9961756310	.9968923459	.9974778895	.9979555777	.9983447335
65+	.9986613376	.9989185801	.9991273271	.9992965137	.9994334746
			R= 7		
25+	0	0	0	.0065575094	.0237709714
30+	.0532370715	.0945112777	.1457757590	.2044945793	.2679376340
35+	.3335373290	.3990918823	.4628506071	.5235197392	.5802215743
40+	.6324310606	.6799058216	.7226191077	.7607005865	.7943868999
45+	.8239821216	.8498272869	.8722777320	.8916868628	.9083950405
50+	.9227224204	.9349647644	.9453914315	.9542449155	.9617414453
55+	.9680722800	.9734054251	.9778875744	.9816461355	.9847912426
60+	.9874176930	.9896067668	.9914279059	.9929402433	.9941939790
65+	.9952316074	.9960890030	.9967963747	.9973790991	.9978584444
70+	.9982521969	.9985751997	.9988398153	.9990563199	.9992332403
			R= 8		
30+	0	0	.0053972176	.0197897980	.0448568755
35+	.0806139976	.1258602556	.1786621699	.2367758112	.2979631088
40+	.3601971259	.4217724499	.4813447316	.5379230957	.5908349964
45+	.6396779080	.6842674822	.7245880297	.7607484580	.7929449772
50+	.8214307471	.8464919980	.8684298350	.8875468267	.9041374923
55+	.9184818762	.9308415103	.9414571776	.9505480006	.9583114772
60+	.9649241700	.9705428234	.9753057416	.9793343014	.9827345132
65+	.9855985645	.9880063058	.9900266505	.9917188732	.9931337984
70+	.9943148776	.9952991560	.9961181349	.9967985351	.9973629690
75+	.9978305292	.9982173020	.9985368125	.9988004103	.9990176009
			R= 9		
35+	0	.0045427970	.0168083489	.0384681985	.0698234093
40+	.1101086198	.1578524638	.2112157878	.2682647236	.3271646618
45+	.3862998265	.4443323769	.5002174994	.5531896658	.6027322720
50+	.6485345924	.6904770619	.7285435899	.7628379264	.7935299499
55+	.8208370138	.8450050533	.8662939300	.8849663999	.9012800851
60+	.9154818717	.9278042221	.9384629648	.9476561994	.9555640232
65+	.9623488446	.9681561003	.9731152367	.9773408470	.9809338868
70+	.9839829100	.9865652831	.9887483519	.9905905409	.9921423750
75+	.9934474186	.9945431296	.9954616305	.9962303994	.9968728853
80+	.9974090537	.9978558665	.9982277041	.9985367336	.9987932295
85+	.9990058519	.9991818873	.9993274553	.9994476873	.9995468789
			R=10		
40+	.0038921833	.0145072286	.0334655387	.0612458901	.0973945355
45+	.1407961468	.1899448816	.2431770989	.2988484811	.3554536978
50+	.4116957294	.4665158044	.5190953360	.5688398434	.6153527349

DIRICHLET 1   PROBABILITY (P=1/B)   B= 4

N	0	1	2	3	4
			R=10		
55+	.6584046742	.6979023899	.7338593176	.7663693896	.7955845431
60+	.8216960299	.8449193215	.8654822404	.8836158821	.8995478815
65+	.9134975994	.9256728503	.9362678399	.9454620345	.9534197325
70+	.9602901500	.9662078738	.9712935648	.9756548207	.9793871309
75+	.9825748717	.9852923043	.9876045493	.9895685186	.9912337931
80+	.9926434395	.9938347613	.9948399850	.9956868811	.9963993228
85+	.9969977858	.9974997936	.9979203116	.9982720940	.9985659902
90+	.9988112110	.9990155624	.9991856481	.9993270449	.9994444549

DIRICHLET 1   PROBABILITY (P=1/B)   B= 5

N	0	1	2	3	4
			R= 1		
5+	.0384000000	.1152000000	.2150400000	.3225600000	.4270694400
10+	.5225472000	.6063636480	.6780026880	.7381156823	.7879072481
15+	.8287692521	.8620793266	.8891009573	.9109429198	.9285514908
20+	.9427194307	.9541024653	.9632381160	.9705641755	.9764355486
25+	.9811389632	.9849054850	.9879209765	.9903347343	.9922665593
30+	.9938125105	.9950495663	.9960393877	.9968313510	.9974649853
35+	.9979719309	.9983775104	.9987019877	.9989615778	.9991692548
			R= 2		
10+	.0116121600	.0425779200	.0936714240	.1608877670	.2383303803
15+	.3202010972	.4017726711	.4796632117	.5517398292	.6168740693
20+	.6746724106	.7252382582	.7689839688	.8064928286	.8384233629
25+	.8654465008	.8882068010	.9073005371	.9232651376	.9365759681
30+	.9476476355	.9568378851	.9644528150	.9707525882	.9759571363
35+	.9802515630	.9837910908	.9867054868	.9891029552	.9910735196
40+	.9926919329	.9940201628	.9951095013	.9960023472	.9967337060
45+	.9973324471	.9978223560	.9982230114	.9985505153	.9988181002
50+	.9990366326	.9992150292	.9993606026	.9994793460	.9995761682
			R= 3		
15+	.0055105290	.0220421161	.0524602363	.0966767212	.1524713944
20+	.2165478259	.2853907302	.3558098740	.4252072515	.4916506291
25+	.5538329153	.6109765626	.6627210328	.7090147426	.7500216540
30+	.7860458217	.8174734703	.8447304174	.8682520708	.8884632562
35+	.9057654563	.9205294594	.9330918376	.9437540445	.9527832348
40+	.9604141504	.9668516114	.9722732900	.9768325531	.9806612332
45+	.9838722438	.9865619901	.9888125556	.9906936579	.9922643822
50+	.9935747029	.9946668131	.9955762768	.9963330252	.9969622137
55+	.9974849555	.9979189502	.9982790183	.9985775559	.9988249212
60+	.9990297605	.9991992850	.9993395033	.9994554181	.9995511909
			R= 4		
20+	.0032038216	.0134560506	.0335504195	.0645859673	.1060662399
25+	.1563544130	.2131989573	.2741775065	.3370064930	.3997206384
30+	.4607501688	.5189282504	.5734568477	.6238520673	.6698831036
35+	.7115133027	.7488478341	.7820897943	.8115049544	.8373944745
40+	.8600745041	.8798614712	.8970619118	.9119658196	.9248426542
45+	.9359393050	.9454794560	.9536639179	.9606716030	.9666608983
50+	.9717712597	.9761249013	.9798284916	.9829747983	.9856442445
55+	.9879063539	.9898210744	.9914399778	.9928073354	.9939610780
60+	.9949336452	.9957527340	.9964419549	.9970214052	.9975081684
65+	.9979167469	.9982594365	.9985466502	.9987871965	.9989885184
70+	.9991568987	.9992976359	.9994151937	.9995133295	.9995952032
			R= 5		
25+	.0020916516	.0090638235	.0233069746	.0462254996	.0781042615
30+	.1182678760	.1653712414	.2177060969	.2734602678	.3309059045

DIRICHLET 1   PROBABILITY (P=1/B)   B= 5

N	0	1	2	3	4
			R= 5		
35+	.3885171362	.4450291733	.4994543960	.5510701009	.5993898386
40+	.6441271227	.6851574421	.7224822331	.7561967981	.7864630127
45+	.8134869376	.8375010225	.8587503746	.8774824816	.8939397793
50+	.9083545029	.9209453281	.9319153849	.9414512999	.9497229902
55+	.9568839890	.9630721352	.9684104937	.9730084131	.9769626452
60+	.9803584776	.9832708429	.9857653785	.9878994244	.9897229482
65+	.9912793940	.9926064553	.9937367721	.9946985573	.9955161556
70+	.9962105403	.9967997525	.9972992889	.9977224427	.9980806017
75+	.9983835097	.9986394935	.9988556610	.9990380733	.9991918927
			R= 6		
30+	.0014719649	.0065187018	.0171348734	.0347309338	.0599372296
35+	.0926240607	.1320495525	.1770580134	.2262755666	.2782738752
40+	.3316909970	.3853098458	.4381006252	.4892356949	.5380852077
45+	.5842006465	.6272918135	.6672012839	.7038790140	.7373587561
50+	.7677371682	.7951559728	.8197871784	.8418211626	.8614573118
55+	.8788968620	.8943375867	.9079699977	.9199747597	.9305210595
60+	.9397657123	.9478528238	.9549138619	.9610680211	.9664227872
65+	.9710746311	.9751097772	.9786050068	.9816284640	.9842404455
70+	.9864941581	.9884364339	.9901083988	.9915460899	.9927810220
75+	.9938407031	.9947491018	.9955270678	.9961927090	.9967617285
80+	.9972477247	.9976624571	.9980160818	.9983173589	.9985738358
85+	.9987920068	.9989774545	.9991349721	.9992686711	.9993820743
			R= 7		
35+	.0010917666	.0049129498	.0131284937	.0270547958	.0474609468
40+	.0745253365	.1079003321	.1468351140	.1903164590	.2372007800
45+	.2863231060	.3365779151	.3869726045	.4366574942	.4849374590
50+	.5312702932	.5752562966	.6166227101	.6552057329	.6909320594
55+	.7238012080	.7538694085	.7812354365	.8060285234	.8283983006
60+	.8485066295	.8665211180	.8826100983	.8969388431	.9096668087
65+	.9209457151	.9309182954	.9397175715	.9474665356	.9542781388
70+	.9602555047	.9654923038	.9700732366	.9740745858	.9775648058
75+	.9806051266	.9832501545	.9855484566	.9875431217	.9892722895
80+	.9907696470	.9920648882	.9931841383	.9941503417	.9949836151
85+	.9957015683	.9963195924	.9968511205	.9973078600	.9977000008
90+	.9980364006	.9983247497	.9985717167	.9987830774	.9989638295
95+	.9991182917	.9992501931	.9993627496	.9994587323	.9995405265
			R= 8		
40+	.0008418647	.0038351612	.0103805030	.0216728705	.0385197355
45+	.0612709210	.0898354892	.1237539424	.1622965983	.2045660848
50+	.2495898450	.2963953857	.3440660027	.3917779049	.4388213703
55+	.4846092410	.5286760712	.5706708906	.6103460298	.6475439078
60+	.6821831753	.7142451751	.7437613362	.7708018520	.7954658020
65+	.8178727428	.8381557058	.8564554890	.8729161011	.8876812069
70+	.9008914263	.9126823462	.9231831199	.9325155399	.9407934907
75+	.9481226965	.9546006965	.9603169913	.9653533131	.9697839831
80+	.9736763269	.9770911237	.9800830713	.9827012550	.9849896074
85+	.9869873542	.9887294394	.9902469270	.9915673774	.9927151969
90+	.9937119604	.9945767083	.9953262172	.9959752468	.9965367632
95+	.9970221415	.9974413469	.9978030988	.9981150168	.9983837514
100+	.9986151005	.9988141130	.9989851809	.9991321204	.9992582444
			R= 9		
45+	.0006688774	.0030768359	.0084137476	.0177530075	.0318915278
50+	.0512707444	.0759664779	.1057277045	.1400435183	.1782208675
55+	.2194605328	.2629236501	.3077850504	.3532725407	.3986930501
60+	.4434475372	.4870369245	.5290613177	.5692145411	.6072756880
65+	.6430990365	.6766033429	.7077612352	.7365891907	.7631383923
70+	.7874866172	.8097312087	.8299831142	.8483619240	.8649918230

DIRICHLET 1 PROBABILITY (P=1/B)   B= 5

N	0	1	2	3	4
			R= 9		
75+	.8799983516	.8935058703	.9056356250	.9165043181	.9262230969
80+	.9348968828	.9426239754	.9494958729	.9555972604	.9610061276
85+	.9657939797	.9700261177	.9737619634	.9770554133	.9799552080
90+	.9825053053	.9847452494	.9867105315	.9884329353	.9899408662
95+	.9912596614	.9924118801	.9934175738	.9942945354	.9950585288
100+	.9957234997	.9963017673	.9968041990	.9972403687	.9976187001
105+	.9979465955	.9982305520	.9984762654	.9986887235	.9988722894
110+	.9990307754	.9991675095	.9992853936	.9993869557	.9994743963
			R=10		
50+	.0005442012	.0025231146	.0069576797	.0148092232	.0268408918
55+	.0435384618	.0650853691	.0913795822	.1220777292	.1566530364
60+	.1944564803	.2347738718	.2768746062	.3200502039	.3636424516
65+	.4070620002	.4497988396	.4914262692	.5315999596	.5700535418
70+	.6065919382	.6410834101	.6734510692	.7036643982	.7317311546
75+	.7576899000	.7816032891	.8035521753	.8236305370	.8419411893
80+	.8585922227	.8736940992	.8873573287	.8996906494	.9107996397
85+	.9207856924	.9297452922	.9377695397	.9449438762	.9513479678
90+	.9570557149	.9621353569	.9666496493	.9706560922	.9742071950
95+	.9773507635	.9801301996	.9825848051	.9847500839	.9866580371
100+	.9883374483	.9898141557	.9911113109	.9922496210	.9932475761
105+	.9941216598	.9948865447	.9955552723	.9961394177	.9966492405
110+	.9970938227	.9974811935	.9978184425	.9981118226	.9983668420
115+	.9985883477	.9987806002	.9989473404	.9990918496	.9992170034

DIRICHLET 1 PROBABILITY (P=1/B)   B= 6

N	0	1	2	3	4
			R= 1		
5+	0	.0154320988	.0540123457	.1140260631	.1890432099
10+	.2718121285	.3562064186	.4378156806	.5138581940	.5828453489
15+	.6442127386	.6980043977	.7446324520	.7847071165	.8189230770
20+	.8479875409	.8725774848	.8933165344	.9107645583	.9254151646
25+	.9376978598	.9479827432	.9565863818	.9637780321	.9697857167
30+	.9748018864	.9789885405	.9824817617	.9853956751	.9878258670
35+	.9898523142	.9915418784	.9929504209	.9941245877	.9951033144
40+	.9959190895	.9965990152	.9971656953	.9976379789	.9980315821
45+	.9983596071	.9986329761	.9988607936	.9990506481	.9992088646
			R= 2		
10+	0	0	.0034382859	.0148992389	.0376619649
15+	.0727487528	.1191555899	.1745763984	.2361476705	.3010076016
20+	.3666333483	.4309956527	.4925893217	.5503917348	.6037874562
25+	.6524832940	.6964275610	.7357401072	.7706552812	.8014775288
30+	.8285481546	.8522213498	.8728475840	.8907626521	.9062809399
35+	.9196917528	.9312578056	.9412151880	.9497742959	.9571213592
40+	.9634202997	.9688147354	.9734300069	.9773751444	.9807447242
45+	.9836205873	.9860734068	.9881641007	.9899450950	.9914614434
50+	.9927518157	.9938493652	.9947824864	.9955754760	.9962491070
55+	.9968211258	.9973066826	.9977187027	.9980682065	.9983645849
60+	.9986158360	.9988287680	.9990091738	.9991619801	.9992913749
			R= 3		
15+	0	0	0	.0013511732	.0064180725
20+	.0176496995	.0368147771	.0646542810	.1008984700	.1444949791
25+	.1939075174	.2473948275	.3032273892	.3598320546	.4158724837
30+	.4702802184	.5222519434	.5712262301	.6168498413	.6589405717
35+	.6974510240	.7324358061	.7640232988	.7923922886	.8177532468
40+	.8403337696	.8603675809	.8780864852	.8937146995	.9074650569

A 8

DIRICHLET 1   PROBABILITY (P=1/B)   B= 6

N	0	1	2	3	4
			R= 3		
45+	.9195366519	.9301135724	.9393644307	.9474424676	.9544860509
50+	.9606194360	.9659536834	.9705876621	.9746090804	.9780955091
55+	.9811153696	.9837288692	.9859888754	.9879417224	.9896279486
60+	.9910829668	.9923376680	.9934189636	.9943502702	.9951519406
65+	.9958416463	.9964347161	.9969444342	.9973823037	.9977582767
70+	.9980809576	.9983577794	.9985951595	.9987986343	.9989729773
75+	.9991223020	.9992501505	.9993595720	.9994531893	.9995332574
			R= 4		
20+	0	0	0	0	.0006851856
25+	.0034259279	.0098971250	.0216322874	.0396998425	.0645580114
30+	.0960635908	.1335790668	.1761231652	.2225242744	.2715527885
35+	.3220220001	.3728562910	.4231304668	.4720862413	.5191321902
40+	.5638328204	.6058913135	.6451293512	.6814663983	.7148999745
45+	.7454878021	.7733322574	.7985672396	.8213473711	.8418393330
50+	.8602150778	.8766466496	.8913023440	.9043439663	.9159249683
55+	.9261892796	.9352706750	.9432925487	.9503679890	.9566000705
60+	.9620822945	.9668991282	.9711266008	.9748329270	.9780791356
65+	.9809196874	.9834030700	.9855723631	.9874657694	.9891171073
70+	.9905562670	.9918096273	.9929004363	.9938491571	.9946737795
75+	.9953901018	.9960119825	.9965515667	.9970194877	.9974250469
80+	.9977763737	.9980805671	.9983438219	.9985715391	.9987684250
85+	.9989385772	.9990855610	.9992124767	.9993220185	.9994165262
			R= 5		
30+	.0004018233	.0020760869	.0061953070	.0139760560	.0264426903
35+	.0442713092	.0677263393	.0966742935	.1306484011	.1689380844
40+	.2106829197	.2549577902	.3008422417	.3474717677	.3940717907
45+	.4399767684	.4846375156	.5276198572	.5685973980	.6073407037
50+	.6437046724	.6776153888	.7090573502	.7380616246	.7646952543
55+	.7890520401	.8112447185	.8313984649	.8496456093	.8661214273
60+	.8809608635	.8942960442	.9062544507	.9169576316	.9265203534
65+	.9350500965	.9426468228	.9494029510	.9554034871	.9607262689
70+	.9654422883	.9696160674	.9733060644	.9765650954	.9794407574
75+	.9819758457	.9842087574	.9861738770	.9879019411	.9894203801
80+	.9907536366	.9919234597	.9929491760	.9938479377	.9946349493
85+	.9953236739	.9959260194	.9964525079	.9969124280	.9973139717
90+	.9976643573	.9979699401	.9982363109	.9984683838	.9986704753
95+	.9988463736	.9989994017	.9991324717	.9992481344	.9993486226
			R= 6		
35+	0	.0002588780	.0013683552	.0041783703	.0096439046
40+	.0186599746	.0319294920	.0498837868	.0726564241	.1000998935
45+	.1318306636	.1672885739	.2057993515	.2466325478	.2890504489
50+	.3323460975	.3758703654	.4190491247	.4613921274	.5024953793
55+	.5420387277	.5797801797	.6155482054	.6492330099	.6807775083
60+	.7101685213	.7374285320	.7626082089	.7857797946	.8070313871
65+	.8264620891	.8441779694	.8602887620	.8749052173	.8881370229
70+	.9000912090	.9108709642	.9205747895	.9292959312	.9371220376
75+	.9441349949	.9504109017	.9560201506	.9610275899	.9654927429
80+	.9694700675	.9730092408	.9761554586	.9789497403	.9814292323
85+	.9836275054	.9855748421	.9872985109	.9888230265	.9901703947
90+	.9913603409	.9924105228	.9933367272	.9941530514	.9948720695
95+	.9955049849	.9960617697	.9965512906	.9969814247	.9973591631
100+	.9976907051	.9979815436	.9982365412	.9984599993	.9986557200
105+	.9988270617	.9989769883	.9991081142	.9992227437	.9993229064
			R= 7		
40+	0	0	.0001781322	.0009574607	.0029743339
45+	.0069848087	.0137496028	.0239293907	.0380089909	.0562566922
50+	.0787162379	.1052242288	.1354441630	.1689088678	.2050647005

A 9

DIRICHLET 1   PROBABILITY (P=1/B)   B= 6

N	0	1	2	3	4
			R= 7		
55+	.2433128425	.2830448510	.3236711347	.3646421173	.4054625794
60+	.4457000792	.4849885284	.5230280236	.5595819455	.5944722083
65+	.6275733864	.6588062884	.6881314098	.7155425721	.7410609562
70+	.7647296561	.7866088210	.8067714041	.8252995098	.8422813055
75+	.8578084540	.8719740133	.8848707509	.8965898171	.9072197267
80+	.9168456023	.9255486350	.9334057249	.9404892690	.9468670659
85+	.9526023139	.9577536819	.9623754339	.9665175952	.9702261464
90+	.9735432363	.9765074061	.9791538188	.9815144882	.9836185052
95+	.9854922573	.9871596397	.9886422571	.9899596146	.9911292974
100+	.9921671388	.9930873774	.9939028027	.9946248891	.9952639207
105+	.9958291049	.9963286774	.9967699971	.9971596341	.9975034489
110+	.9978066642	.9980739310	.9983093877	.9985167134	.9986991770
115+	.9988596805	.9990007981	.9991248123	.9992337449	.9993293861
			R= 8		
45+	0	0	0	.0001286843	.0007006146
50+	.0022056386	.0052504597	.0104776087	.0184841616	.0297555106
55+	.0446217685	.0632380201	.0855854281	.1114880688	.1406398529
60+	.1726363832	.2070076229	.2432484296	.2808451128	.3192970984
65+	.3581334755	.3969246770	.4352898363	.4729005085	.5094814802
70+	.5448093683	.5787096316	.6110525321	.6417484821	.6707431227
75+	.6980123914	.7235577676	.7474018207	.7695841397	.7901576857
80+	.8091855789	.8267383144	.8428913849	.8577232808	.8713138338
85+	.8837428659	.8950891084	.9054293550	.9148378144	.9233856347
90+	.9311405688	.9381667590	.9445246181	.9502707885	.9554581632
95+	.9601359564	.9643498099	.9681419280	.9715512316	.9746135256
100+	.9773616739	.9798257781	.9820333564	.9840095202	.9857771456
105+	.9873570400	.9887681010	.9900274675	.9911506628	.9921517294
110+	.9930433548	.9938369892	.9945429551	.9951705483	.9957281317
115+	.9962232217	.9966625676	.9970522242	.9973976189	.9977036121
120+	.9979745535	.9982143320	.9984264224	.9986139269	.9987796128
125+	.9989259476	.9990551290	.9991691140	.9992696435	.9993582652
			R= 9		
50+	0	0	0	0	.0000965100
55+	.0005308050	.0016889250	.0040646794	.0082018408	.0146312862
60+	.0238151718	.0361053998	.0517191371	.0707307557	.0930774651
65+	.1185749596	.1469392962	.1778116461	.2107832455	.2454186300
70+	.2812759366	.3179236506	.3549536256	.3919905236	.4286980252
75+	.4647822720	.4999930395	.5341231344	.5670064685	.5985152088
80+	.6285563366	.6570678885	.6840150907	.7093865473	.7331905965
85+	.7554519128	.7762084013	.7955084098	.8134082647	.8299701225
90+	.8452601253	.8593468353	.8722999272	.8841891098	.8950832539
95+	.9050496973	.9141537065	.9224580708	.9300228090	.9369049712
100+	.9431585187	.9488342685	.9539798904	.9586399448	.9628559539
105+	.9666664974	.9701073266	.9732114922	.9760094799	.9785293516
110+	.9807968890	.9828357362	.9846675409	.9863120919	.9877874518
115+	.9891100845	.9902949771	.9913557547	.9923047899	.9931533044
120+	.9939114654	.9945884747	.9951926521	.9957315124	.9962118371
125+	.9966397402	.9970207293	.9973597612	.9976612935	.9979293317
130+	.9981674719	.9983789407	.9985666305	.9987331327	.9988807671
135+	.9990116088	.9991275133	.9992301382	.9993209632	.9994013091
			R=10		
60+	.0000745627	.0004134841	.0013271168	.0032228362	.0065632808
65+	.0118176000	.0194151020	.0297076867	.0429443125	.0592581195
70+	.0786649904	.1010712735	.1262879934	.1540489405	.1840303736
75+	.2158705518	.2491878165	.2835964100	.3187196085	.3542000515
80+	.3897073658	.4249433268	.4596448793	.4935853752	.5265743849
85+	.5584564178	.5891088524	.6184393315	.6463828391	.6728986291

DIRICHLET 1   PROBABILITY (P=1/B)   B= 6

N	0	1	2	3	4
			R=10		
90+	.6979671400	.7215869954	.7437721606	.7645493033	.7839553862
95+	.8020355048	.8188409728	.8344276477	.8488544867	.8621823151
100+	.8744727897	.8857875399	.8961874646	.9057321684	.9144795190
105+	.9224853080	.9298030026	.9364835713	.9425753746	.9481241075
110+	.9531727854	.9577617652	.9619287937	.9657090786	.9691353753
115+	.9722380860	.9750453681	.9775832466	.9798757312	.9819449333
120+	.9838111825	.9854931413	.9870079167	.9883711685	.9895972124
125+	.9906991195	.9916888099	.9925771419	.9933739955	.9940883516
130+	.9947283650	.9953014334	.9958142616	.9962729205	.9966829025
135+	.9970491721	.9973762133	.9976680723	.9979283980	.9981604779
140+	.9983672721	.9985514439	.9987153877	.9988612548	.9989909766
145+	.9991062862	.9992087376	.9992997231	.9993804898	.9994521538

DIRICHLET 1   PROBABILITY (P=1/B)   B= 7

N	0	1	2	3	4
			R= 1		
5+	0	0	.0061198990	.0244795961	.0577019051
10+	.1049125547	.1630962010	.2284524404	.2973065453	.3665749237
15+	.4339188263	.4977198231	.5569727114	.6111536514	.6600938106
20+	.7038716281	.7427272334	.7769978107	.8070707585	.8333510369
25+	.8562393434	.8761182816	.8933442601	.9082433889	.9211100811
30+	.9322074209	.9417686268	.9499991441	.9570790458	.9631655314
35+	.9683953851	.9728873129	.9767441086	.9800546296	.9828955745
40+	.9853330675	.9874240600	.9892175629	.9907557252	.9920747745
45+	.9932058342	.9941756326	.9950071160	.9957199786	.9963311199
50+	.9968550382	.9973041703	.9976891830	.9980192242	.9983021381
55+	.9985446513	.9987525307	.9989307209	.9990834610	.9992143851
			R= 2		
10+	0	0	0	0	.0010042130
15+	.0050210648	.0143458994	.0308140811	.0553790755	.0880427148
20+	.1280185009	.1739933610	.2243918397	.2775922459	.3320783075
25+	.3865296941	.4398635648	.4912413010	.5400531370	.5858906559
30+	.6285142441	.6678201370	.7038097949	.7365630049	.7662152135
35+	.7929390508	.8169297001	.8383936238	.8575401144	.8745751587
40+	.8896971496	.9030940450	.9149416352	.9254026450	.9346264472
45+	.9427492138	.9498943697	.9561732452	.9616858485	.9665217022
50+	.9707607005	.9744739583	.9777246318	.9805686973	.9830556808
55+	.9852293330	.9871282499	.9887864380	.9902338272	.9914967333
60+	.9925982736	.9935587391	.9943959269	.9951254356	.9957609290
65+	.9963143684	.9967962196	.9972156350	.9975806146	.9978981485
70+	.9981743413	.9984145224	.9986233428	.9988048600	.9989626125
75+	.9990996851	.9992187664	.9993221992	.9994120240	.9994900177
			R= 3		
20+	0	.0003267581	.0017971694	.0056097358	.0130753241
25+	.0253350544	.0431602208	.0668639724	.0963106256	.1309898842
30+	.1701221559	.2127686089	.2579292564	.3046208416	.3519324191
35+	.3990602720	.4453256682	.4901795331	.5331979306	.5740716713
40+	.6125926678	.6486389732	.6821598410	.7131616690	.7416953192
45+	.7678450478	.7917190957	.8134418774	.8331476345	.8509753903
50+	.8670650271	.8815543121	.8945767111	.9062598444	.9167244574
55+	.9260837981	.9344433111	.9419005708	.9485453944	.9544600841
60+	.9597197594	.9643927480	.9685410118	.9722205904	.9754820460
65+	.9783709023	.9809280682	.9831902426	.9851902961	.9869576291
70+	.9885185035	.9898963506	.9911120527	.9921842020	.9931293361
75+	.9939621522	.9946957013	.9953415638	.9959100078	.9964101315

SOBEL, UPPULURI AND FRANKOWSKI

DIRICHLET 1 PROBABILITY (P=1/B)   B= 7

N	0	1	2	3	4
			R= 3		
80+	.9968499917	.9972367181	.9975766170	.9978752630	.9981375811
85+	.9983679208	.9985701208	.9987475682	.9989032495	.9990397976
			R= 4		
25+	0	0	0	.0001445164	.0008381949
30+	.0027540689	.0067422148	.0136878865	.0243705808	.0393582938
35+	.0589505924	.0831683955	.1117796848	.1443476269	.1802886435
40+	.2189308223	.2595663502	.3014945698	.3440545024	.3866471856
45+	.4287490573	.4699180188	.5097938863	.5480948208	.5846111039
50+	.6191973643	.6517641074	.6822691657	.7107094986	.7371136119
55+	.7615347505	.7840449289	.8047298039	.8236843531	.8410092966
60+	.8568081870	.8711850880	.8842427591	.8960812740	.9067969987
65+	.9164818698	.9252229158	.9331019745	.9401955661	.9465748864
70+	.9523058926	.9574494582	.9620615768	.9661935999	.9698924967
75+	.9732011249	.9761585062	.9788000999	.9811580709	.9832615473
80+	.9851368682	.9868078176	.9882958445	.9896202695	.9907984766
85+	.9918460911	.9927771439	.9936042228	.9943386111	.9949904146
90+	.9955686775	.9960814880	.9965360745	.9969388923	.9972957028
95+	.9976116452	.9978913005	.9981387505	.9983576294	.9985511719
100+	.9987222554	.9988734380	.9990069933	.9991249408	.9992290734
			R= 5		
35+	.0000761259	.0004567555	.0015520365	.0039266250	.0082305754
40+	.0151128778	.0251400782	.0387347410	.0561399979	.0774096781
45+	.1024193861	.1308921381	.1624321409	.1965612576	.2327540959
50+	.2704690703	.3091740180	.3483658997	.3875847779	.4264226734
55+	.4645281087	.5016072143	.5374222383	.5717882161	.6045684402
60+	.6356692505	.6650345458	.6926403179	.7184894165	.7426066850
65+	.7650345474	.7858290870	.8050566220	.8227907643	.8391099331
70+	.8540952831	.8678290078	.8803929712	.8918676279	.9023311893
75+	.9118589988	.9205230833	.9283918489	.9355298967	.9419979341
80+	.9478527628	.9531473270	.9579308073	.9622487492	.9661432161
85+	.9696529594	.9728135989	.9756578086	.9782155046	.9805140299
90+	.9825783365	.9844311606	.9860931915	.9875832313	.9889183470
95+	.9901140133	.9911842466	.9921417293	.9929979267	.9937631941
100+	.9944468767	.9950574011	.9956023600	.9960885895	.9965222403
105+	.9969088429	.9972533663	.9975602729	.9978335673	.9980768413
110+	.9982933149	.9984858727	.9986570977	.9988093020	.9989445537
115+	.9990647018	.9991713990	.9992661214	.9993501872	.9994247727
			R= 6		
40+	0	0	.0000448986	.0002758059	.0009596918
45+	.0024861008	.0053340669	.0100205058	.0170433355	.0268304065
50+	.0397013073	.0558449847	.0753128291	.0980247718	.1237849294
55+	.1523031657	.1832193220	.2161275167	.2505986557	.2861999845
60+	.3225110978	.3591362619	.3957132251	.4319188789	.4674722417
65+	.5021352669	.5357119660	.5680462932	.5990191804	.6285450465
70+	.6565680410	.6830582233	.7080078267	.7314277133	.7533440872
75+	.7737955091	.7928302289	.8105038395	.8268772420	.8420149056
80+	.8559833983	.8688501650	.8806825250	.8915468651	.9015080009
85+	.9106286852	.9189692404	.9265872965	.9335376165	.9398719948
90+	.9456392149	.9508850543	.9556523280	.9599809608	.9639080822
95+	.9674681375	.9706930108	.9736121548	.9762527254	.9786397169
100+	.9807960974	.9827429417	.9844995600	.9860836229	.9875112810
105+	.9887972783	.9899550607	.9909968771	.9919338758	.9927761935
110+	.9935330394	.9942127726	.9948229752	.9953705185	.9958616256
115+	.9963019283	.9966965197	.9970500032	.9973665363	.9976498719
120+	.9979033955	.9981301596	.9983329147	.9985141377	.9986760583
125+	.9988206827	.9989498144	.9990650746	.9991679198	.9992596576

DIRICHLET 1   PROBABILITY (P=1/B)   B= 7

N	0	1	2	3	4
			R= 7		
45+	0	0	0	0	.0000286607
50+	.0001791295	.0006344171	.0016730484	.0036541000	.0069865873
55+	.0120908021	.0193591650	.0291224952	.0416252954	.0570114358
60+	.0753198699	.0964888707	.1203666998	.1467264893	.1752832903
65+	.2057115843	.2376619703	.2707761451	.3046996620	.3390922452
70+	.3736356612	.4080393033	.4420437401	.4754225290	.5079826109
75+	.5395635909	.5700361859	.5993000858	.6272814362	.6539301142
80+	.6792169333	.7031308799	.7256764582	.7468711971	.7667433512
85+	.7853298164	.8026742663	.8188255084	.8338360516	.8477608736
90+	.8606563730	.8725794888	.8835869728	.8937347949	.9030776680
95+	.9116686751	.9195589857	.9267976477	.9334314443	.9395048038
100+	.9450597549	.9501359180	.9547705268	.9589984725	.9628523675
105+	.9663626224	.9695575338	.9724633786	.9751045139	.9775034788
110+	.9796810973	.9816565802	.9834476261	.9850705192	.9865402240
115+	.9878704767	.9890738724	.9901619476	.9911452596	.9920334600
120+	.9928353649	.9935590204	.9942117636	.9948002803	.9953306579
125+	.9958084350	.9962386477	.9966258717	.9969742617	.9972875882
130+	.9975692705	.9978224077	.9980498071	.9982540103	.9984373168
135+	.9986018065	.9987493591	.9988816730	.9990002816	.9991065691
			R= 8		
55+	0	.0000193970	.0001228477	.0004410819	.0011795314
60+	.0026126677	.0050659847	.0088899207	.0144307172	.0220026964
65+	.0318652484	.0442064215	.0591337328	.0766718221	.0967659228
70+	.1192897866	.1440566164	.1708316551	.1993452780	.2293056857
75+	.2604105473	.2923571800	.3248510487	.3576125264	.3903819717
80+	.4229232593	.4550259454	.4865062742	.5172072347	.5469978691
85+	.5757720165	.6034466549	.6299599792	.6552693313	.6793490728
90+	.7021884752	.7237896786	.7441657603	.7633389380	.7813389236
95+	.7982014354	.8139668699	.8286791299	.8423846007	.8551312679
100+	.8669679631	.8779437287	.8881072886	.8975066140	.9061885734
105+	.9141986561	.9215807592	.9283770307	.9346277584	.9403712994
110+	.9456440428	.9504804000	.9549128176	.9589718085	.9626859983
115+	.9660821820	.9691853901	.9720189606	.9746046152	.9769625387
120+	.9791114587	.9810687264	.9828503962	.9844713040	.9859451432
125+	.9872845390	.9885011186	.9896055801	.9906077565	.9915166779
130+	.9923406294	.9930872065	.9937633666	.9943754778	.9949293643
135+	.9954303491	.9958832939	.9962926359	.9966624220	.9969963412
140+	.9972977541	.9975697202	.9978150229	.9980361940	.9982355339
145+	.9984151326	.9985768875	.9987225204	.9988535928	.9989715203
150+	.9990775857	.9991729506	.9992586665	.9993356850	.9994048667
			R= 9		
60+	0	0	0	.0000137315	.0000878818
65+	.0003189997	.0008626768	.0019327052	.0037906913	.0067285071
70+	.0110468334	.0170330735	.0249413559	.0349765124	.0472830336
75+	.0619392297	.0789562370	.0982811239	.1198031545	.1433622206
80+	.1687585108	.1957626141	.2241254127	.2535872848	.2838862957
85+	.3147651913	.3459771195	.3772900911	.4084902519	.4393840815
90+	.4697996528	.4995871001	.5286184387	.5567868758	.5840057353
95+	.6102071083	.6353403235	.6593703154	.6822759560	.7040483980
100+	.7246894703	.7442101525	.7626291485	.7799715705	.7962677417
105+	.8115521169	.8258623226	.8392383097	.8517216145	.8633547207
110+	.8741805132	.8842418173	.8935810145	.9022397254	.9102585541
115+	.9176768859	.9245327306	.9308626069	.9367014609	.9420826142
120+	.9470377371	.9515968415	.9557882927	.9596388337	.9631736216
125+	.9664162726	.9693889140	.9721122412	.9746055784	.9768869411
130+	.9789731012	.9808796507	.9826210669	.9842107753	.9856612121
135+	.9869838845	.9881894288	.9892876672	.9902876610	.9911977627

DIRICHLET 1   PROBABILITY (P=1/B)   B= 7

N	0	1	2	3	4
			R= 9		
140+	.9920256641	.9927784433	.9934626080	.9940841373	.9946485204
145+	.9951607930	.9956255720	.9960470868	.9964292101	.9967754852
150+	.9970891523	.9973731725	.9976302502	.9978628545	.9980732376
155+	.9982634534	.9984353737	.9985907034	.9987309948	.9988576604
160+	.9989719851	.9990751369	.9991681770	.9992520694	.9993276894
			R=10		
70+	.0000100740	.0000650229	.0002381360	.0006499528	.0014698944
75+	.0029105586	.0052159449	.0086457138	.0134578366	.0198917859
80+	.0281539445	.0384063227	.0507591019	.0652670346	.0819293714
85+	.1006927496	.1214563653	.1440787218	.1683852962	.1941765458
90+	.2212357867	.2493365883	.2782494363	.3077475112	.3376115123
95+	.3676335170	.3976199160	.4273934961	.4567947632	.4856826088
100+	.5139344256	.5414457768	.5681297133	.5939158280	.6187491227
105+	.6425887540	.6654067122	.6871864799	.7079217032	.7276149059
110+	.7462762641	.7639224576	.7805756061	.7962622952	.8110126951
115+	.8248597696	.8378385745	.8499856392	.8613384287	.8719348791
120+	.8818130000	.8910105398	.8995647056	.9075119332	.9148877013
125+	.9217263847	.9280611411	.9339238285	.9393449472	.9443536039
130+	.9489774947	.9532429029	.9571747110	.9607964219	.9641301896
135+	.9671968567	.9700159972	.9726059626	.9749839322	.9771659640
140+	.9791670477	.9810011575	.9826813051	.9842195920	.9856272607
145+	.9869147451	.9880917187	.9891671415	.9901493056	.9910458775
150+	.9918639402	.9926100316	.9932901822	.9939099505	.9944744556
155+	.9949884092	.9954561449	.9958816458	.9962685708	.9966202785
160+	.9969398505	.9972301119	.9974936519	.9977328417	.9979498517
165+	.9981466675	.9983251050	.9984868233	.9986333384	.9987660342
170+	.9988861735	.9989949082	.9990932888	.9991822721	.9992627301

DIRICHLET 1   PROBABILITY (P=1/B)   B= 8

N	0	1	2	3	4
			R= 1		
5+	0	0	0	.0024032593	.0108146667
10+	.0281631947	.0557631254	.0933064241	.1393206837	.1917180140
15+	.2482475596	.3067978817	.3655622114	.4231023388	.4783476425
20+	.5305582405	.5792723223	.6242501484	.6654216742	.7028410753
25+	.7366492133	.7670438184	.7942565489	.8185358557	.8401345615
30+	.8593011613	.8762739868	.8912775310	.9045203642	.9161941994
35+	.9264737644	.9355172232	.9434669545	.9504505450	.9565818977
40+	.9619623830	.9666819828	.9708203963	.9744480853	.9776272488
45+	.9804127204	.9828527869	.9849899297	.9868614937	.9885002854
50+	.9899351083	.9911912385	.9922908477	.9932533771	.9940958690
55+	.9948332583	.9954786307	.9960434492	.9965377535	.9969703359
60+	.9973488951	.9976801716	.9979700664	.9982237453	.9984457301
65+	.9986399785	.9988099547	.9989586905	.9990888393	.9992027232
			R= 2		
15+	0	.0002903620	.0016453849	.0052446643	.0124072304
20+	.0242964752	.0417059809	.0649634850	.0939392216	.1281231582
25+	.1667344032	.2088345685	.2534276567	.2995382962	.3462666357
30+	.3928220278	.4385394124	.4828827545	.5254395722	.5659099218
35+	.6040924413	.6398693350	.6731915741	.7040651071	.7325385208
40+	.7586923378	.7826299711	.8044702505	.8243413755	.8423761238
45+	.8587081364	.8734691074	.8867867210	.8987831951	.9095743104
50+	.9192688204	.9279681588	.9357663706	.9427502109	.9489993644
55+	.9545867496	.9595788772	.9640362432	.9680137361	.9715610493
60+	.9747230856	.9775403505	.9800493267	.9822828294	.9842703388

DIRICHLET 1   PROBABILITY (P=1/B)   B= 8

N	0	1	2	3	4
			R= 2		
65+	.9860383099	.9876104594	.9890080301	.9902500332	.9913534703
70+	.9923335349	.9932037967	.9939763678	.9946620536	.9952704889
75+	.9958102608	.9962890192	.9967135766	.9970899972	.9974236773
80+	.9977194171	.9979814855	.9982136774	.9984193657	.9986015471
85+	.9987628836	.9989057390	.9990322124	.9991441665	.9992432549
			R= 3		
20+	0	0	0	0	.0000782233
25+	.0004888955	.0017080785	.0044017778	.0093340872	.0172518933
30+	.0287799384	.0443470834	.0641506893	.0881558253	.1161205885
35+	.1476373214	.1821805045	.2191542675	.2579348747	.2979056883
40+	.3384837567	.3791383011	.4194020364	.4588765778	.4972332561
45+	.5342105851	.5696094628	.6032869940	.6351496294	.6651461363
50+	.6932607660	.7195068567	.7439210180	.7665579658	.7874860282
55+	.8067833041	.8245344350	.8408279366	.8557540287	.8694029022
60+	.8818633634	.8932217981	.9035614047	.9129616497	.9214979059
65+	.9292412375	.9362583027	.9426113487	.9483582765	.9535527604
70+	.9582444055	.9624789323	.9662983802	.9697413197	.9728430703
75+	.9756359175	.9781493267	.9804101500	.9824428260	.9842695698
80+	.9859105523	.9873840699	.9887067029	.9898934631	.9909579309
85+	.9919123827	.9927679081	.9935345177	.9942212427	.9948362257
90+	.9953868040	.9958795857	.9963205196	.9967149577	.9970677134
95+	.9973831138	.9976650467	.9979170046	.9981421231	.9983432163
100+	.9985228094	.9986831670	.9988263197	.9989540876	.9990681014
			R= 4		
30+	0	0	.0000301719	.0001991343	.0007334782
35+	.0019888542	.0044279777	.0085733150	.0149480352	.0240186482
40+	.0361490992	.0515713523	.0703733029	.0925019247	.1177779872
45+	.1459182258	.1765611464	.2092933570	.2436741728	.2792570679
50+	.3156072425	.3523151159	.3890059253	.4253458456	.4610451643
55+	.4958590826	.5295866959	.5620686535	.5931839293	.6228460574
60+	.6509991151	.6776136666	.7026828263	.7262185463	.7482481990
65+	.7688114923	.7879577310	.8057434242	.8222302235	.8374831706
70+	.8515692287	.8645560687	.8765110823	.8875005918	.8975892330
75+	.9068394834	.9153113151	.9230619520	.9301457129	.9366139255
80+	.9425148978	.9478939352	.9527933928	.9572527564	.9613087435
85+	.9649954189	.9683443207	.9713845926	.9741431194	.9766446623
90+	.9789119940	.9809660298	.9828259558	.9845093512	.9860323058
95+	.9874095315	.9886544677	.9897793807	.9907954569	.9917128900
100+	.9925409620	.9932881195	.9939620431	.9945697133	.9951174698
105+	.9956110678	.9960557283	.9964561859	.9968167316	.9971412526
110+	.9974332686	.9976959654	.9979322249	.9981446536	.9983356074
115+	.9985072152	.9986613999	.9987998979	.9989242765	.9990359497
			R= 5		
40+	.0000142756	.0000975502	.0003719102	.0010432030	.0024006868
45+	.0047998091	.0086325419	.0142917957	.0221363516	.0324612990
50+	.0454769643	.0612973836	.0799378498	.1013200741	.1252830160
55+	.1515973400	.1799816386	.2101188837	.2416719569	.2742974833
60+	.3076575272	.3414289739	.3753106242	.4090281636	.4423372511
65+	.4750250173	.5069102668	.5378426713	.5677012132	.5963921062
70+	.6238463854	.6500173214	.6748777826	.6984176385	.7206412725
75+	.7415652513	.7612161799	.7796287591	.7968440507	.8129079468
80+	.8278698378	.8417814650	.8546959446	.8666669483	.8777480235
85+	.8879920379	.8974507345	.9061743809	.9142115015	.9216086796
90+	.9284104196	.9346590587	.9403947204	.9456553019	.9504764894
95+	.9548917952	.9589326126	.9626282835	.9660061764	.9690917710
100+	.9719087480	.9744790804	.9768231274	.9789597267	.9809062865
105+	.9826788745	.9842923053	.9857602235	.9870951843	.9883087297

A15

DIRICHLET 1 PROBABILITY (P=1/B)   B= 8

N	0	1	2	3	4
			R= 5		
110+	.9894114610	.9904131080	.9913225930	.9921480925	.9928970940
115+	.9935764495	.9941924255	.9947507495	.9952566533	.9957149132
120+	.9961298877	.9965055512	.9968455268	.9971531154	.9974313228
125+	.9976828854	.9979102926	.9981158086	.9983014916	.9984692122
130+	.9986206694	.9987574059	.9988808222	.9989921889	.9990926586
			R= 6		
45+	0	0	0	.770778E-05	.0000539545
50+	.0002107597	.0006056832	.0014276855	.0029226050	.0053791540
55+	.0091082522	.0144193233	.0215970578	.0308814129	.0424526234
60+	.0564220002	.0728284572	.0916401122	.1127599552	.1360344341
65+	.1612638300	.1882134142	.2166245644	.2462252173	.2767392304
70+	.3078944013	.3394290336	.3710970500	.4026717324	.4339482210
75+	.4647449329	.4949040732	.5242914107	.5527954794	.5803263537
80+	.6068141240	.6322071817	.6564704038	.6795833086	.7015382375
85+	.7223386058	.7419972507	.7605348978	.7779787562	.7943612490
90+	.8097188778	.8240912198	.8375200502	.8500485820	.8617208162
95+	.8725809901	.8826731167	.8920406045	.9007259478	.9087704815
100+	.9162141884	.9230955561	.9294514727	.9353171581	.9407261240
105+	.9457101582	.9502993294	.9545220078	.9584048993	.9619730901
110+	.9652500988	.9682579353	.9710171635	.9735469666	.9758652146
115+	.9779885321	.9799323657	.9817110513	.9833378790	.9848251569
120+	.9861842729	.9874257536	.9885593210	.9895939470	.9905379044
125+	.9913988159	.9921837003	.9928990157	.9935507008	.9941442127
130+	.9946845638	.9951763544	.9956238052	.9960307857	.9964008421
135+	.9967372228	.9970429014	.9973205992	.9975728052	.9978017949
140+	.9980096480	.9981982640	.9983693777	.9985245722	.9986652921
145+	.9987928548	.9989084614	.9990132067	.9991080879	.9991940134
			R= 7		
55+	0	.456442E-05	.0000325215	.0001293519	.0003785691
60+	.0009087623	.0018943252	.0035495218	.0061169175	.0098520715
65+	.0150066991	.0218123763	.0304664316	.0411211100	.0538765297
70+	.0687774647	.0858136216	.1049228393	.1259965234	.1488866020
75+	.1734133356	.1993734037	.2265477993	.2547091837	.2836284591
80+	.3130804193	.3428484163	.3727280461	.4025299023	.4320814784
85+	.4612283186	.4898345252	.5177827324	.5449736528	.5713252925
90+	.5967719233	.6212628868	.6447612959	.6672426849	.6886936539
95+	.7091105376	.7284981266	.7468684585	.7642396900	.7806350598
100+	.7960819434	.8106110032	.8242554290	.8370502677	.8490318365
105+	.8602372144	.8707038052	.8804689667	.8895696992	.8980423880
110+	.9059225926	.9132448784	.9200426855	.9263482287	.9321924263
115+	.9376048521	.9426137075	.9472458111	.9515266025	.9554801571
120+	.9591292121	.9624951983	.9655982794	.9684573951	.9710903084
125+	.9735136539	.9757429896	.9777928474	.9796767853	.9814074383
130+	.9829965691	.9844551175	.9857932484	.9870203983	.9881453202
135+	.9891761265	.9901203300	.9909848835	.9917762168	.9925002718
140+	.9931625365	.9937680763	.9943215635	.9948273057	.9952892721
145+	.9957111175	.9960962062	.9964476331	.9967682441	.9970606546
150+	.9973272674	.9975702887	.9977917435	.9979934896	.9981772306
155+	.9983445284	.9984968141	.9986353983	.9987614813	.9988761617
160+	.9989804444	.9990752488	.9991614152	.9992397119	.9993108407
			R= 8		
60+	0	0	0	0	.289397E-05
65+	.0000209009	.0000843002	.0002502492	.0006094035	.0012886820
70+	.0024494552	.0042813601	.0069926586	.0107984584	.0159082238
75+	.0225138786	.0307795238	.0408334566	.0527628273	.0666109782
80+	.0823772706	.1000190489	.1194553036	.1405715622	.1632255529
85+	.1872532330	.2124748349	.2387006570	.2657363960	.2933878863

DIRICHLET 1   PROBABILITY (P=1/B)   B= 8

N	0	1	2	3	4
			R= 8		
90+	.3214651664	.3497858447	.3781777706	.4064810479	.4345494448
95+	.4622512670	.4894697681	.5161031692	.5420643606	.5672803517
100+	.5916915311	.6152507886	.6379225475	.6596817453	.6805127946
105+	.7004085521	.7193693135	.7374018519	.7545185093	.7707363497
110+	.7860763779	.8005628253	.8142225048	.8270842317	.8391783088
115+	.8505360731	.8611894987	.8711708539	.8805124065	.8892461739
120+	.8974037131	.9050159474	.9121130246	.9187242047	.9248777720
125+	.9306009692	.9359199508	.9408597532	.9454442780	.9496962881
130+	.9536374129	.9572881633	.9606679515	.9637951183	.9666869633
135+	.9693597797	.9718288904	.9741086871	.9762126696	.9781534866
140+	.9799429759	.9815922050	.9831115108	.9845105389	.9857982819
145+	.9869831163	.9880728384	.9890746992	.9899954370	.9908413099
150+	.9916181253	.9923312695	.9929857347	.9935861453	.9941367823
155+	.9946416064	.9951042805	.9955281897	.9959164610	.9962719818
160+	.9965974165	.9968952228	.9971676668	.9974168367	.9976446562
165+	.9978528965	.9980431880	.9982170304	.9983758031	.9985207743
170+	.9986531090	.9987738777	.9988840634	.9989845681	.9990762199
			R= 9		
70+	0	0	.193380E-05	.0000141168	.0000575740
75+	.0001728703	.0004258744	.0009111551	.0017522461	.0030986232
80+	.0051197740	.0079971147	.0119146955	.0170496543	.0235632728
85+	.0315932943	.0412479466	.0526018886	.0656941105	.0805276590
90+	.0970709600	.1152604376	.1350041107	.1561858442	.1786699618
95+	.2023059618	.2269331261	.2523848569	.2784926235	.3050894435
100+	.3320128553	.3591073722	.3862264284	.4132338464	.4400048645
105+	.4664267742	.4923992168	.5178341916	.5426558272	.5667999617
110+	.5902135759	.6128541186	.6346887575	.6556935840	.6758527978
115+	.6951578894	.7136068383	.7312033380	.7479560568	.7638779435
120+	.7789085783	.7932985783	.8068390643	.8196310763	.8317003142
125+	.8430735514	.8537783769	.8638428675	.8732953178	.8821640078
130+	.8904770039	.8982619904	.9055461291	.9123559435	.9187172249
135+	.9246549588	.9301932679	.9353553709	.9401635539	.9446391535
140+	.9488025498	.9526731673	.9562694832	.9596090411	.9627084701
145+	.9655835073	.9682490236	.9707190520	.9730068174	.9751247679
150+	.9770846071	.9788973266	.9805732385	.9821220081	.9835526862
155+	.9848737404	.9860930861	.9872181169	.9882557330	.9892123702
160+	.9900940262	.9909062871	.9916543520	.9923430569	.9929768973
165+	.9935600496	.9940963918	.9945895224	.9950427793	.9954592568
170+	.9958418218	.9961931294	.9965156375	.9968116201	.9970831802
175+	.9973322620	.9975606617	.9977700385	.9979619242	.9981377325
180+	.9982987674	.9984462315	.9985812333	.9987047942	.9988178552
185+	.9989212824	.9990158734	.9991023619	.9991814227	.9992536766
			R=10		
80+	.134718E-05	.992012E-05	.0000408265	.0001237361	.0003077554
85+	.0006648470	.0012910998	.0023055561	.0038466902	.0060669438
90+	.0091259203	.0131829216	.0183894945	.0248825586	.0327785587
95+	.0421689287	.0531170095	.0656564336	.0797908873	.0954950850
100+	.1127167485	.1313793591	.1513854517	.1726202337	.1949553351
105+	.2182525265	.2423672738	.2671520317	.2924592052	.3181437394
110+	.3440653153	.3700901528	.3960924321	.4219553587	.4475719016
115+	.4728452422	.4976889685	.5220270549	.5457936631	.5689327987
120+	.5913978545	.6131510705	.6341629339	.6544115425	.6738819481
125+	.6925654968	.7104591771	.7275649872	.7438893278	.7594424268
130+	.7742377992	.7882917452	.8016228875	.8142517482	.8262003649
135+	.8374919444	.8481505536	.8582008449	.8676678138	.8765765872
140+	.8849522404	.8928196389	.9002033045	.9071273025	.9136151485
145+	.9196897319	.9253732556	.9306871882	.9356522296	.9402882860

DIRICHLET 1   PROBABILITY (P=1/B)   B= 8

N	0	1	2	3	4
			R=10		
150+	.9446144548	.9486490168	.9524094356	.9559123630	.9591736486
155+	.9622083539	.9650307701	.9676544375	.9700921685	.9723560711
160+	.9744575739	.9764074528	.9782158568	.9798923353	.9814458649
165+	.9828848758	.9842172789	.9854504908	.9865914598	.9876466903
170+	.9886222669	.9895238772	.9903568346	.9911260993	.9918362992
175+	.9924917492	.9930964704	.9936542079	.9941684475	.9946424326
180+	.9950791786	.9954814886	.9958519659	.9961930282	.9965069190
185+	.9967957196	.9970613597	.9973056281	.9975301817	.9977365553
190+	.9979261693	.9981003382	.9982602778	.9984071123	.9985418808
195+	.9986655432	.9987789861	.9988830283	.9989784251	.9990658737

DIRICHLET 1   PROBABILITY (P=1/B)   B= 9

N	0	1	2	3	4
			R= 1		
5+	0	0	0	0	.0009366567
10+	.0046832835	.0133560308	.0286200661	.0513229630	.0814552992
15+	.1183129463	.1607355335	.2073383321	.2566975664	.3074784229
20+	.3585107898	.4088240681	.4576532903	.5044271448	.5487460401
25+	.5903559328	.6291216362	.6650018143	.6980268046	.7282797093
30+	.7558807654	.7809747557	.8037211060	.8242862730	.8428380326
35+	.8595413131	.8745552588	.8880312581	.9001117150	.9109293836
40+	.9206071199	.9292579382	.9369852814	.9438834364	.9500380422
45+	.9555266514	.9604193158	.9647791748	.9686630318	.9721219089
50+	.9752015719	.9779430237	.9803829618	.9825542017	.9844860649
55+	.9862047331	.9877335715	.9890934204	.9903028609	.9913784527
60+	.9923349492	.9931854912	.9939417797	.9946142314	.9952121183
65+	.9957436914	.9962162928	.9966364544	.9970099870	.9973420593
70+	.9976372682	.9978997024	.9981329975	.9983403868	.9985247452
75+	.9986886289	.9988343107	.9989638115	.9990789277	.9991812568
			R= 2		
15+	0	0	0	.0000833117	.0005276406
20+	.0018565134	.0047979832	.0101741114	.0187677194	.0312071112
25+	.0478913990	.0689614019	.0943093610	.1236154632	.1563986693
30+	.1920714706	.2299912884	.2695042283	.3099793258	.3508331210
35+	.3915454497	.4316678660	.4708262812	.5087193436	.5451138998
40+	.5798386449	.6127768230	.6438586220	.6730537131	.7003642346
45+	.7258183964	.7494647982	.7713674848	.7916017234	.8102504574
50+	.8274013780	.8431445442	.8575704826	.8707686996	.8828265427
55+	.8938283541	.9038548664	.9129827940	.9212845844	.9288282959
60+	.9356775727	.9418916981	.9475257031	.9526305181	.9572531529
65+	.9614368969	.9652215302	.9686435400	.9717363376	.9745304715
70+	.9770538356	.9793318686	.9813877447	.9832425540	.9849154725
75+	.9864239215	.9877837158	.9890092022	.9901133872	.9911080555
80+	.9920038784	.9928105149	.9935367030	.9941903444	.9947785813
85+	.9953078673	.9957840315	.9962123373	.9965975355	.9969439135
90+	.9972553386	.9975352991	.9977869400	.9980130963	.9982163233
95+	.9983989230	.9985629693	.9987103302	.9988426878	.9989615562
100+	.9990682987	.9991641421	.9992501904	.9993274368	.9993967749
			R= 3		
25+	0	0	.0000185812	.0001300684	.0005029312
30+	.0014203149	.0032730164	.0065267467	.0116732333	.0191771114
35+	.0294288719	.0427104512	.0591761056	.0788480407	.1016242475
40+	.1272950704	.1555649395	.1860761370	.2184321549	.2522189479
45+	.2870230721	.3224462566	.3581163717	.3936950372	.4288822805
50+	.4634187317	.4970858575	.5297047070	.5611335943	.5912650761

DIRICHLET 1 PROBABILITY (P=1/B)   B= 9

N	0	1	2	3	4
			R= 3		
55+	.6200225197	.6473564918	.6732411448	.6976707268	.7206563019
60+	.7422227361	.7624059767	.7812506378	.7988078866	.8151336188
65+	.8302869037	.8443286764	.8573206519	.8693244377	.8804008191
70+	.8906091949	.9000071427	.9086500938	.9165911003	.9238806798
75+	.9305667234	.9366944549	.9423064331	.9474425854	.9521402691
80+	.9564343503	.9603572987	.9639392920	.9672083271	.9701903346
85+	.9729092957	.9753873584	.9776449516	.9797008971	.9815725173
90+	.9832757389	.9848251925	.9862343060	.9875153943	.9886797430
95+	.9897376872	.9906986860	.9915713908	.9923637103	.9930828701
100+	.9937354685	.9943275280	.9948645431	.9953515244	.9957930393
105+	.9961932496	.9965559460	.9968845803	.9971822942	.9974519464
110+	.9976961371	.9979172312	.9981173781	.9982985316	.9984624666
115+	.9986107956	.9987449825	.9988663568	.9989761249	.9990753817
			R= 4		
35+	0	.625034E-05	.0000462525	.0001887788	.0005617486
40+	.0013613725	.0028491393	.0053369535	.0091635665	.0146668186
45+	.0221562137	.0318894409	.0440551253	.0587627509	.0760395856
50+	.0958336740	.1180215347	.1424190672	.1687942403	.1968803411
55+	.2263888202	.2570210484	.2884785483	.3204714810	.3527253365
60+	.3849858957	.4170226170	.4486306443	.4796316578	.5098737857
65+	.5392307890	.5676007050	.5949041161	.6210821822	.6460945485
70+	.6699172203	.6925404721	.7139668426	.7342092493	.7532892469
75+	.7712354413	.7880820644	.8038677086	.8186342156	.8324257116
80+	.8452877786	.8572667498	.8684091193	.8787610515	.8883679812
85+	.8972742926	.9055230670	.9131558892	.9202127062	.9267317281
90+	.9327493658	.9383001988	.9434169691	.9481305942	.9524701974
95+	.9564631511	.9601351300	.9635101719	.9666107440	.9694578127
100+	.9720709161	.9744682373	.9766666757	.9786819275	.9805285418
105+	.9822200045	.9837687983	.9851864683	.9864836839	.9876702979
110+	.9887554023	.9897473813	.9906539616	.9914822596	.9922388260
115+	.9929296872	.9935603844	.9941360103	.9946612432	.9951403785
120+	.9955773587	.9959758004	.9963390204	.9966700589	.9969717021
125+	.9972465019	.9974967952	.9977247210	.9979322369	.9981211334
130+	.9982930483	.9984494786	.9985917926	.9987212408	.9988389651
135+	.9989460086	.9990433235	.9991317790	.9992121681	.9992852144
			R= 5		
45+	.265624E-05	.0000203645	.0000860913	.0002652137	.0006649308
50+	.0014384519	.0027825983	.0049290171	.0081304418	.0126441181
55+	.0187146565	.0265582969	.0363500479	.0482145594	.0622210158
60+	.0783818754	.0966549562	.1169481834	.1391262388	.1630183784
65+	.1884267578	.2151347225	.2429146433	.2715350030	.3007665504
70+	.3303874324	.3601872889	.3899703521	.4195576288	.4487882674
75+	.4775202263	.5056303590	.5330140302	.5595843684	.5852712494
80+	.6100200935	.6337905450	.6565550936	.6782976825	.6990123404
85+	.7187018652	.7373765774	.7550531589	.7717535833	.7875041432
90+	.8023345740	.8162772729	.8293666092	.8416383214	.8531289942
95+	.8638756101	.8739151690	.8832843683	.8920193384	.9001554264
100+	.9077270233	.9147674286	.9213087472	.9273818155	.9330161506
105+	.9382399210	.9430799338	.9475616370	.9517091334	.9555452041
110+	.9590913400	.9623677795	.9653935502	.9681865154	.9707634211
115+	.9731399467	.9753307544	.9773495406	.9792090858	.9809213040
120+	.9824972914	.9839473736	.9852811514	.9865075441	.9876348327
125+	.9886706997	.9896222678	.9904961365	.9912984171	.9920347652
130+	.9927104123	.9933301945	.9938985809	.9944196986	.9948973579
135+	.9953350745	.9957360912	.9961033976	.9964397491	.9967476839
140+	.9970295396	.9972874678	.9975234487	.9977393038	.9979367084
145+	.9981172023	.9982822008	.9984330042	.9985708066	.9986967046

DIRICHLET 1 PROBABILITY (P=1/B)   B= 9

N	0	1	2	3	4
			R= 5		
150+	.9988117048	.9989167308	.9990126301	.9991001798	.9991800926
			R= 6		
50+	0	0	0	0	.131290E-05
55+	.0000103156	.0000447010	.0001411497	.0003626646	.0008037675
60+	.0015922455	.0028868784	.0048713541	.0077451756	.0117127256
65+	.0169717503	.0237024272	.0320579370	.0421571619	.0540798222
70+	.0678640907	.0835065127	.1009639099	.1201568626	.1409743373
75+	.1632790356	.1869130837	.2117037405	.2374688654	.2640219583
80+	.2911766410	.3187505086	.3465683197	.3744645342	.4022852295
85+	.4298894485	.4571500395	.4839540587	.5102028026	.5358115381
90+	.5607089943	.5848366717	.6081480207	.6306075323	.6521897791
95+	.6728784356	.6926653046	.7115493667	.7295358694	.7466354662
100+	.7628634110	.7782388159	.7927839701	.8065237244	.8194849374
105+	.8316959816	.8431863085	.8539860659	.8641257668	.8736360036
110+	.8825472053	.8908894320	.8986922045	.9059843645	.9127939619
115+	.9191481661	.9250731984	.9305942834	.9357356146	.9405203357
120+	.9449705318	.9491072311	.9529504149	.9565190338	.9598310301
125+	.9629033640	.9657520436	.9683921572	.9708379085	.9731026514
130+	.9751989275	.9771385025	.9789324037	.9805909561	.9821238190
135+	.9835400210	.9848479950	.9860556110	.9871702088	.9881986291
140+	.9891472428	.9900219803	.9908283582	.9915715053	.9922561875
145+	.9928868310	.9934675443	.9940021395	.9944941519	.9949468583
150+	.9953632951	.9957462744	.9960983996	.9964220801	.9967195444
155+	.9969928535	.9972439125	.9974744819	.9976861877	.9978805317
160+	.9980588999	.9982225717	.9983727272	.9985104547	.9986367577
165+	.9987525612	.9988587175	.9989560115	.9990451664	.9991268479
			R= 7		
60+	0	0	0	.721247E-06	.576998E-05
65+	.0000254663	.0000819164	.0002144149	.0004840692	.0009766613
70+	.0018030916	.0030971882	.0050110774	.0077086345	.0113577262
75+	.0161220175	.0221530649	.0295832959	.0385203059	.0490427258
80+	.0611977510	.0750002799	.0904335080	.1074507526	.1259782478
85+	.1459186369	.1671549022	.1895544981	.2129734868	.2372605140
90+	.2622605035	.2878179835	.3137799920	.3399985359	.3663326016
95+	.3926497310	.4188271929	.4447527841	.4703253035	.4954547424
100+	.5200622358	.5440798166	.5674500132	.5901253257	.6120676146
105+	.6332474285	.6536432962	.6732410027	.6920328663	.7100170295
110+	.7271967743	.7435798701	.7591779584	.7740059806	.7880816476
115+	.8014249547	.8140577413	.8260032925	.8372859839	.8479309652
120+	.8579638817	.8674106304	.8762971489	.8846492332	.8924923836
125+	.8998516739	.9067516440	.9132162108	.9192685980	.9249312809
130+	.9302259442	.9351734532	.9397938341	.9441062639	.9481290677
135+	.9518797227	.9553748676	.9586303165	.9616610764	.9644813685
140+	.9671046508	.9695436430	.9718103535	.9739161060	.9758715677
145+	.9776867777	.9793711748	.9809336260	.9823824542	.9837254652
150+	.9849699743	.9861228328	.9871904521	.9881788290	.9890935682
155+	.9899399056	.9907227291	.9914465994	.9921157698	.9927342049
160+	.9933055982	.9938333889	.9943207785	.9947707452	.9951860587
165+	.9955692937	.9959228424	.9962489270	.9965496104	.9968268075
170+	.9970822948	.9973177199	.9975346104	.9977343821	.9979183466
175+	.9980877189	.9982436241	.9983871032	.9985191200	.9986405658
180+	.9987522650	.9988549796	.9989494141	.9990362194	.9991159970
			R= 8		
70+	0	0	.428401E-06	.347481E-05	.0000155551
75+	.0000507620	.0001348161	.0003088412	.0006322679	.0011843095
80+	.0020636852	.0033865288	.0052826636	.0078906055	.0113517637
85+	.0158043439	.0213774262	.0281856222	.0363246097	.0458677395

DIRICHLET 1   PROBABILITY (P=1/B)   B= 9

N	0	1	2	3	4
			R= 8		
90+	.0568638015	.0693359453	.0832816791	.0986738163	.1154622073
95+	.1335760783	.1529267972	.1734109001	.1949132236	.2173100180
100+	.2404719336	.2642668020	.2885621544	.3132274406	.3381359300
105+	.3631662917	.3882038605	.4131416061	.4378808293	.4623316106
110+	.4864130433	.5100532773	.5331894074	.5557672304	.5777408998
115+	.5990725010	.6197315673	.6396945580	.6589443110	.6774694869
120+	.6952640136	.7123265409	.7286599133	.7442706652	.7591685430
125+	.7733660575	.7868780671	.7997213939	.8119144717	.8234770261
130+	.8344297861	.8447942249	.8545923293	.8638463958	.8725788516
135+	.8808120984	.8885683781	.8958696570	.9027375286	.9091931321
140+	.9152570848	.9209494283	.9262895854	.9312963275	.9359877509
145+	.9403812612	.9444935638	.9483406610	.9519378540	.9552997488
150+	.9584402661	.9613726545	.9641095052	.9666627698	.9690437793
155+	.9712632638	.9733313741	.9752577028	.9770513070	.9787207303
160+	.9802740248	.9817187736	.9830621126	.9843107515	.9854709955
165+	.9865487652	.9875496167	.9884787609	.9893410818	.9901411547
170+	.9908832632	.9915714157	.9922093612	.9928006045	.9933484203
175+	.9938558670	.9943257998	.9947608830	.9951636015	.9955362724
180+	.9958810552	.9961999614	.9964948648	.9967675094	.9970195182
185+	.9972524014	.9974675630	.9976663084	.9978498508	.9980193174
190+	.9981757549	.9983201352	.9984533604	.9985762678	.9986896337
195+	.9987941784	.9988905695	.9989794258	.9990613208	.9991367856
			R= 9		
80+	0	.270210E-06	.221573E-05	.0000100312	.0000331153
85+	.0000889865	.0002062801	.0004273494	.0008100347	.0014282844
90+	.0023714739	.0037424337	.0056543466	.0082267802	.0115811811
95+	.0158361781	.0211030178	.0274814103	.0350559995	.0438936002
100+	.0540412762	.0655252707	.0783507485	.0925022728	.1079449097
105+	.1246258395	.1424763523	.1614141017	.1813455066	.2021681980
110+	.2237734276	.2460483677	.2688782500	.2921483059	.3157454825
115+	.3395599233	.3634862097	.3874243692	.4112806610	.4349681553
120+	.4584071238	.4815252640	.5042577759	.5265473138	.5483438335
125+	.5696043531	.5902926464	.6103788841	.6298392375	.6486554571
130+	.6668144369	.6843077745	.7011313325	.7172848096	.7327713256
135+	.7475970234	.7617706919	.7753034109	.7882082205	.8004998140
140+	.8121942555	.8233087220	.8338612691	.8438706191	.8533559714
145+	.8623368333	.8708328700	.8788637733	.8864491462	.8936084039
150+	.9003606882	.9067247953	.9127191149	.9183615807	.9236696292
155+	.9286601682	.9333495520	.9377535638	.9418874036	.9457656815
160+	.9494024157	.9528110336	.9560043772	.9589947105	.9617937300
165+	.9644125762	.9668618480	.9691516170	.9712914442	.9732903963
170+	.9751570632	.9768995759	.9785256242	.9800424749	.9814569895
175+	.9827756423	.9840045377	.9851494273	.9862157273	.9872085343
180+	.9881326417	.9889925547	.9897925058	.9905364689	.9912281730
185+	.9918711160	.9924685769	.9930236287	.9935391493	.9940178332
190+	.9944622022	.9948746151	.9952572773	.9956122506	.9959414612
195+	.9962467080	.9965296707	.9967919169	.9970349092	.9972600117
200+	.9974684962	.9976615480	.9978402716	.9980056958	.9981587785
205+	.9983004115	.9984314246	.9985525897	.9986646251	.9987681984
210+	.9988639301	.9989523969	.9990341348	.9991096413	.9991793785
			R=10		
90+	.178749E-06	.147874E-05	.675639E-05	.0000225162	.0000610918
95+	.0001430108	.0002992159	.0005728125	.0010200711	.0017105097
100+	.0027259909	.0041588813	.0061094091	.0086824226	.0119837891
105+	.0161166774	.0211779561	.0272549044	.0344223882	.0427406098
110+	.0522534871	.0629876794	.0749522373	.0881388270	.1025224568
115+	.1180626238	.1347047892	.1523820930	.1710172229	.1905243582

DIRICHLET 1 PROBABILITY (P=1/B)   B= 9

N	0	1	2	3	4
			R=10		
120+	.2108111216	.2317804801	.2533325498	.2753662690	.2977809136
125+	.3204774395	.3433596450	.3663351485	.3893161899	.4122202606
130+	.4349705743	.4574963936	.4797332249	.5016228990	.5231135514
135+	.5441595188	.5647211644	.5847646477	.6042616485	.6231890585
140+	.6415286489	.6592667227	.6763937601	.6929040621	.7087953984
145+	.7240686630	.7387275419	.7527781950	.7662289538	.7790900374
150+	.7913732858	.8030919136	.8142602815	.8248936869	.8350081735
155+	.8446203578	.8537472729	.8624062291	.8706146890	.8783901571
160+	.8857500831	.8927117772	.8992923369	.9055085841	.9113770124
165+	.9169137420	.9221344833	.9270545073	.9316886221	.9360511554
170+	.9401559417	.9440163144	.9476451009	.9510546222	.9542566944
175+	.9572626337	.9600832627	.9627289194	.9652094665	.9675343035
180+	.9697123788	.9717522025	.9736618608	.9754490299	.9771209906
185+	.9786846429	.9801465214	.9815128095	.9827893543	.9839816812
190+	.9850950083	.9861342603	.9871040819	.9880088520	.9888526957
195+	.9896394975	.9903729132	.9910563817	.9916931364	.9922862158
200+	.9928384743	.9933525916	.9938310830	.9942763080	.9946904792
205+	.9950756708	.9954338261	.9957667654	.9960761931	.9963637044
210+	.9966307918	.9968788515	.9971091887	.9973230233	.9975214954
215+	.9977056698	.9978765409	.9980350368	.9981820239	.9983183104
220+	.9984446499	.9985617454	.9986702520	.9987707802	.9988638988
225+	.9989501376	.9990299899	.9991039148	.9991723398	.9992356625

DIRICHLET 1 PROBABILITY (P=1/B)   B=10

N	0	1	2	3	4
			R= 1		
10+	.0003628800	.0019958400	.0061871040	.0142702560	.0273158646
15+	.0459502243	.0703098108	.1000944296	.1346726200	.1732015476
20+	.2147373232	.2583238657	.3030569819	.3481253463	.3928324120
25+	.4366039318	.4789854654	.5196335295	.5583032101	.5948342777
30+	.6291371893	.6611798504	.6909756324	.7185728740	.7440459200
35+	.7674876384	.7890032941	.8087056244	.8267109551	.8431361970
40+	.8580965754	.8717039615	.8840656890	.8952837595	.9054543524
45+	.9146675717	.9230073735	.9305516279	.9373722794	.9435355776
50+	.9491023529	.9541283226	.9586644113	.9627570767	.9664486332
55+	.9697775664	.9727788367	.9754841673	.9779223169	.9801193347
60+	.9820987983	.9838820345	.9854883226	.9869350827	.9882380477
65+	.9894114212	.9904680214	.9914194129	.9922760263	.9930472674
70+	.9937416156	.9943667139	.9949294503	.9954360316	.9958920494
75+	.9963025413	.9966720445	.9970046460	.9973040262	.9975734993
80+	.9978160500	.9980343655	.9982308654	.9984077280	.9985669145
85+	.9987101904	.9988391453	.9989552100	.9990596723	.9991536917
			R= 2		
20+	.0000237588	.0001663117	.0006402999	.0017952789	.0040996878
25+	.0080917790	.0143149797	.0232540758	.0352843380	.0506397187
30+	.0694009381	.0915006739	.1167412834	.1448201231	.1753580693
35+	.2079278217	.2420796496	.2773632189	.3133449277	.3496207397
40+	.3858248782	.4216349437	.4567741049	.4910110107	.5241580214
45+	.5560682787	.5866320464	.6157726624	.6434423675	.6696182004
50+	.6942980929	.7174972525	.7392448801	.7595812465	.7785551281
55+	.7962215895	.8126400934	.8278729086	.8419837882	.8550368870
60+	.8670958889	.8782233156	.8884799913	.8979246384	.9066135851
65+	.9146005644	.9219365888	.9286698867	.9348458887	.9405072521
70+	.9456939168	.9504431841	.9547898124	.9587661254	.9624021278
75+	.9657256270	.9687623552	.9715360929	.9740687903	.9763806856

DIRICHLET 1   PROBABILITY (P=1/B)   B=10

N	0	1	2	3	4
			R= 2		
80+	.9784904199	.9804151474	.9821706405	.9837713904	.9852307012
85+	.9865607792	.9877728169	.9888770715	.9898829377	.9907990167
90+	.9916331791	.9923926246	.9930839358	.9937131296	.9942857036
95+	.9948066793	.9952806420	.9957117775	.9961039057	.9964605120
100+	.9967847756	.9970795958	.9973476160	.9975912458	.9978126814
105+	.9980139239	.9981967963	.9983629593	.9985139253	.9986510713
110+	.9987756513	.9988888064	.9989915756	.9990849041	.9991696522
			R= 3		
30+	.438680E-05	.0000339977	.0001441502	.0004427858	.0011021113
35+	.0023593563	.0045057577	.0078659882	.0127720800	.0195363327
40+	.0284270631	.0396498538	.0533356184	.0695356221	.0882227416
45+	.1092977237	.1325989979	.1579146124	.1849950344	.2135657997
50+	.2433392689	.2740250005	.3053384744	.3370080745	.3687803710
55+	.4004238304	.4317311383	.4625203421	.4926350302	.5219437527
60+	.5503388720	.5777350081	.6040672197	.6292890354	.6533704287
65+	.6762958064	.6980620644	.7186767496	.7381563500	.7565247311
70+	.7738117229	.7900518590	.8052832646	.8195466851	.8328846474
75+	.8453407432	.8569590225	.8677834876	.8778576755	.8872243192
80+	.8959250770	.9040003223	.9114889836	.9184284295	.9248543901
85+	.9308009095	.9363003249	.9413832664	.9460786747	.9504138321
90+	.9544144048	.9581044939	.9615066923	.9646421460	.9675306189
95+	.9701905585	.9726391634	.9748924497	.9769653169	.9788716127
100+	.9806241961	.9822349975	.9837150778	.9850746845	.9863233046
105+	.9874697160	.9885220354	.9894877643	.9903738317	.9911866346
110+	.9919320762	.9926156018	.9932422318	.9938165933	.9943429492
115+	.9948252256	.9952670368	.9956717092	.9960423035	.9963816345
120+	.9966922907	.9969766513	.9972369029	.9974750546	.9976929515
125+	.9978922883	.9980746205	.9982413763	.9983938661	.9985332921
130+	.9986607573	.9987772730	.9988837669	.9989810890	.9990700188
			R= 4		
40+	.128686E-05	.0000105523	.0000472742	.0001531854	.0004015146
45+	.0009034822	.0018101772	.0033090481	.0056153878	.0089600646
50+	.0135752193	.0196797243	.0274659606	.0370890602	.0486592837
55+	.0622377576	.0778354265	.0954148184	.1148940663	.1361525665
60+	.1590376673	.1833718390	.2089598649	.2355956951	.2630687028
65+	.2911691770	.3196929617	.3484452193	.3772433390	.4059190501
70+	.4343198195	.4623096262	.4897692094	.5165958853	.5427030228
75+	.5680192600	.5924875329	.6160639803	.6387167754	.6604249276
80+	.6811770886	.7009703902	.7198093307	.7377047278	.7546727435
85+	.7707339892	.7859127111	.8002360575	.8137334249	.8264358795
90+	.8383756509	.8495856914	.8600992979	.8699497882	.8791702300
95+	.8877932133	.8958506650	.9033736982	.9103924940	.9169362098
100+	.9230329126	.9287095315	.9339918290	.9389043865	.9434706021
105+	.9477126994	.9516517444	.9553076696	.9586993031	.9618444027
110+	.9647596932	.9674609049	.9699628158	.9722792927	.9744233337
115+	.9764071113	.9782420135	.9799386857	.9815070709	.9829564490
120+	.9842954748	.9855322147	.9866741819	.9877283702	.9887012858
125+	.9895989788	.9904270713	.9911907858	.9918949712	.9925441274
130+	.9931424286	.9936937454	.9942016654	.9946695129	.9951003667
135+	.9954970774	.9958622836	.9961984268	.9965077651	.9967923869
140+	.9970542226	.9972950562	.9975165363	.9977201855	.9979074096
145+	.9980795069	.9982376753	.9983830206	.9985165626	.9986392420
150+	.9987519265	.9988554156	.9989504468	.9990376993	.9991177992
			R= 5		
50+	.491205E-06	.417524E-05	.0000193850	.0000650691	.0001765672
55+	.0004110186	.0008512114	.0016069644	.0028135975	.0046275520
60+	.0072196506	.0107667685	.0154428048	.0214098221	.0288100865

SOBEL, UPPULURI AND FRANKOWSKI

DIRICHLET 1   PROBABILITY (P=1/8)   B=10

N	0	1	2	3	4
			R= 5		
65+	.0377595504	.0483431045	.0606117223	.0745814509	.0902340671
70+	.1075191390	.1263571800	.1466435797	.1682530040	.1910439940
75+	.2148635336	.2395514052	.2649441977	.2908788763	.3171958615
80+	.3437415961	.3703706063	.3969470782	.4233459896	.4494538418
85+	.4751690416	.5004019863	.5250749013	.5491214798	.5724863673
90+	.5951245320	.6170005539	.6380878642	.6583679590	.6778296074
95+	.6964680722	.7142843533	.7312844665	.7474787622	.7628812898
100+	.7775092107	.7913822606	.8045222605	.8169526774	.8286982303
105+	.8397845418	.8502378304	.8600846423	.8693516194	.8780652992
110+	.8862519460	.8939374080	.9011469995	.9079054042	.9142365987
115+	.9201637916	.9257093788	.9308949110	.9357410727	.9402676707
120+	.9444936315	.9484370045	.9521149717	.9555438630	.9587391738
125+	.9617155872	.9644869980	.9670665391	.9694666084	.9716988982
130+	.9737744236	.9757035521	.9774960332	.9791610272	.9807071343
135+	.9821424229	.9834744568	.9847103222	.9858566538	.9869196596
140+	.9879051451	.9888185362	.9896649018	.9904489746	.9911751714
145+	.9918476123	.9924701387	.9930463312	.9935795254	.9940728275
150+	.9945291295	.9949511223	.9953413093	.9957020184	.9960354137
155+	.9963435064	.9966281651	.9968911251	.9971339976	.9973582784
160+	.9975653553	.9977565159	.9979329545	.9980957784	.9982460140
165+	.9983846128	.9985124562	.9986303607	.9987390826	.9988393220
170+	.9989317273	.9990168982	.9990953899	.9991677160	.9992343515
			R= 6		
60+	.222254E-06	.193679E-05	.922049E-05	.0000317356	.0000882885
65+	.0002106516	.0004469903	.0008642532	.0015490420	.0026067094
70+	.0041586941	.0063383266	.0092855048	.0131407264	.0180389837
75+	.0241039812	.0314430530	.0401430520	.0502673695	.0618541406
80+	.0749156025	.0894385038	.1053854159	.1226967710	.1412934420
85+	.1610796826	.1819462598	.2037736349	.2264350693	.2497995599
90+	.2737345330	.2981082483	.3227918858	.3476613037	.3725984704
95+	.3974925851	.4222409069	.4467493191	.4709326599	.4947148468
100+	.5180288296	.5408163987	.5630278774	.5846217245	.6055640688
105+	.6258281965	.6453940092	.6642474662	.6823800245	.6997880859
110+	.7164724595	.7324378458	.7476923464	.7622470039	.7761153719
115+	.7893131185	.8018576624	.8137678405	.8250636087	.8357657720
120+	.8458957441	.8554753342	.8645265599	.8730714829	.8811320677
125+	.8887300593	.8958868804	.9026235441	.9089605817	.9149179845
130+	.9205151567	.9257708793	.9307032832	.9353298301	.9396673015
135+	.9437317923	.9475387112	.9511027848	.9544380655	.9575579429
140+	.9604751576	.9632018174	.9657494151	.9681288473	.9703504348
145+	.9724239434	.9743586050	.9761631393	.9778457756	.9794142739
150+	.9808759465	.9822376790	.9835059508	.9846868552	.9857861189
155+	.9868091209	.9877609110	.9886462272	.9894695128	.9902349328
160+	.9909463892	.9916075365	.9922217953	.9927923661	.9933222428
165+	.9938142242	.9942709260	.9946947920	.9950881047	.9954529948
170+	.9957914509	.9961053284	.9963963578	.9966661526	.9969162170
175+	.9971479523	.9973626644	.9975615692	.9977457990	.9979164078
180+	.9980743763	.9982206170	.9983559787	.9984812507	.9985971667
185+	.9987044089	.9988036110	.9988953617	.9989802082	.9990586582
			R= 7		
70+	.113266E-06	.100524E-05	.487539E-05	.0000170981	.0000484699
75+	.0001178378	.0002547522	.0005017433	.0009158440	.0015690859
80+	.0025478255	.0039509036	.0058867733	.0084698263	.0118162115
85+	.0160394567	.0212461921	.0275322351	.0349792379	.0436520379
90+	.0535967873	.0648398771	.0773876270	.0912266711	.1063249442
95+	.1226331586	.1400866553	.1586075134	.1781068120	.1984869476
100+	.2196439252	.2414695559	.2638535105	.2866851885	.3098553807

DIRICHLET 1 PROBABILITY (P=1/B) B=10

N	0	1	2	3	4
			R= 7		
105+	.3332577095	.3567898448	.3803544973	.4038602006	.4272218950
110+	.4503613307	.4732073093	.4956957835	.5177698341	.5393795449
115+	.5604817929	.5810399708	.6010236575	.6204082504	.6391745715
120+	.6573084582	.6748003467	.6916448569	.7078403840	.7233887019
125+	.7382945827	.7525654340	.7662109573	.7792428289	.7916744023
130+	.8035204347	.8147968355	.8255204380	.8357087919	.8453799775
135+	.8545524383	.8632448337	.8714759080	.8792643758	.8866288222
140+	.8935876176	.9001588435	.9063602309	.9122091087	.9177223611
145+	.9229163936	.9278071067	.9324098753	.9367395350	.9408103728
150+	.9446361227	.9482299646	.9516045272	.9547718936	.9577436091
155+	.9605306909	.9631436401	.9655924546	.9678866433	.9700352408
160+	.9720468234	.9739295250	.9756910533	.9773387066	.9788793903
165+	.9803196334	.9816656046	.9829231286	.9840977017	.9851945077
170+	.9862184320	.9871740774	.9880657771	.9888976093	.9896734097
175+	.9903967848	.9910711236	.9916996098	.9922852326	.9928307975
180+	.9933389368	.9938121193	.9942526594	.9946627261	.9950443514
185+	.9953994386	.9957297694	.9960370113	.9963227248	.9965883693
190+	.9968353094	.9970648210	.9972780964	.9974762495	.9976603207
195+	.9978312816	.9979900394	.9981374405	.9982742748	.9984012793
200+	.9985191413	.9986285014	.9987299571	.9988240652	.9989113443
205+	.9989922778	.9990673158	.9991368775	.9992013529	.9992611055
			R= 8		
80+	.630264E-07	.567238E-06	.279081E-05	.993102E-05	.0000285700
85+	.0000704923	.0001546665	.0003091409	.0005725959	.0009953210
90+	.0016394531	.0025783960	.0038954294	.0056815935	.0080329977
95+	.0110477430	.0148226599	.0194500649	.0250147142	.0315911039
100+	.0392412274	.0480128606	.0579384070	.0690343019	.0813009452
105+	.0947231126	.1092707778	.1249002718	.1415557020	.1591705524
110+	.1776693950	.1969696461	.2169833110	.2376186719	.2587818787
115+	.2803784169	.3023144319	.3244978974	.3468396224	.3692540973
120+	.3916601828	.4139816493	.4361475778	.4580926342	.4797572288
125+	.5010875765	.5220356703	.5425591798	.5626212906	.5821904922
130+	.6012403286	.6197491184	.6376996554	.6550788952	.6718776354
135+	.6880901946	.7037140946	.7187497500	.7332001677	.7470706588
140+	.7603685651	.7731030009	.7852846109	.7969253449	.8080382494
145+	.8186372763	.8287371069	.8383529928	.8475006117	.8561959375
150+	.8644551244	.8722944036	.8797299925	.8867780147	.8934504307
155+	.8997749775	.9057551178	.9114099963	.9167544038	.9218027476
160+	.9265690279	.9310668188	.9353092552	.9393090225	.9430783511
165+	.9466290134	.9499723239	.9531191418	.9560798753	.9588644882
170+	.9614825078	.9639430340	.9662547495	.9684259308	.9704644603
175+	.9723778380	.9741731944	.9758573034	.9774365955	.9789171703
180+	.9803048102	.9816049930	.9828229047	.9839634524	.9850312764
185+	.9860307625	.9869660535	.9878410611	.9886594764	.9894247816
190+	.9901402593	.9908090033	.9914339281	.9920177780	.9925631361
195+	.9930724328	.9935479543	.9939918496	.9944061390	.9947927204
200+	.9951533765	.9954897811	.9958035054	.9960960235	.9963687185
205+	.9966228873	.9968597459	.9970804337	.9972860188	.9974775014
210+	.9976558182	.9978218463	.9979764067	.9981202675	.9982541476
215+	.9983787191	.9984946108	.9986024102	.9987026667	.9987958935
220+	.9988825704	.9989631451	.9990380360	.9991076338	.9991723028
			R= 9		
90+	.375237E-07	.341466E-06	.169926E-05	.611761E-05	.0000178089
95+	.0000444693	.0000987493	.0001997656	.0003744791	.0006587657
100+	.0010980325	.0017472817	.0026705753	.0039399143	.0056335939
105+	.0078341379	.0106259397	.0140927512	.0183151589	.0233681767
110+	.0293190634	.0362254524	.0441338510	.0530785454	.0630809194

DIRICHLET 1 PROBABILITY (P=1/B)   B=10

N	0	1	2	3	4
			R= 9		
115+	.0741491775	.0862784439	.0994511986	.1136380010	.1287984486
120+	.1448823162	.1618308216	.1795779685	.1980519203	.2171763664
125+	.2368718471	.2570570092	.2776497732	.2985683958	.3197324183
130+	.3410634953	.3624861032	.3839281298	.4053213494	.4266017904
135+	.4477100029	.4685912361	.4891955332	.5094777552	.5293975424
140+	.5489192223	.5680116742	.5866481566	.6048061072	.6224669208
145+	.6396157114	.6562410644	.6723347836	.6878916363	.7029091005
150+	.7173871172	.7313278494	.7447354501	.7576158409	.7699765015
155+	.7818262718	.7931751662	.8040342006	.8144152325	.8243308129
160+	.8337940515	.8428184934	.8514180066	.8596066815	.8673987398
165+	.8748084541	.8818500754	.8885377701	.8948855646	.9009072969
170+	.9066165754	.9120267434	.9171508493	.9220016221	.9265914514
175+	.9309323711	.9350360478	.9389137717	.9425764504	.9460346061
180+	.9492983744	.9523775055	.9552813667	.9580189470	.9605988629
185+	.9630293645	.9653183440	.9674733434	.9695015642	.9714098764
190+	.9732048288	.9748926592	.9764793047	.9779704123	.9793713495
195+	.9806872150	.9819228489	.9830828435	.9841715532	.9851931047
200+	.9861514071	.9870501612	.9878928689	.9886828428	.9894232144
205+	.9901169431	.9907668245	.9913754981	.9919454554	.9924790470
210+	.9929784897	.9934458739	.9938831692	.9942922317	.9946748094
215+	.9950325478	.9953669961	.9956796116	.9959717655	.9962447468
220+	.9964997676	.9967379668	.9969604148	.9971681167	.9973620165
225+	.9975430004	.9977119001	.9978694963	.9980165209	.9981536607
230+	.9982815596	.9984008209	.9985120106	.9986156584	.9987122609
235+	.9988022832	.9988861607	.9989643009	.9990370854	.9991048712
			R=10		
100+	.235707E-07	.216422E-06	.108703E-05	.395097E-05	.0000116140
105+	.0000292883	.0000656896	.0001342260	.0002541592	.0004516131
110+	.0007603110	.0012219469	.0018861296	.0028098756	.0040566660
115+	.0056951147	.0077973235	.0104370152	.0136875451	.0176198926
120+	.0223007266	.0277906255	.0341425188	.0414003984	.0495983295
125+	.0587597760	.0688972367	.0800121807	.0920952541	.1051267295
130+	.1190771589	.1339081921	.1495735210	.1660199112	.1831882856
135+	.2010148263	.2194320683	.2383699601	.2577568709	.2775205306
140+	.2975888900	.3178908943	.3383571657	.3589205929	.3795168291
145+	.4000847008	.4205665326	.4409083914	.4610602597	.4809761410
150+	.5006141072	.5199362934	.5389088480	.5575018445	.5756891607
155+	.5934483325	.6107603865	.6276096574	.6439835933	.6598725546
160+	.6752696080	.6901703202	.7045725529	.7184762611	.7318832974
165+	.7447972225	.7572231239	.7691674432	.7806378125	.7916429010
170+	.8021922707	.8122962426	.8219657725	.8312123368	.8400478272
175+	.8484844548	.8565346634	.8642110501	.8715262949	.8784930968
180+	.8851241176	.8914319319	.8974289832	.9031275454	.9085396901
185+	.9136772580	.9185518350	.9231747322	.9275569697	.9317092632
190+	.9356420139	.9393653011	.9428888772	.9462221644	.9493742535
195+	.9523539047	.9551695493	.9578292929	.9603409198	.9627118982
200+	.9649493861	.9670602380	.9690510121	.9709279783	.9726971258
205+	.9743641716	.9759345689	.9774135161	.9788059650	.9801166300
210+	.9813499964	.9825103296	.9836016831	.9846279075	.9855926582
215+	.9864994043	.9873514359	.9881518724	.9889036696	.9896096274
220+	.9902723969	.9908944870	.9914782715	.9920259953	.9925397805
225+	.9930216327	.9934734465	.9938970111	.9942940157	.9946660543
230+	.9950146310	.9953411642	.9956469913	.9959333731	.9962014974
235+	.9964524835	.9966873852	.9969071950	.9971128470	.9973052201
240+	.9974851413	.9976533884	.9978106928	.9979577422	.9980951827
245+	.9982236215	.9983436291	.9984557413	.9985604612	.9986582609
250+	.9987495839	.9988348460	.9989144376	.9989887246	.9990580503

TABLE B

(Forward, p < 1/b)

This table gives the same I - value as Table A for $p = 1/j$, $j = b + 1(1)10$.

Starting and Stopping Rule: Start with the line of n-values for which I first
crosses .75 and stop with the line of n-values for which I first crosses
.99 . One extra line (marked with a *) gives the line where I crosses the
value .999 .

DIRICHLET 1 PROBABILITY, B= 1

N	0	1	2	3	4
			P=1/ 2, R= 1		
0+	0	.5000000000	.7500000000	.8750000000	.9375000000
5+	.9687500000	.9843750000	.9921875000	.9960937500	.9980468750
* 10+	.9990234375	.9995117188	.9997558594	.9998779297	.9999389648
			P=1/ 2, R= 2		
5+	.8125000000	.8906250000	.9375000000	.9648437500	.9804687500
* 10+	.9892578125	.9941406250	.9968261719	.9982910156	.9990844727
			P=1/ 2, R= 3		
5+	.5000000000	.6562500000	.7734375000	.8554687500	.9101562500
10+	.9453125000	.9672851563	.9807128906	.9887695313	.9935302734
* 15+	.9963073730	.9979095459	.9988250732	.9993438721	.9996356964
			P=1/ 2, R= 4		
10+	.8281250000	.8867187500	.9270019531	.9538574219	.9713134766
15+	.9824218750	.9893646240	.9936370850	.9962310791	.9977874756
* 20+	.9987115860	.9992551804	.9995722771	.9997558594	.9998614192
			P=1/ 2, R= 5		
10+	.6230468750	.7255859375	.8061523438	.8665771484	.9102172852
15+	.9407653809	.9615936279	.9754791260	.9845581055	.9903945923
* 20+	.9940910339	.9964013100	.9978282452	.9987002611	.9992280602
			P=1/ 2, R= 6		
10+	.3769531250	.5000000000	.6127929688	.7094726563	.7880249023
15+	.8491210938	.8949432373	.9282684326	.9518737793	.9682159424
20+	.9793052673	.9866981506	.9915497303	.9946889877	.9966946244
* 25+	.9979613423	.9987530410	.9992431402	.9995438829	.9997269437
			P=1/ 2, R= 7		
15+	.6963806152	.7727508545	.8338470459	.8810577393	.9164657593
20+	.9423408508	.9608230591	.9737606049	.9826551676	.9886720777
25+	.9926833510	.9953223467	.9970376939	.9981404170	.9988421500
* 30+	.9992845468	.9995610449	.9997324736	.9998379683	.9999024372
			P=1/ 2, R= 8		
15+	.5000000000	.5981903076	.6854705811	.7596588135	.8203582764
20+	.8684120178	.9053764343	.9330997467	.9534301758	.9680426717
25+	.9783573747	.9855203629	.9904213548	.9937295243	.9959349707
* 30+	.9973885603	.9983365536	.9989487992	.9993406364	.9995893023
			P=1/ 2, R= 9		
20+	.7482776642	.8083448410	.8568606377	.8949801922	.9242051840
25+	.9461239278	.9622406512	.9738805071	.9821509309	.9879402276
30+	.9919375991	.9946630797	.9964998167	.9977243079	.9985324722
* 35+	.9990608873	.9994033785	.9996235515	.9997640066	.9998529616
			P=1/ 2, R=10		
20+	.5880985260	.6681880951	.7382664680	.7975635529	.8462718725
25+	.8852385283	.9156812280	.9389609396	.9564207233	.9692858271
30+	.9786130274	.9852753133	.9899691965	.9932345066	.9954794073
* 35+	.9970059397	.9980334135	.9987183960	.9991709737	.9994674902
			P=1/ 3, R= 1		
0+	0	.3333333333	.5555555556	.7037037037	.8024691358
5+	.8683127572	.9122085048	.9414723365	.9609815577	.9739877051
10+	.9826584701	.9884389801	.9922926534	.9948617689	.9965745126
* 15+	.9977163417	.9984775612	.9989850408	.9993233605	.9995489070
			P=1/ 3, R= 2		
5+	.5390946502	.6488340192	.7366255144	.8049077884	.8569323782
10+	.8959508205	.9248533704	.9460485736	.9614632669	.9725961009
15+	.9805889048	.9862980504	.9903578873	.9932336052	.9952635236
* 20+	.9966919847	.9976944136	.9983961138	.9988861901	.9992277585
			P=1/ 3, R= 3		
10+	.7008586090	.7658893461	.8188773542	.8612677607	.8946662627
15+	.9206428754	.9406248852	.9558492736	.9673521449	.9759792983
20+	.9824073734	.9871689105	.9906774115	.9932503123	.9951289383

DIRICHLET 1 PROBABILITY, B= 1

N	0	1	2	3	4
			P=1/ 3, R= 3		
* 25+	.9964952117	.9974852649	.9982003033	.9987151310	.9990847508
			P=1/ 3, R= 4		
15+	.7907598119	.8340541664	.8695777393	.8983349174	.9213406599
20+	.9395535394	.9538381507	.9649484040	.9735247398	.9800999307
25+	.9851095999	.9889048038	.9917649575	.9939100728	.9955117588
* 30+	.9967027562	.9975849764	.9982361390	.9987151551	.9990664336
			P=1/ 3, R= 5		
15+	.5959352165	.6608767483	.7186025543	.7689276160	.8120633832
20+	.8484891421	.8788439412	.9038420110	.9242108087	.9406487857
25+	.9537991674	.9642359782	.9724589201	.9788942659	.9838995348
30+	.9877702762	.9907477695	.9930268385	.9947632720	.9960805663
* 35+	.9970758554	.9978249978	.9983868545	.9988068282	.9991197498
			P=1/ 3, R= 6		
20+	.7027861064	.7513537850	.7938505037	.8305143395	.8617464959
25+	.8880472592	.9099645619	.9280550340	.9428563294	.9548689749
30+	.9645458282	.9722873108	.9784407971	.9833028109	.9871229646
35+	.9901088318	.9924311730	.9942291146	.9956150279	.9966789613
* 40+	.9974925575	.9981124403	.9985830920	.9989392609	.9992079497
			P=1/ 3, R= 7		
25+	.7784607455	.8149895834	.8466479095	.8737836177	.8968078549
30+	.9161615616	.9322896504	.9456222039	.9565617350	.9654754269
35+	.9726912728	.9784971258	.9831418082	.9868375770	.9897633940
40+	.9920685831	.9938765746	.9952885298	.9963867172	.9972375651
* 45+	.9978943600	.9983995868	.9987869274	.9990829438	.9993084800
			P=1/ 3, R= 8		
25+	.6297361912	.6793110426	.7245372229	.7652407851	.8014217293
30+	.8332171045	.8608652569	.8846733880	.9049896600	.9221803517
35+	.9366120434	.9486384532	.9585913441	.9667748321	.9734624138
40+	.9788960738	.9832869102	.9868167984	.9896407088	.9918893783
* 50+	.9982559774	.9986639242	.9989791557	.9992220008	.9994085340
			P=1/ 3, R= 9		
30+	.7139844475	.7537286665	.7894408633	.8211850382	.8491199121
35+	.8734733920	.8945196091	.9125592238	.9279032639	.9408604533
40+	.9517277735	.9607838736	.9682848858	.9744621900	.9795216963
45+	.9836442569	.9869868737	.9896844241	.9918516868	.9935854969
* 55+	.9985691347	.9988965600	.9991512240	.9993487186	.9995014478
			P=1/ 3, R=10		
35+	.7787654148	.8103347405	.8383963634	.8631173169	.8847126325
40+	.9034285728	.9195283064	.9332801621	.9449484033	.9547863322
45+	.9630314536	.9699023880	.9755972166	.9802929524	.9841458639
50+	.9872924082	.9898505744	.9919214708	.9935910306	.9949317378
* 60+	.9988339648	.9990957228	.9993004310	.9994600777	.9995842473
			P=1/ 4, R= 1		
5+	.7626953125	.8220214844	.8665161133	.8998870850	.9249153137
10+	.9436864853	.9577648640	.9683236480	.9762427360	.9821820520
15+	.9866365390	.9899774042	.9924830532	.9943622899	.9957717174
* 25+	.9992474565	.9994355924	.9995766943	.9996825207	.9997618905
			P=1/ 4, R= 2		
10+	.7559747696	.8029026985	.8416182399	.8732945919	.8990316279
15+	.9198192339	.9365235602	.9498870212	.9605360292	.9689925944
20+	.9756873751	.9809727284	.9851349440	.9884052563	.9909694785
* 30+	.9980355970	.9984820522	.9988280550	.9990959282	.9993031113
			P=1/ 4, R= 3		
15+	.7639121888	.8028889501	.8362976026	.8646949573	.8886552253
20+	.9087395675	.9254765194	.9393505717	.9507966648	.9601988127
25+	.9678914791	.9741626746	.9792580209	.9833852515	.9867187838
30+	.9894041293	.9915619962	.9932920102	.9946760214	.9957809981

B   3

DIRICHLET 1 PROBABILITY,   B= 1

N	0	1	2	3	4
			P=1/ 4, R= 3		
* 40+	.9989842849	.9992021776	.9993739776	.9995092701	.9996156892
			P=1/ 4, R= 4		
20+	.7748439523	.8083178561	.8376075220	.8630432844	.8849816295
25+	.9037859253	.9198123137	.9333999040	.9448644332	.9544946378
30+	.9625506743	.9692640380	.9748385276	.9794518982	.9832579290
35+	.9863886963	.9889569038	.9910581645	.9927731641	.9941696637
* 45+	.9984450791	.9987586354	.9990102096	.9992117548	.9993729910
			P=1/ 4, R= 5		
20+	.5851584975	.6325798612	.6765143599	.7167876504	.7533515589
25+	.7862590766	.8156407887	.8416836700	.8646127285	.8846756547
30+	.9021304004	.9172354689	.9302426112	.9413915903	.9509066673
35+	.9589944827	.9658430361	.9716215030	.9764806684	.9805537923
40+	.9839577602	.9867944001	.9891518778	.9911061027	.9927220965
* 50+	.9978916334	.9982941762	.9986217776	.9988879537	.9991038823
			P=1/ 4, R= 6		
25+	.6217214884	.6628558854	.7010521112	.7362100009	.7683106828
30+	.7974019258	.8235840444	.8469969006	.8678083282	.8862041437
35+	.9023797746	.9165334516	.9288608478	.9395510116	.9487834258
40+	.9567260174	.9635339531	.9693490649	.9742997681	.9785013517
45+	.9820565379	.9850562263	.9875803543	.9896988188	.9914724171
* 60+	.9990438713	.9992230648	.9993694596	.9994888869	.9995861789
			P=1/ 4, R= 7		
30+	.6519457110	.6883097647	.7221283346	.7533454761	.7819611891
35+	.8080219278	.8316113895	.8528419050	.8718466407	.8887727334
40+	.9037754065	.9170130592	.9286432827	.9388197282	.9476897382
45+	.9553926416	.9620586157	.9678080183	.9727511023	.9769880314
50+	.9806091278	.9836952895	.9863185270	.9885425761	.9904235539
* 65+	.9988078424	.9990222148	.9991990720	.9993447617	.9994646032
			P=1/ 4, R= 8		
35+	.6777182345	.7102941578	.7406234658	.7686780756	.7944702169
40+	.8180458460	.8394782361	.8588619419	.8763072771	.8919353899
45+	.9058739770	.9182536431	.9292048863	.9388556693	.9473295275
50+	.9547441535	.9612103971	.9668316202	.9717033469	.9759131542
55+	.9795407541	.9826582228	.9853303388	.9876149980	.9895636779
60+	.9912219295	.9926298791	.9938227252	.9948312223	.9956821418
* 70+	.9985786318	.9988247350	.9990295005	.9991996134	.9993407298
			P=1/ 4, R= 9		
40+	.7001677003	.7296372367	.7570974866	.7825386004	.8059807696
45+	.8274694247	.8470705627	.8648663328	.8809509712	.8954271457
50+	.9084027412	.9199880943	.9302936700	.9394281575	.9474969549
55+	.9546010047	.9608359421	.9662915122	.9710512189	.9751921637
60+	.9787850422	.9818942641	.9845781678	.9868893071	.9888747859
65+	.9905766249	.9920321451	.9932743563	.9943323412	.9952316284
* 75+	.9983617348	.9986357073	.9988653607	.9990575706	.9992182032
			P=1/ 4, R=10		
45+	.7200261493	.7468869681	.7719328668	.7951662333	.8166124178
50+	.8363160997	.8543377601	.8707503436	.8856361752	.8990841708
55+	.9111873668	.9220407763	.9317395677	.9403775539	.9480459701
60+	.9548325185	.9608206494	.9660890531	.9707113318	.9747558256
65+	.9782855657	.9813583305	.9840267841	.9863386772	.9883370932
70+	.9900607270	.9915441823	.9928182790	.9939103619	.9948446047
* 80+	.9981607054	.9984585921	.9987099340	.9989216809	.9990998059
			P=1/ 5, R= 1		
5+	.6723200000	.7378560000	.7902848000	.8322278400	.8657822720
10+	.8926258176	.9141006541	.9312805233	.9450244186	.9560195349
15+	.9648156279	.9718525023	.9774820019	.9819856015	.9855884812
20+	.9884707850	.9907766280	.9926213024	.9940970419	.9952776335

DIRICHLET 1 PROBABILITY, B= 1

N	0	1	2	3	4
		P=1/ 5,	R= 1		
* 30+	.9987620600	.9990096480	.9992077184	.9993661747	.9994929398
		P=1/ 5,	R= 2		
10+	.6241903616	.6778774528	.7251220931	.7663537791	.8020879070
15+	.8328742326	.8592625116	.8817805098	.9009208082	.9171337669
20+	.9308247097	.9423539248	.9520384654	.9601550328	.9669434346
25+	.9726102744	.9773326409	.9812616498	.9845257495	.9872337433
30+	.9894775097	.9913344197	.9928694654	.9941371160	.9951829277
* 40+	.9985378492	.9988036948	.9990216882	.9992003364	.9993466579
		P=1/ 5,	R= 3		
15+	.6019767907	.6481562791	.6903775256	.7286581225	.7631106596
20+	.7939152811	.8212971668	.8455085184	.8668145078	.8854826128
25+	.9017747772	.9159418766	.9282200295	.9388283535	.9479678327
30+	.9558210148	.9625523138	.9683087350	.9732208811	.9774041281
35+	.9809598880	.9839768964	.9865324800	.9886937736	.9905188659
* 50+	.9987145850	.9989331323	.9991151064	.9992665089	.9993923807
		P=1/ 5,	R= 4		
25+	.7660067408	.7931603481	.8177166538	.8398173289	.8596195338
30+	.8772891936	.8929955579	.9069069090	.9191872742	.9299939956
35+	.9394760221	.9477727953	.9550136155	.9613173884	.9667926654
40+	.9715379055	.9756418970	.9791842896	.9822361971	.9848608375
45+	.9871141873	.9890456300	.9906985857	.9921111114	.9933164667
* 60+	.9989865415	.9991504228	.9992883091	.9994042270	.9995015981
		P=1/ 5,	R= 5		
30+	.7447667453	.7712712349	.7956160995	.8178742614	.8381368640
35+	.8565082903	.8731018367	.8880360284	.9014315458	.9134087143
40+	.9240855045	.9335759847	.9419891672	.9494281917	.9559897928
45+	.9617640017	.9668340388	.9712763571	.9751608028	.9785508645
50+	.9815039849	.9840719157	.9863010982	.9882330564	.9899047916
55+	.9913491709	.9925953020	.9936688918	.9945925842	.9953862755
* 65+	.9982552482	.9985208637	.9987470654	.9989395163	.9991030996
		P=1/ 5,	R= 6		
35+	.7279083060	.7536283029	.7775230096	.7996256134	.8199867999
40+	.8386711828	.8557540471	.8713184346	.8854525812	.8982477033
45+	.9097961212	.9201896973	.9295185656	.9378701239	.9453282597
50+	.9519727806	.9578790215	.9631176003	.9677542999	.9718500512
55+	.9754609993	.9786386336	.9814299673	.9838777522	.9860207186
60+	.9878938300	.9895285454	.9909530831	.9921926808	.9932698484
* 75+	.9987678770	.9989496396	.9991052899	.9992384573	.9993522880
		P=1/ 5,	R= 7		
40+	.7141086302	.7390211407	.7623677220	.7841578645	.8044168078
45+	.8231829869	.8405056138	.8564424305	.8710576575	.8844201508
50+	.8966017725	.9076759742	.9177165836	.9267967870	.9349882896
55+	.9423606419	.9489807134	.9549122974	.9602158314	.9649482156
60+	.9691627162	.9729089389	.9762328602	.9791769048	.9817800600
65+	.9840780177	.9861033363	.9878856167	.9894516861	.9908257856
* 85+	.9991402062	.9992638441	.9993701727	.9994615366	.9995399757
		P=1/ 5,	R= 8		
45+	.7025432642	.7266712088	.7494380898	.7708389579	.7908826978
50+	.8095901884	.8269925052	.8431291990	.8580466759	.8717966982
55+	.8844350164	.8960201415	.9066122559	.9162722642	.9250609776
60+	.9330384252	.9402632834	.9467924145	.9526805037	.9579797839
65+	.9627398391	.9670074748	.9708266471	.9742384410	.9772810900
70+	.9799900292	.9823979750	.9845350270	.9864287838	.9881044717
75+	.9895850795	.9908914981	.9920426612	.9930556847	.9939460039
* 90+	.9987421372	.9989151624	.9990651176	.9991949612	.9993072909
		P=1/ 5,	R= 9		
50+	.6926683722	.7160527355	.7382406894	.7592183914	.7789840483

DIRICHLET 1 PROBABILITY, B= 1

N	0	1	2	3	4
		P=1/ 5,	R= 9		
55+	.7975465782	.8149242659	.8311434410	.8462372040	.8602442160
60+	.8732075684	.8851737397	.8961916485	.9063118017	.9155855421
65+	.9240643904	.9317994802	.9388410791	.9452381927	.9510382424
70+	.9562868119	.9610274554	.9653015593	.9691482528	.9726043590
75+	.9757043816	.9784805211	.9809627165	.9831787055	.9851541013
80+	.9869124818	.9884754867	.9898629212	.9910928632	.9921817718
* 95+	.9982513910	.9984819898	.9986832397	.9988587115	.9990115670
		P=1/ 5,	R=10		
55+	.6841088951	.7067964317	.7284219985	.7489662870	.7684204704
60+	.7867852195	.8040696893	.8202904994	.8354707292	.8496389437
65+	.8628282634	.8750754888	.8864202871	.8969044455	.9065711949
70+	.9154646044	.9236290459	.9311087278	.9379472941	.9441874858
75+	.9498708605	.9550375647	.9597261560	.9639734681	.9678145156
80+	.9712824327	.9744084425	.9772218514	.9797500653	.9820186249
85+	.9840512543	.9858699227	.9874949147	.9889449075	.9902370531
* 105+	.9988241282	.9989771629	.9991109500	.9992278090	.9993297949
		P=1/ 6,	R= 1		
5+	.5981224280	.6651020233	.7209183528	.7674319606	.8061933005
10+	.8384944171	.8654120143	.8878433452	.9065361210	.9221134342
15+	.9350945285	.9459121071	.9549267559	.9624389632	.9686991360
20+	.9739159467	.9782632889	.9818860741	.9849050617	.9874208848
25+	.9895174040	.9912645033	.9927204194	.9939336829	.9949447357
* 35+	.9983070022	.9985891685	.9988243071	.9990202559	.9991835466
		P=1/ 6,	R= 2		
15+	.7403781139	.7728308497	.8016777259	.8272192309	.8497558530
20+	.8695797335	.8869691023	.9021848001	.9154683458	.9270411318
25+	.9371044239	.9458399206	.9534106844	.9599623069	.9656242029
30+	.9705109584	.9747236786	.9783512988	.9814718323	.9841535408
35+	.9864560178	.9884311818	.9901241796	.9915742009	.9928152100
* 50+	.9987912670	.9989744084	.9991300785	.9992623473	.9993746909
		P=1/ 6,	R= 3		
20+	.6713409284	.7043807292	.7348121247	.7627075706	.7881676998
25+	.8113132718	.8322784638	.8512053733	.8682395918	.8835267110
30+	.8972096263	.9094265150	.9203093756	.9299830295	.9385644966
35+	.9461626706	.9528782285	.9588037207	.9640237972	.9686155311
40+	.9726488110	.9761867757	.9792862747	.9819983364	.9843686342
45+	.9864379418	.9882425705	.9898147849	.9911831937	.9923731145
* 60+	.9985128004	.9987222152	.9989026431	.9990580116	.9991917304
		P=1/ 6,	R= 4		
30+	.7603804731	.7831853320	.8042255292	.8235728369	.8413078690
35+	.8575173069	.8722915342	.8857226499	.8979028284	.9089229899
40+	.9188717467	.9278345908	.9358932883	.9431254527	.9496042666
45+	.9553983279	.9605715969	.9651834258	.9692886523	.9729377426
50+	.9761769712	.9790486278	.9815912404	.9838398094	.9858260454
55+	.9875786065	.9891233318	.9904834674	.9916798829	.9927312783
* 70+	.9984249430	.9986341404	.9988161627	.9989744429	.9991119959
		P=1/ 6,	R= 5		
35+	.7156847250	.7393234887	.7614848296	.7821911330	.8014764155
40+	.8193841779	.8359654394	.8512769646	.8653796852	.8783373131
45+	.8902151387	.9010790036	.9109944358	.9200259341	.9282363872
50+	.9356866131	.9424350061	.9485372764	.9540462704	.9590118602
55+	.9634808911	.9674971770	.9711015361	.9743318580	.9772231955
60+	.9798078759	.9821156264	.9841737079	.9860070564	.9876384258
65+	.9890885319	.9903761945	.9915184758	.9925308150	.9934271571
* 80+	.9984579372	.9986531733	.9988243219	.9989742597	.9991055344
		P=1/ 6,	R= 6		
40+	.6761220788	.6999990954	.7226601527	.7440962880	.7643101876

DIRICHLET 1 PROBABILITY, B= 1

N	0	1	2	3	4
			P=1/ 6, R= 6		
45+	.7833147085	.8011314469	.8177893730	.8333235501	.8477739475
50+	.8611843541	.8736013972	.8850736654	.8956509339	.9053834900
55+	.9143215517	.9225147749	.9300118419	.9368601243	.9431054132
60+	.9487917103	.9539610712	.9586534971	.9629068655	.9667568974
65+	.9702371521	.9733790487	.9762119063	.9787630012	.9810576369
70+	.9831192236	.9849693655	.9866279523	.9881132539	.9894420164
75+	.9906295577	.9916898625	.9926356743	.9934785852	.9942291224
* 90+	.9985592303	.9987344723	.9988889977	.9990251695	.9991450936
			P=1/ 6, R= 7		
50+	.7494309656	.7680565303	.7856473415	.8022183955	.8177904852
55+	.8323893193	.8460446914	.8587897053	.8706600614	.8816934052
60+	.8919287399	.9014059016	.9101650966	.9182464966	.9256898915
65+	.9325343924	.9388181857	.9445783296	.9498505924	.9546693272
70+	.9590673788	.9630760196	.9667249106	.9700420842	.9730539458
75+	.9757852909	.9782593354	.9804977566	.9825207429	.9843470499
80+	.9859940620	.9874778567	.9888132719	.9900139741	.9910925270
* 100+	.9986938835	.9988474157	.9989834399	.9991038779	.9992104511
			P=1/ 6, R= 8		
55+	.7176841940	.7368017149	.7550088777	.7723056823	.7886980788
60+	.8041972999	.8188192065	.8325836557	.8455138959	.8576359960
65+	.8689783119	.8795709920	.8894455243	.8986343252	.9071703697
70+	.9150868626	.9224169486	.9291934605	.9354487021	.9412142658
75+	.9465208791	.9513982811	.9558751235	.9599788957	.9637358702
80+	.9671710668	.9703082327	.9731698367	.9757770759	.9781498922
85+	.9803069980	.9822659082	.9840429786	.9856534487	.9871114874
90+	.9884302420	.9896218877	.9906976789	.9916680004	.9925424180
* 110+	.9988400854	.9989724727	.9990902210	.9991948862	.9992878670
			P=1/ 6, R= 9		
60+	.6879531418	.7073271682	.7259091746	.7436882548	.7606591949
65+	.7768219951	.7921813812	.8067463164	.8205295177	.8335469856
70+	.8458175496	.8573624351	.8682048540	.8783696218	.8878828018
75+	.8967713792	.9050629625	.9127855156	.9199671169	.9266357467
80+	.9328191006	.9385444283	.9438383957	.9487269692	.9532353203
85+	.9573877490	.9612076238	.9647173379	.9679382780	.9708908065
90+	.9735942533	.9760669181	.9783260797	.9803880129	.9822680108
95+	.9839804120	.9855386315	.9869551947	.9882417736	.9894092248
100+	.9904676284	.9914263273	.9922939670	.9930785348	.9937873987
* 120+	.9989849040	.9990978196	.9991985776	.9992884349	.9993685251
			P=1/ 6, R=10		
70+	.7503798295	.7662861162	.7814655027	.7959220612	.8096633213
75+	.8226999014	.8350451477	.8467147835	.8577265722	.8680999963
80+	.8778559547	.8870164790	.8956044706	.9036434581	.9111573766
85+	.9181703672	.9247065975	.9307901019	.9364446412	.9416935807
90+	.9465597850	.9510655297	.9552324278	.9590813698	.9626324769
95+	.9659050659	.9689176236	.9716877916	.9742323588	.9765672612
100+	.9787075885	.9806675951	.9824607172	.9840995921	.9855960826
105+	.9869613019	.9882056425	.9893388049	.9903698284	.9913071226
* 125+	.9984489711	.9986141209	.9987623322	.9988952619	.9990144145
			P=1/ 7, R= 1		
5+	.5373356340	.6034305434	.6600833229	.7086428482	.7502652985
10+	.7859416844	.8165214438	.8427326661	.8651994281	.8844566526
15+	.9009628451	.9151110101	.9272380087	.9376325788	.9465422104
20+	.9541790375	.9607248893	.9663356194	.9711448166	.9752669857
25+	.9788002734	.9818288058	.9844246907	.9866497349	.9885569156
30+	.9901916420	.9915928360	.9927938594	.9938233081	.9947056926
* 45+	.9990286081	.9991673784	.9992863243	.9993882780	.9994756669

DIRICHLET 1 PROBABILITY, B= 1

N	0	1	2	3	4
			P=1/ 7,　R= 2		
15+	.6533699579	.6887403704	.7210790332	.7505303154	.7772592102
20+	.8014424959	.8232620019	.8428995572	.8605332804	.8763349284
25+	.8904680795	.9030869643	.9143357988	.9243484977	.9332486744
30+	.9411498517	.9481558217	.9543611095	.9598515023	.9647046174
35+	.9689904853	.9727721335	.9761061579	.9790432747	.9816288447
40+	.9839033687	.9859029502	.9876597254	.9892022598	.9905559123
* 60+	.9989417572	.9990791914	.9991989554	.9993032931	.9993941679
			P=1/ 7,　R= 3		
25+	.7138036915	.7390414612	.7624765331	.7841707139	.8041961116
30+	.8226321920	.8395632862	.8550765056	.8692600204	.8822016607
35+	.8939877974	.9047024671	.9144267051	.9232380555	.9312102297
40+	.9384128890	.9449115290	.9507674463	.9560377719	.9607755559
45+	.9650298925	.9688460749	.9722657708	.9753272128	.9780653970
50+	.9805122849	.9826970064	.9846460581	.9863834986	.9879311346
55+	.9893087007	.9905340290	.9916232098	.9925907418	.9934496733
* 70+	.9983576889	.9985550408	.9987291028	.9988825614	.9990178027
			P=1/ 7,　R= 4		
35+	.7564828695	.7761264306	.7944944358	.8116276172	.8275719655
40+	.8423774318	.8560967828	.8687846037	.8804964384	.8912880574
45+	.9012148429	.9103312786	.9186905352	.9263441403	.9333417221
50+	.9397308185	.9455567423	.9508624943	.9556887177	.9600736864
55+	.9640533218	.9676612331	.9709287754	.9738851231	.9765573544
60+	.9789705428	.9811478556	.9831106549	.9848786000	.9864697506
65+	.9879006683	.9891865159	.9903411546	.9913772366	.9923062948
* 85+	.9988342202	.9989674484	.9990857716	.9991908136	.9992840273
			P=1/ 7,　R= 5		
40+	.6943227687	.7154734348	.7355624845	.7545942158	.7725802476
45+	.7895385062	.8054922686	.8204692700	.8345008793	.8476213452
50+	.8598671133	.8712762140	.8818877181	.8917412575	.9008766090
55+	.9093333343	.9171504754	.9243662979	.9310180804	.9371419437
60+	.9427727166	.9479438346	.9526872662	.9570334646	.9610113411
65+	.9646482567	.9679700298	.9710009564	.9737638419	.9762800411
70+	.9785695059	.9806508376	.9825413435	.9842570968	.9858129973
75+	.9872228337	.9884993455	.9896542847	.9906984764	.9916418773
* 95+	.9985863346	.9987395361	.9988765610	.9989990625	.9991085320
			P=1/ 7,　R= 6		
50+	.7374094320	.7549033865	.7715280761	.7872937393	.8022148133
55+	.8163093555	.8295984954	.8421059211	.8538574035	.8648803573
60+	.8752034411	.8848561947	.8938687147	.9022713649	.9100945220
65+	.9173683533	.9241226252	.9303865402	.9361885996	.9415564914
70+	.9465169985	.9510959281	.9553180580	.9592070988	.9627856700
75+	.9660752882	.9690963661	.9718682203	.9744090866	.9767361423
80+	.9788655330	.9808124045	.9825909375	.9842143851	.9856951120
85+	.9870446352	.9882736653	.9893921477	.9904093042	.9913336736
* 110+	.9990756285	.9991728473	.9992601083	.9993383986	.9994086114
			P=1/ 7,　R= 7		
55+	.6871093850	.7055665237	.7232853768	.7402597402	.7564879778
60+	.7719726035	.7867198660	.8007393415	.8140435377	.8266475130
65+	.8385685143	.8498256342	.8604394900	.8704319258	.8798257363
70+	.8886444156	.8969119275	.9046524990	.9118904360	.9186499593
75+	.9249550608	.9308293790	.9362960914	.9413778241	.9460965759
80+	.9504736568	.9545296391	.9582843199	.9617566938	.9649649355
85+	.9679263892	.9706575672	.9731741527	.9754910091	.9776221941
90+	.9795809769	.9813798591	.9830305980	.9845442323	.9859311091
95+	.9872009119	.9883626897	.9894248866	.9903953710	.9912814655
* 120+	.9989758447	.9990793587	.9991727142	.9992568721	.9993327067

DIRICHLET 1 PROBABILITY, B= 1

N	0	1	2	3	4
			P=1/ 7, R= 8		
65+	.7278735024	.7436870755	.7588497267	.7733625501	.7872296037
70+	.8004576227	.8130557360	.8250351919	.8364090929	.8471921419
75+	.8574004015	.8670510671	.8761622545	.8847528027	.8928420914
80+	.9004498749	.9075961295	.9143009166	.9205842599	.9264660362
85+	.9319658789	.9371030947	.9418965908	.9463648139	.9505256989
90+	.9543966268	.9579943911	.9613351723	.9644345188	.9673073350
95+	.9699678742	.9724297367	.9747058729	.9768085891	.9787495579
100+	.9805398305	.9821898512	.9837094752	.9851079863	.9863941170
105+	.9875760692	.9886615356	.9896577211	.9905713656	.9914087653
* 130+	.9989050826	.9990118845	.9991085969	.9991961353	.9992753368
			P=1/ 7, R= 9		
75+	.7616979676	.7753697439	.7884670758	.8009949585	.8129603648
80+	.8243720400	.8352403021	.8455768489	.8553945729	.8647073853
85+	.8735300497	.8818780253	.8897673209	.8972143595	.9042358530
90+	.9108486881	.9170698222	.9229161892	.9284046154	.9335517444
95+	.9383739716	.9428873863	.9471077221	.9510503150	.9547300685
100+	.9581614241	.9613583393	.9643342696	.9671021561	.9696744175
105+	.9720629460	.9742791065	.9763337392	.9782371652	.9799991938
110+	.9816291326	.9831357987	.9845275319	.9858122087	.9869972575
115+	.9880896744	.9890967039	.9900225350	.9908749591	.9916587464
* 140+	.9988600350	.9989677292	.9990655910	.9991544804	.9992351864
			P=1/ 7, R=10		
80+	.7229349268	.7374259429	.7513994228	.7648533408	.7777878026
85+	.7902048858	.8021084806	.8135041299	.8243988715	.8348010840
90+	.8447203367	.8541672441	.8631533267	.8716908785	.8797928409
95+	.8874726842	.8947442967	.9016218809	.9081198583	.9142527806
100+	.9200352503	.9254818466	.9306070598	.9354252327	.9399505074
105+	.9441967803	.9481776611	.9519064390	.9553960533	.9586590690
110+	.9617076585	.9645535834	.9672081856	.9696823779	.9719866395
115+	.9741310135	.9761251079	.9779780982	.9796987320	.9812953359
120+	.9827758231	.9841477031	.9854180921	.9865937238	.9876809622
125+	.9886858135	.9896139398	.9904706717	.9912610225	.9919897012
* 150+	.9988369100	.9989436491	.9990409384	.9991295777	.9992103027
			P=1/ 8, R= 1		
10+	.7369244238	.7698088709	.7985827620	.8237599167	.8457899272
15+	.8650661863	.8819329130	.8966912989	.9096048865	.9209042757
20+	.9307912412	.9394423361	.9470120441	.9536355386	.9594310962
25+	.9645022092	.9689394331	.9728220039	.9762192534	.9791918468
30+	.9817928659	.9840687577	.9860601630	.9878026426	.9893273123
35+	.9906613982	.9918287235	.9928501330	.9937438664	.9945258831
* 50+	.9987399068	.9988974185	.9990352412	.9991558360	.9992613565
			P=1/ 8, R= 2		
20+	.7330519304	.7577693443	.7804784683	.8012951652	.8203377119
25+	.8377243849	.8535716130	.8679925905	.8810962672	.8929866405
30+	.9037622912	.9135161131	.9223351936	.9303008148	.9374885433
35+	.9439683894	.9498050155	.9550579790	.9597819983	.9640272318
40+	.9678395632	.9712608863	.9743293854	.9770798084	.9795437290
45+	.9817497974	.9837239780	.9854897728	.9870684318	.9884791483
50+	.9897392415	.9908643247	.9918684614	.9927643089	.9935632498
* 70+	.9990406957	.9991497075	.9992464556	.9993323024	.9994084617
			P=1/ 8, R= 3		
30+	.7421275294	.7623318746	.7812299044	.7988680656	.8152971593
35+	.8305710823	.8447457456	.8578781544	.8700256325	.8812451782
40+	.8915929349	.9011237634	.9098909038	.9179457140	.9253374758
45+	.9321132574	.9383178249	.9439935941	.9491806164	.9539165933
50+	.9582369127	.9621747038	.9657609064	.9690243508	.9719918455
55+	.9746882711	.9771366763	.9793583774	.9813730563	.9831988591

DIRICHLET 1   PROBABILITY,   B= 1

N	0	1	2	3	4
			P=1/ 8, R= 3		
60+	.9848524918	.9863493146	.9877034318	.9889277795	.9900342084
* 85+	.9989879943	.9990951627	.9991911677	.9992771486	.9993541316
			P=1/ 8, R= 4		
40+	.7536228456	.7708691067	.7871509388	.8024934344	.8169249694
45+	.8304765327	.8431811233	.8550732110	.8661882589	.8765623036
50+	.8862315898	.8952322551	.9036000612	.9113701669	.9185769399
55+	.9252538031	.9314331116	.9371460572	.9424225972	.9472914046
60+	.9517798364	.9559139183	.9597183429	.9632164790	.9664303915
65+	.9693808686	.9720874555	.9745684935	.9768411630	.9789215297
70+	.9808245925	.9825643327	.9841537644	.9856049847	.9869292232
75+	.9881368915	.9892376308	.9902403590	.9911533160	.9919841068
* 100+	.9990667744	.9991603433	.9992447223	.9993207913	.9993893484
			P=1/ 8, R= 5		
50+	.7653655120	.7804737717	.7948185822	.8084162670	.8212855045
55+	.8334469339	.8449227926	.8557365825	.8659127668	.8754764956
60+	.8844533592	.8928691689	.9007497625	.9081208351	.9150077906
65+	.9214356157	.9274287723	.9330111077	.9382057809	.9430352037
70+	.9475209944	.9516839442	.9555439928	.9591202142	.9624308105
75+	.9654931121	.9683235845	.9709378403	.9733506552	.9755759878
80+	.9776270026	.9795160953	.9812549192	.9828544143	.9843248362
85+	.9856757863	.9869162421	.9880545871	.9890986415	.9900556914
* 110+	.9986746705	.9988002212	.9989141849	.9990175963	.9991114019
			P=1/ 8, R= 6		
55+	.6996712103	.7163931758	.7324593779	.7478690285	.7626244958
60+	.7767309957	.7901962912	.8030304009	.8152453211	.8268547603
65+	.8378738891	.8483191049	.8582078134	.8675582252	.8763891696
70+	.8847199239	.8925700577	.8999592935	.9069073809	.9134339851
75+	.9195585883	.9253004037	.9306783013	.9357107437	.9404157326
80+	.9448107645	.9489127943	.9527382069	.9563027960	.9596217482
85+	.9627096342	.9655804032	.9682473831	.9707232836	.9730202033
90+	.9751496394	.9771224992	.9789491151	.9806392596	.9822021632
95+	.9836465319	.9849805670	.9862119839	.9873480329	.9883955190
100+	.9893608220	.9902499169	.9910683936	.9918214765	.9925140438
* 125+	.9989237639	.9990220805	.9991116625	.9991932591	.9992675585
			P=1/ 8, R= 7		
65+	.7184999940	.7334217309	.7477839027	.7615868915	.7748333082
70+	.7875277909	.7996768075	.8112884638	.8223723175	.8329392004
75+	.8430010485	.8525707410	.8616619488	.8702889929	.8784667117
80+	.8862103394	.8935353925	.9004575677	.9069926476	.9131564162
85+	.9189645827	.9244327141	.9295761753	.9344100762	.9389492272
90+	.9432080992	.9472007917	.9509410051	.9544420189	.9577166740
95+	.9607773601	.9636360066	.9663040766	.9687925650	.9711119985
100+	.9732724386	.9752834865	.9771542903	.9788935532	.9805095436
105+	.9820101062	.9834026737	.9846942800	.9858915735	.9870000307
110+	.9880279706	.9889785688	.9898578722	.9906708130	.9914220235
* 135+	.9986452514	.9987635979	.9988719304	.9989710629	.9990617466
			P=1/ 8, R= 8		
75+	.7351955333	.7486712227	.7616586625	.7741590733	.7861753132
80+	.7977117381	.8087740632	.8193692294	.8295052717	.8391911937
85+	.8484368465	.8572528135	.8656503011	.8736410354	.8812371655
90+	.8884511732	.8952957889	.9017839143	.9079285506	.9137427342
95+	.9192394766	.9244317121	.9293322489	.9339537274	.9383085821
100+	.9424090091	.9462669378	.9498940064	.9533015419	.9565005433
105+	.9595016684	.9623152231	.9649511544	.9674190451	.9697281112
110+	.9718872011	.9739047973	.9757890187	.9775476254	.9791880239
115+	.9807172738	.9821420958	.9834688796	.9847036932	.9858522923
120+	.9869201305	.9879123695	.9888338898	.9896893010	.9904829530

B 10

DIRICHLET 1 PROBABILITY, B= 1

N	0	1	2	3	4
			P=1/ 8, R= 8		
* 150+	.9989512793	.9990405978	.9991225505	.9991977210	.9992666487
			P=1/ 8, R= 9		
85+	.7502017854	.7624811680	.7743276237	.7857429584	.7967302180
90+	.8072935864	.8174382848	.8271704728	.8364971530	.8454260777
95+	.8539656597	.8621248869	.8699132400	.8773406161	.8844172550
100+	.8911536709	.8975605882	.9036488819	.9094295224	.9149135249
105+	.9201119022	.9250356230	.9296955730	.9341025206	.9382670862
110+	.9421997143	.9459106502	.9494099186	.9527073061	.9558123460
115+	.9587343057	.9614821767	.9640646666	.9664901932	.9687668807
120+	.9709025572	.9729047538	.9747807058	.9765373538	.9781813472
125+	.9797190479	.9811565352	.9824996113	.9837538074	.9849243903
130+	.9860163701	.9870345070	.9879833195	.9888670925	.9896898851
135+	.9904555393	.9911676882	.9918297641	.9924450071	.9930164733
* 160+	.9987603511	.9988621249	.9989558332	.9990420876	.9991214554
			P=1/ 8, R=10		
90+	.7016599021	.7148641126	.7276858841	.7401214577	.7521684196
95+	.7638256269	.7750931310	.7859721005	.7964647429	.8065742271
100+	.8163046056	.8256607387	.8346482199	.8432733027	.8515428301
105+	.8594641670	.8670451339	.8742939450	.8812191485	.8878295700
110+	.8941342595	.9001424414	.9058634675	.9113067739	.9164818404
115+	.9213981536	.9260651726	.9304922981	.9346888442	.9386640128
120+	.9424268713	.9459863320	.9493511348	.9525298311	.9555307715
125+	.9583620934	.9610317128	.9635473156	.9659163525	.9681460344
130+	.9702433289	.9722149590	.9740674025	.9758068921	.9774394172
135+	.9789707257	.9804063274	.9817514975	.9830112808	.9841904966
140+	.9852937437	.9863254060	.9872896586	.9881904735	.9890316263
145+	.9898167022	.9905491029	.9912320533	.9918686075	.9924616563
* 175+	.9990607191	.9991364361	.9992062589	.9992706268	.9993299482
			P=1/ 9, R= 1		
10+	.6920538523	.7262700910	.7566845253	.7837195781	.8077507360
15+	.8291117654	.8480993470	.8649771973	.8799797310	.8933153164
20+	.9051691701	.9157059290	.9250719369	.9333972772	.9407975798
25+	.9473756265	.9532227791	.9584202481	.9630402205	.9671468627
30+	.9707972113	.9740419656	.9769261916	.9794899481	.9817688428
35+	.9837945269	.9855951350	.9871956756	.9886183783	.9898830029
40+	.9910071137	.9920063233	.9928945096	.9936840085	.9943857854
* 55+	.9984632215	.9986339747	.9987857553	.9989206714	.9990405968
			P=1/ 9, R= 2		
20+	.6680920955	.6944339927	.7190197634	.7419144493	.7631903191
25+	.7829244592	.8011968111	.8180885853	.8336809923	.8480542399
30+	.8612867535	.8734545822	.8846309581	.8948859840	.9042864245
35+	.9128905863	.9207732426	.9279756751	.9345506752	.9405626422
40+	.9460426823	.9510387302	.9555906850	.9597355544	.9635076048
45+	.9669385138	.9700575219	.9728915836	.9754655141	.9778021318
50+	.9799223960	.9818455381	.9835891870	.9851694875	.9866012128
55+	.9878978696	.9890717976	.9901342617	.9910955388	.9919649980
* 75+	.9984880100	.9986398162	.9987765543	.9988996985	.9990105816
			P=1/ 9, R= 3		
35+	.7622353243	.7789753530	.7947306740	.8095356742	.8234267854
40+	.8364418806	.8486197474	.8599996344	.8706208622	.8805224947
45+	.8897430625	.8983203349	.9062911334	.9136911834	.9205549980
50+	.9269157906	.9328054134	.9382543162	.9432915240	.9479446311
55+	.9522398068	.9562018138	.9598540342	.9632185040	.9663159523
60+	.9691658462	.9717864384	.9741948169	.9764069571	.9784377744
65+	.9803011766	.9820101169	.9835766455	.9850119606	.9863264579
70+	.9875297789	.9886308569	.9896379620	.9905587439	.9914002722
* 95+	.9988579004	.9989650300	.9990622824	.9991505489	.9992306425

DIRICHLET 1 PROBABILITY, B= 1

N	0	1	2	3	4
			P=1/ 9, R= 4		
45+	.7514345456	.7668021586	.7814152893	.7952903831	.8084460276
50+	.8209025798	.8326818255	.8438066686	.8543008516	.8641887041
55+	.8734949182	.8822443503	.8904618463	.8981720894	.9053994688
60+	.9121679669	.9185010646	.9244216617	.9299520123	.9351136728
65+	.9399274619	.9444134302	.9485908398	.9524781516	.9560930192
70+	.9594522902	.9625720112	.9654674385	.9681530522	.9706425735
75+	.9729489844	.9750845502	.9770608423	.9788887642	.9805785764
80+	.9821399234	.9835818600	.9849128784	.9861409347	.9872734756
85+	.9883174639	.9892794043	.9901653682	.9909810175	.9917316280
* 110+	.9987484345	.9988590464	.9989600992	.9990523962	.9991366754
			P=1/ 9, R= 5		
55+	.7455344743	.7597523014	.7733625290	.7863735643	.7987956226
60+	.8106404944	.8219213247	.8326524069	.8428489908	.8525271043
65+	.8617033897	.8703949533	.8786192285	.8863938519	.8937365519
70+	.9006650483	.9071969640	.9133497471	.9191406017	.9245864295
75+	.9297037788	.9345088017	.9390172182	.9432442875	.9472047849
80+	.9509129840	.9543826439	.9576270013	.9606587654	.9634901175
85+	.9661327129	.9685976852	.9708956540	.9730367334	.9750305427
90+	.9768862188	.9786124292	.9802173860	.9817088610	.9830942010
95+	.9843803439	.9855738342	.9866808397	.9877071674	.9886582796
100+	.9895393099	.9903550786	.9911101086	.9918086397	.9924546437
* 125+	.9987390470	.9988451475	.9989425513	.9990319481	.9991139753
			P=1/ 9, R= 6		
65+	.7424116796	.7556663140	.7684139406	.7806589726	.7924072925
70+	.8036660991	.8144437601	.8247496717	.8345941245	.8439881775
75+	.8529435388	.8614724544	.8695876041	.8773020057	.8846289259
80+	.8915817991	.8981741530	.9044195409	.9103314809	.9159234014
85+	.9212085921	.9262001611	.9309109971	.9353537367	.9395407364
90+	.9434840482	.9471954005	.9506861814	.9539674264	.9570498080
95+	.9599436295	.9626588199	.9652049326	.9675911445	.9698262582
100+	.9719187050	.9738765500	.9757074976	.9774188988	.9790177589
105+	.9805107461	.9819042009	.9832041454	.9844162941	.9855460638
110+	.9865985843	.9875787092	.9884910267	.9893398705	.9901293300
* 140+	.9987922297	.9988897409	.9989796041	.9990623979	.9991386588
			P=1/ 9, R= 7		
75+	.7410015222	.7534395241	.7654431830	.7770147853	.7881578098
80+	.7988768227	.8091773756	.8190659064	.8285496436	.8376365144
85+	.8463350574	.8546543390	.8626038748	.8701935551	.8774335753
90+	.8843343709	.8909065573	.8971608732	.9031081297	.9087591626
95+	.9141247899	.9192157721	.9240427774	.9286163502	.9329468829
100+	.9370445913	.9409194928	.9445813880	.9480398447	.9513041840
105+	.9543834701	.9572865008	.9600218008	.9625976169	.9650219143
110+	.9673023754	.9694463986	.9714610998	.9733533139	.9751295980
115+	.9767962349	.9783592379	.9798243558	.9811970790	.9824826451
120+	.9836860462	.9848120356	.9858651348	.9868496413	.9877696361
125+	.9886289908	.9894313762	.9901802693	.9908789611	.9915305643
* 155+	.9988806232	.9989678600	.9990485056	.9991230390	.9991919060
			P=1/ 9, R= 8		
85+	.7407098924	.7524460219	.7638025016	.7747804319	.7853818901
90+	.7956098551	.8054681346	.8149612927	.8240945794	.8328738628
95+	.8413055628	.8493965880	.8571542751	.8645863309	.8717007775
100+	.8785059003	.8850101993	.8912223430	.8971511258	.9028054279
105+	.9081941786	.9133263221	.9182107864	.9228564547	.9272721394
110+	.9314665588	.9354483162	.9392258809	.9428075719	.9462015432
115+	.9494157715	.9524580452	.9553359555	.9580568889	.9606280211
120+	.9630563127	.9653485053	.9675111198	.9695504548	.9714725866
125+	.9732833699	.9749884389	.9765932097	.9781028830	.9795224472

B 12

DIRICHLET 1   PROBABILITY,   B= 1

N	0	1	2	3	4
			P=1/ 9, R= 8		
130+	.9808566824	.9821101645	.9832872695	.9843921788	.9854288837
135+	.9864011915	.9873127300	.9881669538	.9889671498	.9897164428
140+	.9904178011	.9910740429	.9916878412	.9922617299	.9927981096
* 170+	.9989857676	.9990624271	.9991334772	.9991993123	.9992603001
			P=1/ 9, R= 9		
95+	.7411791254	.7523042851	.7630923188	.7735436473	.7836595010
100+	.7934418651	.8028934245	.8120175106	.8208180476	.8292995007
105+	.8374668260	.8453254207	.8528810764	.8601399331	.8671084355
110+	.8737932915	.8802014323	.8863399749	.8922161867	.8978374517
115+	.9032112397	.9083450765	.9132465175	.9179231217	.9223824292
120+	.9266319394	.9306790920	.9345312490	.9381956791	.9416795431
125+	.9449898812	.9481336022	.9511174729	.9539481104	.9566319740
130+	.9591753599	.9615843957	.9638650367	.9660230626	.9680640755
135+	.9699934986	.9718165756	.9735383705	.9751637687	.9766974777
140+	.9781440294	.9795077818	.9807929219	.9820034685	.9831432753
145+	.9842160347	.9852252811	.9861743953	.9870666082	.9879050051
150+	.9886925299	.9894319899	.9901260596	.9907772854	.9913880904
* 180+	.9986791954	.9987750754	.9988642432	.9989471475	.9990242091
			P=1/ 9, R=10		
105+	.7421813647	.7527686382	.7630527252	.7730336531	.7827121286
110+	.7920894961	.8011676955	.8099492218	.8184370833	.8266347615
115+	.8345461715	.8421756235	.8495277850	.8566076441	.8634204750
120+	.8699718032	.8762673739	.8823131204	.8881151347	.8936796396
125+	.8990129622	.9041215088	.9090117414	.9136901560	.9181632620
130+	.9224375634	.9265195408	.9304156358	.9341322359	.9376756610
135+	.9410521515	.9442678568	.9473288255	.9502409972	.9530101940
140+	.9556421144	.9581423272	.9605162666	.9627692283	.9649063661
145+	.9669326894	.9688530611	.9706721966	.9723946631	.9740248793
150+	.9755671155	.9770254949	.9784039943	.9797064460	.9809365393
155+	.9820978227	.9831937067	.9842274656	.9852022414	.9861210457
160+	.9869867635	.9878021563	.9885698653	.9892924149	.9899722163
165+	.9906115707	.9912126732	.9917776159	.9923083919	.9928068985
* 195+	.9988472100	.9989289408	.9990050869	.9990760126	.9991420598
			P=1/10, R= 1		
10+	.6513215599	.6861894039	.7175704635	.7458134172	.7712320755
15+	.7941088679	.8146979811	.8332281830	.8499053647	.8649148282
20+	.8784233454	.8905810109	.9015229098	.9113706188	.9202335569
25+	.9282102012	.9353891811	.9418502630	.9476652367	.9528987130
30+	.9576088417	.9618479576	.9656631618	.9690968456	.9721871611
35+	.9749684450	.9774716005	.9797244404	.9817519964	.9835767967
40+	.9852191171	.9866972054	.9880274848	.9892247363	.9903022627
* 65+	.9989388834	.9990449950	.9991404955	.9992264460	.9993038014
			P=1/10, R= 2		
25+	.7287940935	.7487357043	.7674010520	.7848459731	.8011278994
30+	.8163049808	.8304353669	.8435766260	.8557852795	.8671164362
35+	.8776235086	.8873580023	.8963693621	.9047048699	.9124095826
40+	.9195263040	.9260955853	.9321557473	.9377429211	.9428911026
45+	.9476322186	.9519962004	.9560110636	.9597029922	.9630964244
50+	.9662141403	.9690773488	.9717057741	.9741177409	.9763302566
55+	.9783590918	.9802188573	.9819230788	.9834842675	.9849139876
60+	.9862229210	.9874209279	.9885171042	.9895198359	.9904368503
* 85+	.9986525935	.9987744334	.9988853795	.9989863919	.9990783481
			P=1/10, R= 3		
35+	.6937497401	.7121371169	.7296592055	.7463302211	.7621676860
40+	.7771918757	.7914253185	.8048923452	.8176186854	.8296311090
45+	.8409571083	.8516246193	.8616617774	.8710967061	.8799573347
50+	.8882712437	.8960655333	.9033667149	.9102006208	.9165923328

B 13

DIRICHLET 1 PROBABILITY, B= 1

N	0	1	2	3	4
			P=1/10, R= 3		
55+	.9225661252	.9281454218	.9333527654	.9382097967	.9427372438
60+	.9469549182	.9508817185	.9545356394	.9579337859	.9610923909
65+	.9640268368	.9667516795	.9692806741	.9716268029	.9738023042
70+	.9758187016	.9776868346	.9794168881	.9810184234	.9825004075
75+	.9838712427	.9851387959	.9863104261	.9873930124	.9883929802
80+	.9893163272	.9901686475	.9909551557	.9916807095	.9923498313
* 105+	.9987440990	.9988498224	.9989468033	.9990357486	.9991173098
			P=1/10, R= 4		
50+	.7497060940	.7635626090	.7768129014	.7894682828	.8015415166
55+	.8130465982	.8239985509	.8344132380	.8443071907	.8536974513
60+	.8626014306	.8710367793	.8790212733	.8865727099	.8937088175
65+	.9004471748	.9068051410	.9127997949	.9184478828	.9237657748
70+	.9287694277	.9334743551	.9378956031	.9420477316	.9459448008
75+	.9496003614	.9530274496	.9562385842	.9592457684	.9620604928
80+	.9646937415	.9671560001	.9694572648	.9716070539	.9736144195
85+	.9754879607	.9772358375	.9788657853	.9803851295	.9818008008
90+	.9831193505	.9843469657	.9854894848	.9865524127	.9875409357
95+	.9884599362	.9893140073	.9901074669	.9908443714	.9915285291
* 125+	.9989580297	.9990411432	.9991177699	.9991884023	.9992534972
			P=1/10, R= 5		
60+	.7290417419	.7423977107	.7552616176	.7676375832	.7795310958
65+	.7909488680	.8018986987	.8123893429	.8224303881	.8320321376
70+	.8412055013	.8499618939	.8583131401	.8662713864	.8738490209
75+	.8810585989	.8879127751	.8944242426	.9006056767	.9064696859
80+	.9120287666	.9172952641	.9222813377	.9269989304	.9314597427
85+	.9356752104	.9396564854	.9434144206	.9469595571	.9503021144
90+	.9534519830	.9564187197	.9592115443	.9618393384	.9643106458
95+	.9666336748	.9688163009	.9708660716	.9727902111	.9745956271
100+	.9762889173	.9778763769	.9793640065	.9807575208	.9820623569
105+	.9832836835	.9844264098	.9854951951	.9864944574	.9874283833
110+	.9883009370	.9891158692	.9898767265	.9905868599	.9912494339
* 140+	.9987879943	.9988797246	.9989646922	.9990433796	.9991162363
			P=1/10, R= 6		
70+	.7127784092	.7256211184	.7380551960	.7500809904	.7617000300
75+	.7729149291	.7837292960	.7941476440	.8041753038	.8138183411
80+	.8230834756	.8319780047	.8405097306	.8486868913	.8565180952
85+	.8640122600	.8711785550	.8780263481	.8845651553	.8908045955
90+	.8967543474	.9024241110	.9078235718	.9129623691	.9178500660
95+	.9224961240	.9269098791	.9311005213	.9350770763	.9388483898
100+	.9424231135	.9458096939	.9490163622	.9520511266	.9549217660
105+	.9576358251	.9602006110	.9626231909	.9649103913	.9670687979
110+	.9691047564	.9710243745	.9728335240	.9745378442	.9761427458
115+	.9776534146	.9790748166	.9804117028	.9816686146	.9828498892
120+	.9839596657	.9850018905	.9859803239	.9868985461	.9877599630
125+	.9885678128	.9893251720	.9900349615	.9906999528	.9913227740
* 155+	.9986779332	.9987735556	.9988624654	.9989451164	.9990219333
			P=1/10, R= 7		
85+	.7578449260	.7684616594	.7787333490	.7886626489	.7982528995
90+	.8075080691	.8164326970	.8250318384	.8333110117	.8412761474
95+	.8489335393	.8562897978	.8633518059	.8701266774	.8766217173
100+	.8828443846	.8888022575	.8945030011	.8999543372	.9051640162
105+	.9101397911	.9148893945	.9194205162	.9237407836	.9278577444
110+	.9317788498	.9355114404	.9390627338	.9424398128	.9456496160
115+	.9486989290	.9515943775	.9543424214	.9569493496	.9594212761
120+	.9617641374	.9639836902	.9660855102	.9680749916	.9699573471
125+	.9717376087	.9734206291	.9750110834	.9765134712	.9779321193
130+	.9792711848	.9805346579	.9817263656	.9828499758	.9839090006

DIRICHLET 1 PROBABILITY, B= 1

N	0	1	2	3	4
			P=1/10, R= 7		
135+	.9849068004	.9858465887	.9867314355	.9875642722	.9883478958
140+	.9890849733	.9897780463	.9904295348	.9910417424	.9916168601
* 175+	.9990381988	.9991058473	.9991688799	.9992275996	.9992822910
			P=1/10, R= 8		
95+	.7450117927	.7554039673	.7654925504	.7752784759	.7847632961
100+	.7939491382	.8028386628	.8114350223	.8197418202	.8277630719
105+	.8355031663	.8429668288	.8501590854	.8570852284	.8637507840
110+	.8701614800	.8763232170	.8822420393	.8879241088	.8933756792
115+	.8986030729	.9036126585	.9084108304	.9130039895	.9173985255
120+	.9216008006	.9256171342	.9294537898	.9331169619	.9366127649
125+	.9399472231	.9431262616	.9461556984	.9490412369	.9517884603
130+	.9544028262	.9568896621	.9592541616	.9615013820	.9636362414
135+	.9656635173	.9675878456	.9694137200	.9711454915	.9727873696
140+	.9743434222	.9758175773	.9772136242	.9785352153	.9797858680
145+	.9809689672	.9820877675	.9831453961	.9841448550	.9850890248
150+	.9859806667	.9868224267	.9876168376	.9883663232	.9890732011
155+	.9897396859	.9903678929	.9909598410	.9915174561	.9920425744
* 190+	.9990124822	.9990795735	.9991422528	.9992007986	.9992554729
			P=1/10, R= 9		
105+	.7339144269	.7440733009	.7539626537	.7635822968	.7729325900
110+	.7820144094	.7908291165	.7993785265	.8076648778	.8156908009
115+	.8234592887	.8309736671	.8382375663	.8452548927	.8520298024
120+	.8585666747	.8648700873	.8709447920	.8767956917	.8824278188
125+	.8878463134	.8930564043	.8980633901	.9028726209	.9074894825
130+	.9119193803	.9161677249	.9202399186	.9241413429	.9278773468
135+	.9314532363	.9348742644	.9381456225	.9412724322	.9442597382
140+	.9471125013	.9498355934	.9524337918	.9549117750	.9572741191
145+	.9595252939	.9616696613	.9637114719	.9656548643	.9675038634
150+	.9692623795	.9709342082	.9725230301	.9740324108	.9754658021
155+	.9768265420	.9781178564	.9793428600	.9805045581	.9816058479
160+	.9826495206	.9836382631	.9845746604	.9854611976	.9863002621
165+	.9870941463	.9878450495	.9885550807	.9892262612	.9898605267
170+	.9904597304	.9910256449	.9915599655	.9920643120	.9925402316
* 205+	.9990101621	.9990753594	.9991364078	.9991935601	.9992470547
			P=1/10, R=10		
115+	.7241952776	.7341216787	.7438068775	.7532499464	.7624504410
120+	.7714083772	.7801242069	.7885987950	.7968333947	.8048296244
125+	.8125894438	.8201151308	.8274092581	.8344746713	.8413144663
130+	.8479319679	.8543307091	.8605144107	.8664869615	.8722523996
135+	.8778148944	.8831787285	.8883482821	.8933280162	.8981224578
140+	.9027361858	.9071738174	.9114399950	.9155393747	.9194766147
145+	.9232563651	.9268832580	.9303618983	.9336968557	.9368926566
150+	.9399537772	.9428846375	.9456895945	.9483729381	.9509388854
155+	.9533915770	.9557350735	.9579733518	.9601103026	.9621497282
160+	.9640953402	.9659507582	.9677195087	.9694050239	.9710106412
165+	.9725396033	.9739950576	.9753800568	.9766975592	.9779504294
170+	.9791414391	.9802732683	.9813485059	.9823696519	.9833391179
175+	.9842592293	.9851322265	.9859602670	.9867454268	.9874897026
180+	.9881950133	.9888632024	.9894960397	.9900952232	.9906623813
* 220+	.9990257754	.9990881792	.9991467269	.9992016463	.9992531524

DIRICHLET 1   PROBABILITY,   B= 2

N	0	1	2	3	4
			P=1/ 3, R= 1		
5+	.7407407407	.8257887517	.8834019204	.9221155312	.9480262155
10+	.9653338753	.9768836051	.9845871884	.9897241651	.9931492343
* 15+	.9954327532	.9969551455	.9979700893	.9986467236	.9990978149
			P=1/ 3, R= 2		
10+	.7937814358	.8504575296	.8923925704	.9230413160	.9452363166
15+	.9611946053	.9726024428	.9807181520	.9864680957	.9905273750
* 25+	.9989307433	.9992607606	.9994895727	.9996479811	.9997574981
			P=1/ 3, R= 3		
10+	.4386526444	.5499161713	.6463709048	.7265152670	.7911270175
15+	.8420782154	.8815934203	.9118452019	.9347659927	.9519842302
20+	.9648252754	.9743421013	.9813565470	.9865013131	.9902581493
* 30+	.9986988347	.9990769285	.9993463749	.9995379813	.9996739600
			P=1/ 3, R= 4		
15+	.5942983671	.6745662695	.7423351169	.7981995862	.8434022755
20+	.8794406328	.9078280719	.9299648311	.9470795521	.9602129906
25+	.9702248664	.9778120277	.9835309385	.9878205745	.9910236960
* 35+	.9986465656	.9990211363	.9992935515	.9994911861	.9996342359
			P=1/ 3, R= 5		
20+	.7017777954	.7601557582	.8089259488	.8490344785	.8815946457
25+	.9077399986	.9285385026	.9449486699	.9578026330	.9678054434
30+	.9755434016	.9814967997	.9860542295	.9895267839	.9921612361
* 40+	.9987044123	.9990487913	.9993031975	.9994906548	.9996284439
			P=1/ 3, R= 6		
25+	.7779854262	.8209128862	.8566125569	.8859649978	.9098626902
30+	.9291524255	.9446038264	.9568954525	.9666121204	.9742489429
35+	.9802190468	.9848629747	.9884585123	.9912301822	.9933579786
* 45+	.9988200813	.9991232172	.9993499530	.9995191052	.9996449859
			P=1/ 3, R= 7		
25+	.5714497291	.6384045874	.6980704554	.7502156837	.7950558740
30+	.8330919137	.8649826960	.8914527037	.9132294231	.9310039969
35+	.9454088517	.9570071131	.9662898318	.9736781248	.9795281933
40+	.9841378245	.9877534547	.9905772002	.9927734986	.9944751593
* 50+	.9989596356	.9992192379	.9994153820	.9995632007	.9996743268
			P=1/ 3, R= 8		
30+	.6725043660	.7252255316	.7713201054	.8110732544	.8449568078
35+	.8735437192	.8974457161	.9172705803	.9335948150	.9469477294
40+	.9578036258	.9665795083	.9736363851	.9792827708	.9837794069
45+	.9873445241	.9901591958	.9923724927	.9941062662	.9954594668
* 55+	.9991028210	.9993211045	.9994874157	.9996138122	.9997096423
			P=1/ 3, R= 9		
35+	.7495203486	.7905133498	.8259485986	.8562665875	.8819720342
40+	.9035906784	.9216394827	.9366073671	.9489438448	.9590533544
45+	.9672935566	.9739762735	.9793701010	.9837039890	.9871712935
50+	.9899339579	.9921265963	.9938603337	.9952263199	.9962988774
* 55+	.9971382728	.9977931215	.9983024488	.9986974376	.9990028957
			P=1/ 3, R=10		
35+	.5716661661	.6294810440	.6821785796	.7294658440	.7713299108
40+	.8079612367	.8396865768	.8669144039	.8900930036	.9096799148
45+	.9261207847	.9398356291	.9512106892	.9605943750	.9682960945
50+	.9745870450	.9797022751	.9838435055	.9871823412	.9898636134
55+	.9920086744	.9937185271	.9950767187	.9961519588	.9970004502
* 60+	.9976679313	.9981914465	.9986008625	.9989201556	.9991684947
			P=1/ 4, R= 1		
5+	.5566406250	.6596679688	.7408447266	.8036804199	.8517837524
10+	.8883495331	.9160180092	.9368914366	.9526075423	.9644251391
15+	.9733035956	.9799700673	.9849737358	.9887283945	.9915453422
* 25+	.9984949429	.9988711997	.9991533961	.9993650452	.9995237830

DIRICHLET 1   PROBABILITY,   B= 2

N	0	1	2	3	4
			P=1/ 4,	R= 2	
10+	.5446643829	.6250925064	.6944669485	.7530589104	.8017558828
15+	.8417289220	.8742220472	.9004301704	.9214363620	.9381864140
20+	.9514853765	.9620060151	.9703029090	.9768284537	.9819486726
25+	.9859577673	.9890909436	.9915354023	.9934395738	.9949207657
* 35+	.9989264244	.9991736270	.9993643272	.9995113259	.9996245549
			P=1/ 4,	R= 3	
15+	.5575523134	.6246911692	.6844703578	.7367580023	.7818332510
20+	.8202284589	.8526095116	.8796910871	.9021805396	.9207435558
25+	.9359854576	.9484431913	.9585842445	.9668097754	.9734600742
30+	.9788210996	.9831312883	.9865881495	.9893543711	.9915633045
* 40+	.9979686085	.9984043765	.9987479669	.9990185466	.9992313819
			P=1/ 4,	R= 4	
20+	.5758117860	.6340281092	.6866519683	.7335219725	.7747464721
25+	.8106189816	.8415482575	.8680039679	.8904767366	.9094503505
30+	.9253836569	.9386998148	.9497808974	.9589662269	.9665531897
35+	.9727996024	.9779269576	.9821240794	.9855508763	.9883419855
40+	.9906101868	.9924495194	.9939380740	.9951404540	.9961099230
* 50+	.9990036153	.9992091665	.9993729658	.9995033369	.9996069822
			P=1/ 4,	R= 5	
25+	.5952296306	.6468735101	.6939508104	.7363333778	.7740777121
30+	.8073743560	.8365040570	.8618018800	.8836292235	.9023529831
35+	.9183307905	.9319011685	.9433775140	.9530449525	.9611592717
40+	.9679473000	.9736082403	.9783155898	.9822193763	.9854485188
45+	.9881131812	.9903070312	.9921093516	.9935869737	.9947960206
* 55+	.9985575818	.9988405063	.9990690234	.9992533469	.9994018302
			P=1/ 4,	R= 6	
30+	.6144805892	.6609825067	.7035922454	.7422205458	.7769097680
35+	.8078009841	.8351047244	.8590762660	.8799956509	.8981521685
40+	.9138327977	.9273139922	.9388561886	.9487004552	.9570667748
45+	.9641535334	.9701378669	.9751765940	.9794075198	.9829509537
50+	.9859113223	.9883787951	.9904308635	.9921338368	.9935442309
* 60+	.9980877668	.9984461441	.9987389277	.9989777788	.9991723608
			P=1/ 4,	R= 7	
35+	.6330807454	.6753911745	.7142923391	.7497281759	.7817395541
40+	.8104415201	.8360027769	.8586280106	.8785432560	.8959842113
45+	.9111872403	.9243827146	.9357903181	.9456159488	.9540498827
50+	.9612659075	.9674211802	.9726566080	.9770975875	.9808549794
55+	.9840262185	.9866964866	.9889398965	.9908206460	.9923941197
* 70+	.9991260922	.9992875961	.9994199054	.9995281586	.9996166194
			P=1/ 4,	R= 8	
40+	.6508515371	.6896437200	.7253942652	.7580732873	.7877241448
45+	.8144469626	.8383836620	.8597049091	.8785991363	.8952636041
50+	.9098973591	.9226958748	.9338471378	.9435289357	.9519071218
55+	.9591346519	.9653512156	.9706833147	.9752446641	.9791368167
60+	.9824499344	.9852636426	.9876479238	.9896640133	.9913652746
* 75+	.9989152545	.9991086449	.9992684025	.9994002029	.9995087991
			P=1/ 4,	R= 9	
45+	.6677402401	.7035146388	.7365401451	.7668069453	.7943617000
50+	.8192951858	.8417310303	.8618158181	.8797106756	.8955843279
55+	.9096075358	.9219487766	.9327710096	.9422293605	.9504695666
60+	.9576270353	.9638263892	.9691813846	.9737951094	.9777603851
65+	.9811603073	.9840688768	.9865516811	.9886665961	.9904644857
* 80+	.9987045119	.9989279240	.9991138464	.9992683633	.9993966126
			P=1/ 4,	R=10	
50+	.6837497959	.7168926495	.7475277268	.7756604124	.8013403371
55+	.8246518540	.8457053495	.8646295774	.8815650913	.8966587715
60+	.9100593870	.9219140966	.9323657820	.9415510913	.9495990819

DIRICHLET 1 PROBABILITY,   B= 2

N	0	1	2	3	4
			P=1/ 4, R=10		
65+	.9566303545	.9627565825	.9680803522	.9726952420	.9766860797
70+	.9801293277	.9830935554	.9856399662	.9878229527	.9896906614
75+	.9912855510	.9926449333	.9938014893	.9947837549	.9956165738
* 85+	.9984988721	.9987498838	.9989601476	.9991360419	.9992829919
			P=1/ 5, R= 1		
10+	.7912982528	.8318292787	.8647378289	.8913549066	.9128227114
15+	.9301014408	.9439871156	.9551332703	.9640727629	.9712378984
20+	.9769781315	.9815751929	.9852557669	.9882019811	.9905600054
* 35+	.9991887208	.9993509732	.9994807765	.9995846200	.9996676952
			P=1/ 5, R= 2		
15+	.6818914830	.7293392779	.7707642535	.8066149344	.8374158924
20+	.8637171802	.8860606282	.9049587962	.9208830588	.9342580424
25+	.9454603109	.9548197530	.9626225782	.9691151601	.9745082216
30+	.9789810324	.9826854200	.9857494804	.9882809331	.9903701054
* 45+	.9989329006	.9991288935	.9992891739	.9994201870	.9995272282
			P=1/ 5, R= 3		
20+	.6138258486	.6611798056	.7042108939	.7429340851	.7774878707
25+	.8080959786	.8350360404	.8586149760	.8791503889	.8969570679
30+	.9123376511	.9255765713	.9369365091	.9466567044	.9549526018
35+	.9620164177	.9680183096	.9731079127	.9774160690	.9810566244
40+	.9841282100	.9867159516	.9888930697	.9907223542	.9922575034
* 55+	.9989939223	.9991675415	.9993115825	.9994310065	.9995299583
			P=1/ 5, R= 4		
30+	.7621331408	.7914537786	.8177424605	.8411855459	.8619895798
35+	.8803707544	.8965467827	.9107308529	.9231273223	.9339288251
40+	.9433145010	.9514490894	.9584826740	.9645509017	.9697755336
45+	.9742652174	.9781163936	.9814142717	.9842338281	.9866407902
50+	.9886925846	.9904392325	.9919241827	.9931850786	.9942544561
* 60+	.9979731768	.9983009076	.9985766591	.9988084811	.9990032140
			P=1/ 5, R= 5		
35+	.7238521693	.7542925478	.7820773908	.8072982457	.8300765845
40+	.8505545504	.8688871788	.8852360367	.8997641639	.9126321696
45+	.9239953250	.9340014935	.9427897483	.9504895440	.9572203224
50+	.9630914524	.9682024191	.9726431923	.9764947179	.9798294881
55+	.9827121560	.9852001669	.9873443873	.9891897178	.9907756778
* 70+	.9984840833	.9987198118	.9989196557	.9990889370	.9992322139
			P=1/ 5, R= 6		
40+	.6915665306	.7223969932	.7509275568	.7771848674	.8012282073
45+	.8231421022	.8430297116	.8610070743	.8771982248	.8917311447
50+	.9047344902	.9163350151	.9266556045	.9358138314	.9439209518
55+	.9510812626	.9573917506	.9629419728	.9678141160	.9720831900
60+	.9758173195	.9790781021	.9819210117	.9843958237	.9865470516
65+	.9884143796	.9900330853	.9914344459	.9926461221	.9936925200
* 80+	.9988990310	.9990649428	.9992064055	.9993269251	.9994295218
			P=1/ 5, R= 7		
45+	.6639856040	.6947773608	.7235839668	.7503886749	.7752066541
50+	.7980793478	.8190691623	.8382546111	.8557259868	.8715815906
55+	.8859245183	.8988599795	.9104931104	.9209227332	.9302625095
60+	.9385949361	.9460156306	.9526103611	.9584592750	.9636367891
65+	.9682116080	.9722468394	.9758001851	.9789241842	.9816664931
70+	.9840701887	.9861740829	.9880130405	.9896182933	.9910177465
* 85+	.9982804662	.9985277255	.9987403712	.9989230908	.9990799637
			P=1/ 5, R= 8		
50+	.6401422717	.6706418254	.6994276019	.7264543102	.7517063262
55+	.7751935004	.7969470211	.8170154665	.8354611394	.8523567449
60+	.8677824421	.8818232798	.8945670097	.9061022562	.9165170188
65+	.9258974725	.9343270345	.9418856629	.9486493528	.9546898014

DIRICHLET 1 PROBABILITY, B= 2

N	0	1	2	3	4
			P=1/ 5, R= 8		
70+	.9600742120	.9648652114	.9691208604	.9728947341	.9762360590
75+	.9791898901	.9817973175	.9840956913	.9861188585	.9878974043
80+	.9894588923	.9908281012	.9920272535	.9930762347	.9939928017
* 95+	.9988087919	.9989764937	.9991212081	.9992459849	.9993534854
			P=1/ 5, R= 9		
60+	.7542758104	.7765399173	.7972409847	.8164186134	.8341240572
65+	.8504176217	.8653663248	.8790418313	.8915186614	.9028726651
70+	.9131797458	.9225148139	.9309509488	.9385587453	.9454058213
75+	.9515564653	.9570714018	.9620076561	.9664185007	.9703534677
80+	.9738584144	.9769756295	.9797439702	.9821990214	.9843732707
85+	.9862962924	.9879949359	.9894935149	.9908139943	.9919761720
* 100+	.9982892524	.9985206026	.9987216059	.9988960956	.9990474435
			P=1/ 5, R=10		
65+	.7351175438	.7577024262	.7788432211	.7985613556	.8168904013
70+	.8338738345	.8495629624	.8640150400	.8772915876	.8894569141
75+	.9005768409	.9107176173	.9199450161	.9283235926	.9359160919
80+	.9427829891	.9489821428	.9545685502	.9595941864	.9641079149
85+	.9681554584	.9717794177	.9750193285	.9779117504	.9804903772
90+	.9827861659	.9848274759	.9866402172	.9882480008	.9896722916
95+	.9909325589	.9920464233	.9930298001	.9938970360	.9946610392
* 110+	.9988374723	.9989924370	.9991273484	.9992447088	.9993467225
			P=1/ 6, R= 1		
10+	.6943303641	.7423850485	.7833940371	.8182104731	.8476523557
15+	.8724727152	.8933466530	.9108684710	.9255545660	.9378493651
20+	.9481326221	.9567270636	.9639058054	.9698992283	.9749011728
25+	.9790744101	.9825554081	.9854584398	.9878790997	.9898972941
30+	.9915797746	.9929822763	.9941513175	.9951257116	.9959378355
* 40+	.9986393348	.9988660972	.9990550710	.9992125524	.9993437892
			P=1/ 6, R= 2		
20+	.7496097879	.7815065426	.8098328375	.8348684403	.8569039152
25+	.8762285564	.8931220187	.9078488240	.9206550531	.9317666559
30+	.9413889291	.9497068082	.9568857054	.9630726899	.9683978629
35+	.9729758185	.9769071176	.9802797214	.9831703527	.9856457662
40+	.9877639158	.9895750160	.9911224996	.9924438738	.9935714828
* 55+	.9989401984	.9991021055	.9992394835	.9993560109	.9994548218
			P=1/ 6, R= 3		
25+	.6464119530	.6827578182	.7162767167	.7469987739	.7750029992
30+	.8004045113	.8233436342	.8439768378	.8624694011	.8789896232
35+	.8937043798	.9067758221	.9183590232	.9286003945	.9376367147
40+	.9455946378	.9525905652	.9587307898	.9641118339	.9688209229
45+	.9729365432	.9765290515	.9796613038	.9823892854	.9847627260
50+	.9868256899	.9886171332	.9901714255	.9915188327	.9926859603
* 65+	.9986135882	.9988113643	.9989813313	.9991273310	.9992526859
			P=1/ 6, R= 4		
35+	.7274752070	.7542729046	.7789643259	.8016193116	.8223256665
40+	.8411839887	.8583032649	.8737971969	.8877812040	.9003700306
45+	.9116758885	.9218070559	.9308668665	.9389530202	.9461571577
50+	.9525646476	.9582545398	.9632996483	.9677667295	.9717167316
55+	.9752050918	.9782820638	.9809930635	.9833790191	.9854767209
60+	.9873191610	.9889358614	.9903531855	.9915946321	.9926811095
* 75+	.9984630205	.9986704111	.9988503425	.9990063667	.9991415895
			P=1/ 6, R= 5		
40+	.6600295985	.6889838582	.7161782303	.7416035047	.7652747252
45+	.7872269781	.8075114648	.8261919243	.8433414425	.8590396611
50+	.8733703793	.8864195314	.8982735132	.9090178232	.9187359877
55+	.9275087308	.9354133592	.9425233259	.9489079476	.9546322453
60+	.9597568870	.9643382114	.9684283146	.9720751842	.9753228696

DIRICHLET 1   PROBABILITY,   B= 2

N	0	1	2	3	4
			P=1/ 6, R= 5		
65+	.9782116765	.9807783775	.9830564315	.9850762073	.9868652050
70+	.9884482741	.9898478235	.9910840240	.9921749994	.9931370072
* 85+	.9984408846	.9986417579	.9988173211	.9989706770	.9991045608
			P=1/ 6, R= 6		
50+	.7339349223	.7565288777	.7776422036	.7973057385	.8155614856
55+	.8324602903	.8480597413	.8624223077	.8756137103	.8877015208
60+	.8987539769	.9088389969	.9180233753	.9263721419	.9339480632
65+	.9408112697	.9470189903	.9526253773	.9576814080	.9622348493
70+	.9663302732	.9700091124	.9733097490	.9762676261	.9789153772
75+	.9812829680	.9833978459	.9852850924	.9869675778	.9884661133
80+	.9897995999	.9909851731	.9920383409	.9929731159	.9938021402
* 95+	.9985013638	.9986870362	.9988502674	.9989936881	.9991196324
			P=1/ 6, R= 7		
55+	.6825449184	.7066990908	.7295534950	.7511023690	.7713533342
60+	.7903254848	.8080475553	.8245561964	.8398943820	.8541099605
65+	.8672543569	.8793814271	.8905464604	.9008053236	.9102137382
70+	.9188266793	.9266978865	.9338794733	.9404216253	.9463723748
75+	.9517774428	.9566801370	.9611212978	.9651392854	.9687699977
80+	.9720469167	.9750011742	.9776616342	.9800549876	.9822058546
85+	.9841368932	.9858689106	.9874209757	.9888105305	.9900535007
* 105+	.9986094595	.9987760449	.9989231974	.9990531107	.9991677413
			P=1/ 6, R= 8		
65+	.7479921284	.7674036863	.7856727753	.8028197839	.8188720424
70+	.8338625933	.8478290534	.8608125761	.8728569168	.8840076032
75+	.8943112064	.9038147114	.9125649782	.9206082898	.9279899781
80+	.9347541209	.9409433024	.9465984298	.9517585998	.9564610069
85+	.9607408889	.9646315026	.9681641267	.9713680845	.9742707849
90+	.9768977773	.9792728163	.9814179352	.9833535248	.9850984162
95+	.9866699654	.9880841388	.9893555983	.9904977855	.9915230029
* 115+	.9987408870	.9988873817	.9990173021	.9991324599	.9992344775
			P=1/ 6, R= 9		
70+	.7062003534	.7268932133	.7465464378	.7651598666	.7827414484
75+	.7993061919	.8148751500	.8294744543	.8431344092	.8558886547
80+	.8677734004	.8788267360	.8890880145	.8985973102	.9073949461
85+	.9155210872	.9230153968	.9299167491	.9362629938	.9420907682
90+	.9474353500	.9523305484	.9568086263	.9609002514	.9646344709
95+	.9680387070	.9711387687	.9739588786	.9765217106	.9788484368
100+	.9809587814	.9828710801	.9846023432	.9861683208	.9875835696
105+	.9888615198	.9900145423	.9910540140	.9919903817	.9928332248
* 125+	.9988797744	.9990068043	.9991198404	.9992203706	.9993097314
			P=1/ 6, R=10		
80+	.7641867502	.7811086028	.7971013106	.8121816038	.8263706995
85+	.8396935703	.8521782612	.8638552578	.8747569096	.8849169092
90+	.8943698267	.9031506981	.9112946657	.9188366680	.9258111748
95+	.9322519654	.9381919454	.9436629990	.9486958725	.9533200864
100+	.9575638719	.9614541296	.9650164062	.9682748882	.9712524085
105+	.9739704646	.9764492464	.9787076712	.9807634250	.9826330085
110+	.9843317861	.9858740378	.9872730123	.9885409807	.9896892908
115+	.9907284199	.9916680281	.9925170090	.9932835397	.9939751289
* 135+	.9990162901	.9991253276	.9992226322	.9993094214	.9993867920
			P=1/ 7, R= 1		
10+	.6064549818	.6577368969	.7031039103	.7429978405	.7779125798
15+	.8083537435	.8348134868	.8577556363	.8776077427	.8947576959
20+	.9095532715	.9223034903	.9332810329	.9427252005	.9508450908
25+	.9578227751	.9638163460	.9689627631	.9733804567	.9771716790
30+	.9804246038	.9832151861	.9856088004	.9876616744	.9894221411
35+	.9909317273	.9922260973	.9933358708	.9942873293	.9951030252

DIRICHLET 1   PROBABILITY,   B= 2

N	0	1	2	3	4
		P=1/ 7,	R= 1		
* 50+	.9991011926	.9992295866	.9993396406	.9994339741	.9995148323
		P=1/ 7,	R= 2		
20+	.6318087454	.6688912512	.7030440915	.7343252267	.7628372400
25+	.7887141448	.8121105954	.8331932606	.8521340811	.8691051283
30+	.8842747931	.8978050616	.9098496611	.9205528929	.9300489959
35+	.9384619131	.9459053579	.9524830949	.9582893705	.9634094413
40+	.9679201599	.9718905898	.9753826253	.9784516015	.9811468826
45+	.9835124207	.9855872814	.9874061323	.9889996941	.9903951547
* 65+	.9989466554	.9990844086	.9992043014	.9993086253	.9993993816
		P=1/ 7,	R= 3		
30+	.6676005826	.6970010753	.7244137627	.7498649348	.7734038764
35+	.7950976606	.8150265825	.8332802317	.8499541734	.8651471958
40+	.8789590688	.8914887560	.9028330209	.9130853728	.9223352969
45+	.9306677238	.9381626933	.9448951761	.9509350212	.9563470016
50+	.9611909334	.9655218517	.9693902248	.9728421955	.9759198378
55+	.9786614221	.9811016812	.9832720730	.9852010356	.9869142331
60+	.9884347886	.9897835055	.9909790736	.9920382631	.9929761029
* 80+	.9990997970	.9992103406	.9993074924	.9993928481	.9994678174
		P=1/ 7,	R= 4		
40+	.7018191355	.7261222154	.7489070759	.7701973875	.7900296152
45+	.8084504451	.8255144670	.8412821223	.8558179144	.8691888703
50+	.8814632394	.8927094095	.9029950200	.9123862498	.9209472607
55+	.9287397740	.9358227622	.9422522385	.9480811278	.9533592046
60+	.9581330865	.9624462713	.9663392084	.9698493972	.9730115047
65+	.9758574972	.9784167809	.9807163484	.9827809265	.9846331238
70+	.9862935767	.9877810902	.9891127744	.9903041752	.9913693979
* 90+	.9987335045	.9988801180	.9990100733	.9991252200	.9992272086
		P=1/ 7,	R= 5		
50+	.7327849247	.7533459269	.7726799692	.7908109929	.8077705241
55+	.8235962135	.8383305083	.8520194616	.8647116782	.8764573931
60+	.8873076774	.8973137616	.9065264690	.9149957481	.9227702944
65+	.9298972516	.9364219821	.9423878986	.9478363477	.9528065379
70+	.9573355047	.9614581073	.9652070504	.9686129264	.9717042740
75+	.9745076493	.9770477058	.9793472817	.9814274907	.9833078157
80+	.9850062031	.9865391565	.9879218293	.9891681148	.9902907330
* 100+	.9984127551	.9985873363	.9987431515	.9988821613	.9990061288
		P=1/ 7,	R= 6		
60+	.7604259268	.7780773682	.7947057339	.8103348623	.8249932701
65+	.8387132495	.8515300439	.8634811059	.8746054351	.8849429947
70+	.8945342036	.9034194989	.9116389645	.9192320203	.9262371664
75+	.9326917765	.9386319361	.9440923190	.9491060985	.9537048876
80+	.9579187049	.9617759618	.9653034681	.9685264521	.9714685930
85+	.9741520636	.9765975799	.9788244568	.9808506681	.9826929091
90+	.9843666612	.9858862568	.9872649445	.9885149531	.9896475550
95+	.9906731272	.9916012103	.9924405663	.9931992320	.9938845719
* 115+	.9989431586	.9990559545	.9991569930	.9992474647	.9993284439
		P=1/ 7,	R= 7		
65+	.6949193963	.7147700530	.7337036592	.7517181403	.7688179773
70+	.7850134200	.8003197243	.8147564226	.8283466364	.8411164348
75+	.8530942436	.8643103064	.8747961975	.8845843873	.8937078571
80+	.9021997625	.9100931417	.9174206662	.9242144307	.9305057781
85+	.9363251577	.9417020117	.9466646883	.9512403771	.9554550652
90+	.9593335098	.9628992271	.9661744928	.9691803536	.9719366486
95+	.9744620367	.9767740304	.9788890345	.9808223882	.9825884093
100+	.9842004403	.9856708961	.9870113111	.9882323871	.9893440398
105+	.9903554453	.9912750846	.9921107875	.9928697745	.9935586967
* 125+	.9988020899	.9989250766	.9990357661	.9991353487	.9992249040

DIRICHLET 1 PROBABILITY, B= 2

N	0	1	2	3	4
			P=1/ 7, R= 8		
75+	.7282204749	.7455653136	.7620961121	.7778183076	.7927417017
80+	.8068799071	.8202498162	.8328710987	.8447657300	.8559575537
85+	.8664718794	.8763351168	.8855744450	.8942175173	.9022921993
90+	.9098263398	.9168475717	.9233831411	.9294597633	.9351035021
95+	.9403396719	.9451927585	.9496863588	.9538431354	.9576847859
100+	.9612320236	.9645045691	.9675211516	.9702995169	.9728564425
105+	.9752077585	.9773683716	.9793522943	.9811726758	.9828418349
110+	.9843712948	.9857718192	.9870534475	.9882255320	.9892967731
115+	.9902752554	.9911684823	.9919834101	.9927264810	.9934036546
* 135+	.9986940069	.9988234985	.9989405160	.9990462188	.9991416637
			P=1/ 7, R= 9		
85+	.7572751454	.7725149678	.7870307529	.8008321372	.8139317037
90+	.8263445828	.8380880733	.8491812853	.8596448087	.8695004053
95+	.8787707284	.8874790671	.8956491162	.9033047708	.9104699444
100+	.9171684098	.9234236608	.9292587936	.9346964063	.9397585161
105+	.9444664903	.9488409924	.9529019402	.9566684747	.9601589393
110+	.9633908678	.9663809791	.9691451798	.9716985710	.9740554609
115+	.9762293807	.9782331044	.9800786709	.9817774078	.9833399580
120+	.9847763060	.9860958060	.9873072105	.9884186980	.9894379021
125+	.9903719392	.9912274362	.9920105575	.9927270318	.9933821770
* 145+	.9986169484	.9987498276	.9988703286	.9989795612	.9990785405
			P=1/ 7, R=10		
90+	.7056232497	.7223823747	.7384783261	.7539081318	.7686724574
95+	.7827752666	.7962234827	.8090266573	.8211966494	.8327473167
100+	.8436942232	.8540543631	.8638459029	.8730879425	.8818002954
105+	.8900032884	.8977175795	.9049639955	.9117633858	.9181364950
110+	.9241038498	.9296856624	.9349017471	.9397714505	.9443135927
115+	.9485464211	.9524875718	.9561540422	.9595621698	.9627276194
120+	.9656653754	.9683897404	.9709143384	.9732521218	.9754153822
125+	.9774157639	.9792642801	.9809713308	.9825467223	.9839996883
130+	.9853389117	.9865725464	.9877082405	.9887531592	.9897140072
135+	.9905970519	.9914081455	.9921527468	.9928359430	.9934624700
* 155+	.9985676701	.9987013971	.9988230397	.9989336468	.9990341807
			P=1/ 8, R= 1		
15+	.7434958335	.7738884217	.8008995445	.8248474831	.8460368340
20+	.8647536944	.8812630811	.8958078948	.9086089321	.9198655838
25+	.9297569619	.9384432737	.9460673135	.9527559861	.9586218030
30+	.9637643139	.9682714519	.9722207784	.9756806245	.9787111290
35+	.9813651748	.9836892307	.9857241039	.9875056112	.9890651750
40+	.9904303520	.9916253009	.9926711955	.9935865890	.9943877350
* 55+	.9987075083	.9988690530	.9990104088	.9991340982	.9992423288
			P=1/ 8, R= 2		
25+	.6950148998	.7227898587	.7484726990	.7721402182	.7938838061
30+	.8138047193	.8320102019	.8486103547	.8637156548	.8774350319
35+	.8898744119	.9011356522	.9113157955	.9205065841	.9287941813
40+	.9362590561	.9429759949	.9490142082	.9544375085	.9593045388
45+	.9636690345	.9675801070	.9710825370	.9742170715	.9770207179
50+	.9795270293	.9817663805	.9837662299	.9855513679	.9871441495
55+	.9885647117	.9898311753	.9909598321	.9919653164	.9928607633
* 75+	.9989521045	.9990719007	.9991781220	.9992722902	.9993557587
			P=1/ 8, R= 3		
35+	.6824580908	.7071010272	.7302718927	.7519912871	.7722919885
40+	.7912165186	.8088149658	.8251430674	.8402605428	.8542296647
45+	.8671140499	.8789776517	.8898839322	.8998951954	.9090720607
50+	.9174730580	.9251543277	.9321694090	.9385691036	.9444014000
55+	.9497114496	.9545415826	.9589313580	.9629176367	.9665346763
60+	.9698142375	.9727857024	.9754761970	.9779107186	.9801122631

DIRICHLET 1 PROBABILITY, B= 2

N	0	1	2	3	4
		P=1/ 8,	R= 3		
65+	.9821019517	.9838991554	.9855216155	.9869855596	.9883058139
70+	.9894959080	.9905681760	.9915338509	.9924031531	.9931853736
* 90+	.9988461690	.9989694804	.9990798047	.9991784853	.9992667299
		P=1/ 8,	R= 4		
45+	.6821807391	.7042558445	.7251977262	.7450111451	.7637096312
50+	.7813141386	.7978517907	.8133547270	.8278590546	.8414039079
55+	.8540306140	.8657819594	.8767015547	.8868332891	.8962208690
60+	.9049074319	.9129352301	.9203453751	.9271776362	.9334702876
65+	.9392599962	.9445817449	.9494687873	.9539526268	.9580630182
70+	.9618279862	.9652738588	.9684253117	.9713054218	.9739357272
75+	.9763362924	.9785257765	.9805215033	.9823395326	.9839947312
80+	.9855008427	.9868705566	.9881155752	.9892466781	.9902737843
* 105+	.9989023115	.9990135713	.9991137609	.9992039569	.9992851344
		P=1/ 8,	R= 5		
55+	.6872328685	.7072301539	.7263032508	.7444522977	.7616837768
60+	.7780096995	.7934468271	.8080159375	.8217411411	.8346492508
65+	.8467692071	.8581315589	.8687679986	.8787109508	.8879932109
70+	.8966476313	.9047068525	.9122030742	.9191678645	.9256320027
75+	.9316253526	.9371767629	.9423139917	.9470636519	.9514511755
80+	.9555007930	.9592355267	.9626771958	.9658464304	.9687626943
85+	.9714443131	.9739085085	.9761714362	.9782482264	.9801530271
90+	.9818990481	.9834986065	.9849631715	.9863034099	.9875292303
95+	.9886498268	.9896737213	.9906088048	.9914623767	.9922411833
* 120+	.9990321384	.9991261210	.9992111684	.9992881084	.9993576946
		P=1/ 8,	R= 6		
65+	.6948559805	.7131393582	.7306395280	.7473556091	.7632914258
70+	.7784549734	.7928579021	.8065150221	.8194438361	.8316641009
75+	.8431974207	.8540668726	.8642966656	.8739118317	.8829379493
80+	.8914008967	.8993266359	.9067410228	.9136696445	.9201376790
85+	.9261697787	.9317899724	.9370215862	.9418871810	.9464085039
90+	.9506064532	.9545010547	.9581114474	.9614558789	.9645517074
95+	.9674154101	.9700625976	.9725080316	.9747656475	.9768485785
100+	.9787691835	.9805390751	.9821691500	.9836696199	.9850500424
105+	.9863193523	.9874858929	.9885574469	.9895412664	.9904441023
* 130+	.9986704921	.9987935642	.9989055276	.9990073544	.9990999345
		P=1/ 8,	R= 7		
75+	.7037568783	.7205962018	.7367539040	.7522293014	.7670252890
80+	.7811479689	.7946062885	.8074116905	.8195777792	.8311200040
85+	.8420553623	.8524021232	.8621795718	.8714077753	.8801073696
90+	.8882993670	.8960049835	.9032454861	.9100420571	.9164156765
95+	.9223870201	.9279763726	.9332035539	.9380878587	.9426480069
100+	.9469021045	.9508676139	.9545613324	.9579993775	.9611971797
105+	.9641694801	.9669303328	.9694931122	.9718705234	.9740746154
110+	.9761167973	.9780078561	.9797579763	.9813767607	.9828732520
115+	.9842559552	.9855328603	.9867114652	.9877987984	.9888014420
120+	.9897255535	.9905768886	.9913608218	.9920823676	.9927462008
* 145+	.9989273840	.9990237300	.9991116469	.9991918486	.9992649912
		P=1/ 8,	R= 8		
85+	.7132625032	.7288624063	.7438576670	.7582481783	.7720366149
90+	.7852281627	.7978302537	.8098523092	.8213054921	.8322024709
95+	.8425571969	.8523846947	.8617008671	.8705223140	.8788661668
100+	.8867499363	.8941913755	.9012083556	.9078187555	.9140403637
105+	.9198907921	.9253874009	.9305472342	.9353869644	.9399228469
110+	.9441706814	.9481457816	.9518629515	.9553364672	.9585800649
115+	.9616069330	.9644297087	.9670604789	.9695107831	.9717916205
120+	.9739134584	.9758862430	.9777194126	.9794219105	.9810022010
125+	.9824682846	.9838277149	.9850876156	.9862546975	.9873352760

DIRICHLET 1   PROBABILITY,   B= 2

N	0	1	2	3	4
		P=1/ 8,	R= 8		
130+	.9883352888	.9892603129	.9901155818	.9909060022	.9916361706
* 155+	.9986597701	.9987755524	.9988816232	.9989787679	.9990677111
		P=1/ 8,	R= 9		
95+	.7229975021	.7375171294	.7514927739	.7649249735	.7778164680
100+	.7901719954	.8019980908	.8133028907	.8240959449	.8343880352
105+	.8441910040	.8535175919	.8623812857	.8707961767	.8787768288
110+	.8863381585	.8934953237	.9002636237	.9066584083	.9126949968
115+	.9183886055	.9237542838	.9288068590	.9335608873	.9380306128
120+	.9422299327	.9461723682	.9498710409	.9533386544	.9565874802
125+	.9596293476	.9624756373	.9651372782	.9676247471	.9699480710
130+	.9721168311	.9741401695	.9760267967	.9777850009	.9794226589
135+	.9809472466	.9823658516	.9836851859	.9849115986	.9860510898
140+	.9871093239	.9880916432	.9890030819	.9898483793	.9906319936
* 170+	.9989601007	.9990479903	.9991286690	.9992027074	.9992706326
		P=1/ 8,	R=10		
105+	.7327428042	.7463088580	.7593802315	.7719580611	.7840452536
110+	.7956463268	.8067672533	.8174153079	.8275989201	.8373275329
115+	.8466114671	.8554617935	.8638902121	.8719089391	.8795306021
120+	.8867681417	.8936347229	.9001436524	.9063083041	.9121420519
125+	.9176582090	.9228699739	.9277903828	.9324322678	.9368082206
130+	.9409305613	.9448113118	.9484621745	.9518945136	.9551193414
135+	.9581473074	.9609886903	.9636533934	.9661509416	.9684904814
140+	.9706807824	.9727302404	.9746468824	.9764383727	.9781120197
145+	.9796747847	.9811332899	.9824938285	.9837623746	.9849445934
150+	.9860458521	.9870712306	.9880255326	.9889132964	.9897388063
155+	.9905061029	.9912189946	.9918810675	.9924956965	.9930660546
* 180+	.9987692889	.9988699531	.9989626445	.9990479699	.9991264922
		P=1/ 9,	R= 1		
15+	.6812821320	.7141331616	.7439034250	.7708087078	.7950689351
20+	.8169014643	.8365165101	.8541141587	.8698825539	.8839969368
25+	.8966193019	.9078984852	.9179705505	.9269593722	.9349773385
30+	.9421261216	.9484974749	.9541740284	.9592300646	.9637322610
35+	.9677403903	.9713079762	.9744829004	.9773079615	.9798213875
40+	.9820573020	.9840461491	.9858150767	.9873882840	.9887873339
45+	.9900314342	.9911376904	.9921213319	.9929959154	.9937735061
* 65+	.9990535905	.9991587381	.9992522047	.9993352877	.9994091404
		P=1/ 9,	R= 2		
30+	.7371029985	.7588344153	.7790359896	.7977722525	.8151129666
35+	.8311311792	.8459016122	.8594993494	.8719987835	.8834727871
40+	.8939920747	.9036247273	.9124358511	.9204873502	.9278377894
45+	.9345423331	.9406527428	.9462174218	.9512814960	.9558869217
50+	.9600726135	.9638745853	.9673261007	.9704578275	.9732979938
55+	.9758725430	.9782052864	.9803180506	.9822308194	.9839618706
60+	.9855279042	.9869441652	.9882245582	.9893817553	.9904272967
* 85+	.9989566718	.9990626157	.9991578979	.9992435796	.9993206177
		P=1/ 9,	R= 3		
40+	.6934436915	.7146338041	.7346907452	.7536303686	.7714755612
45+	.7882549827	.8040019235	.8187532808	.8325486502	.8454295292
50+	.8574386243	.8686192559	.8790148534	.8886685304	.8976227342
55+	.9059189599	.9135975223	.9206973793	.9272559976	.9333092583
60+	.9388913934	.9440349504	.9487707804	.9531280440	.9571342344
65+	.9608152120	.9641952490	.9672970829	.9701419738	.9727497676
70+	.9751389609	.9773267679	.9793291871	.9811610688	.9828361810
75+	.9843672743	.9857661444	.9870436930	.9882099865	.9892743114
80+	.9902452282	.9911306213	.9919377476	.9926732811	.9933433562
* 100+	.9986067687	.9987385517	.9988580617	.9989664199	.9990646475

DIRICHLET 1 PROBABILITY, B= 2

N	0	1	2	3	4
			P=1/ 9, R= 4		
55+	.7587163387	.7745283134	.7895066747	.8036715778	.8170458983
60+	.8296547049	.8415247800	.8526841890	.8631618961	.8729874249
65+	.8821905608	.8908010930	.8988485926	.9063622250	.9133705921
70+	.9199016024	.9259823660	.9316391111	.9368971213	.9417806887
75+	.9463130830	.9505165331	.9544122199	.9580202790	.9613598106
80+	.9644488969	.9673046245	.9699431111	.9723795359	.9746281726
85+	.9767024242	.9786148596	.9803772505	.9820006086	.9834952235
90+	.9848706997	.9861359931	.9872994467	.9883688259	.9893513517
95+	.9902537337	.9910822011	.9918425330	.9925400870	.9931798261
* 120+	.9990166858	.9991053215	.9991861131	.9992597393	.9993268218
			P=1/ 9, R= 5		
65+	.7375712191	.7530920774	.7679081053	.7820287005	.7954659558
70+	.8082342824	.8203500557	.8318312855	.8426973120	.8529685267
75+	.8626661186	.8718118448	.8804278254	.8885363607	.8961597701
80+	.9033202514	.9100397585	.9163398972	.9222418363	.9277662347
85+	.9329331798	.9377621400	.9422719263	.9464806649	.9504057770
90+	.9540639671	.9574712172	.9606427870	.9635932188	.9663363464
95+	.9688853080	.9712525612	.9734499014	.9754884807	.9773788297
100+	.9791308796	.9807539851	.9822569483	.9836480421	.9849350346
105+	.9861252125	.9872254046	.9882420051	.9891809963	.9900479701
* 135+	.9989628008	.9990520126	.9991337163	.9992085270	.9992770114
			P=1/ 9, R= 6		
75+	.7220228660	.7371142690	.7516009483	.7654862430	.7787759191
80+	.7914779005	.8036020117	.8151597316	.8261639602	.8366288004
85+	.8465693533	.8560015295	.8649418747	.8734074103	.8814154892
90+	.8889836648	.8961295740	.9028708334	.9092249471	.9152092263
95+	.9208407197	.9261361538	.9311118818	.9357838418	.9401675222
100+	.9442779338	.9481295887	.9517364843	.9551120931	.9582693559
105+	.9612206796	.9639779386	.9665524786	.9689551232	.9711961829
110+	.9732854656	.9752322890	.9770454936	.9787334577	.9803041123
115+	.9817649566	.9831230747	.9843851517	.9855574902	.9866460272
120+	.9876563497	.9885937117	.9894630498	.9902689987	.9910159066
* 150+	.9989690928	.9990542894	.9991326142	.9992046054	.9992707602
			P=1/ 9, R= 7		
85+	.7103831956	.7249996195	.7390895411	.7526528370	.7656914989
90+	.7782094408	.7902123091	.8017072987	.8127029756	.8232091068
95+	.8332364992	.8427968467	.8519025877	.8605667713	.8688029336
100+	.8766249836	.8840470993	.8910836313	.8977490180	.9040577068
105+	.9100240858	.9156624216	.9209868059	.9260111076	.9307489330
110+	.9352135902	.9394180607	.9433749745	.9470965913	.9505947848
115+	.9538810314	.9569664019	.9598615570	.9625767445	.9651218000
120+	.9675061487	.9697388100	.9718284029	.9737831532	.9756109019
125+	.9773191144	.9789148905	.9804049750	.9817957692	.9830933422
130+	.9843034427	.9854315113	.9864826922	.9874618455	.9883735593
135+	.9892221612	.9900117306	.9907461095	.9914289146	.9920635472
* 165+	.9990149301	.9990935738	.9991661018	.9992329744	.9992946190
			P=1/ 9, R= 8		
95+	.7015756146	.7157129179	.7293856423	.7425916797	.7553307479
100+	.7676042485	.7794151260	.7907677284	.8016676707	.8121217030
105+	.8221375826	.8317239517	.8408902213	.8496464605	.8580032934
110+	.8659718012	.8735634321	.8807899175	.8876631944	.8941953350
115+	.9003984816	.9062847887	.9118663697	.9171552503	.9221633264
120+	.9269023268	.9313837817	.9356189941	.9396190162	.9433946294
125+	.9469563274	.9503143029	.9534784369	.9564582908	.9592631009
130+	.9619017749	.9643828905	.9667146954	.9689051094	.9709617271
135+	.9728918224	.9747023538	.9763999704	.9779910184	.9794815494
140+	.9808773277	.9821838387	.9834062979	.9845496596	.9856186260

DIRICHLET 1   PROBABILITY,   B= 2

N	0	1	2	3	4
		P=1/ 9, R= 8			
145+	.9866176564	.9875509763	.9884225869	.9892362737	.9899956161
150+	.9907039958	.9913646059	.9919804596	.9925543979	.9930890986
* 175+	.9986337012	.9987382794	.9988350898	.9989246884	.9990075932
		P=1/ 9, R= 9			
110+	.7590719266	.7706533452	.7818184890	.7925709403	.8029153260
115+	.8128572165	.8224030274	.8315599241	.8403357317	.8487388482
120+	.8567781621	.8644629757	.8718029320	.8788079470	.8854881468
125+	.8918538091	.8979153097	.9036830734	.9091675289	.9143790690
130+	.9193280135	.9240245768	.9284788390	.9327007203	.9366999588
135+	.9404860910	.9440684359	.9474560805	.9506578691	.9536823935
140+	.9565379863	.9592327156	.9617743811	.9641705126	.9664283685
145+	.9685549374	.9705569387	.9724408260	.9742127902	.9758787636
150+	.9774444252	.9789152054	.9802962925	.9815926386	.9828089661
155+	.9839497747	.9850193479	.9860217607	.9869608865	.9878404041
160+	.9886638054	.9894344022	.9901553334	.9908295723	.9914599333
* 190+	.9987686431	.9988604044	.9989455369	.9990245016	.9990977286
		P=1/ 9, R=10			
120+	.7521805425	.7635143633	.7744676050	.7850424348	.7952419703
125+	.8050701975	.8145318916	.8236325391	.8323782630	.8407757504
130+	.8488321832	.8565551729	.8639526973	.8710330422	.8778047458
135+	.8842765468	.8904573363	.8963561127	.9019819412	.9073439150
140+	.9124511215	.9173126100	.9219373640	.9263342751	.9305121205
145+	.9344795428	.9382450319	.9418169101	.9452033181	.9484122045
150+	.9514513159	.9543281892	.9570501457	.9596242864	.9620574887
155+	.9643564042	.9665274579	.9685768480	.9705105469	.9723343029
160+	.9740536426	.9756738738	.9772000891	.9786371698	.9799897909
165+	.9812624249	.9824593476	.9835846433	.9846422098	.9856357644
170+	.9865688498	.9874448392	.9882669429	.9890382133	.9897615516
175+	.9904397127	.9910753116	.9916708286	.9922286151	.9927508985
* 205+	.9989051432	.9989848184	.9990588773	.9991277005	.9991916442
		P=1/10, R= 1			
20+	.7683759059	.7903853938	.8104245172	.8286441957	.8451894803
25+	.8601982956	.8738006768	.8861183776	.8972647547	.9073448511
30+	.9164556235	.9246862671	.9321186052	.9388275165	.9448813824
35+	.9503425381	.9552677195	.9597084957	.9637116846	.9673197470
40+	.9705711569	.9735007489	.9761400402	.9785175291	.9806589706
45+	.9825876290	.9843245105	.9858885750	.9872969299	.9885650068
50+	.9897067221	.9907346226	.9916600185	.9924931033	.9932430622
* 70+	.9987470071	.9988722899	.9989850477	.9990865324	.9991778708
		P=1/10, R= 2			
35+	.7667446082	.7843501491	.8008030102	.8161532354	.8304528072
40+	.8437547520	.8561123932	.8675787370	.8782059743	.8880450829
45+	.8971455181	.9055549774	.9133192292	.9204819954	.9270848787
50+	.9331673273	.9387666300	.9439179357	.9486542934	.9530067064
55+	.9570041994	.9606738933	.9640410869	.9671293422	.9699605720
60+	.9725551282	.9749318899	.9771083505	.9791007020	.9809239174
65+	.9825918302	.9841172099	.9855118353	.9867865624	.9879513908
70+	.9890155246	.9899874310	.9908748946	.9916850687	.9924245232
* 95+	.9989605190	.9990554621	.9991418121	.9992203381	.9992917412
		P=1/10, R= 3			
45+	.7018920455	.7204667608	.7381426478	.7549319724	.7708513381
50+	.7859209821	.8001641296	.8136064079	.8262753185	.8381997657
55+	.8494096386	.8599354451	.8698079918	.8790581090	.8877164152
60+	.8958131188	.9033778521	.9104395363	.9170262714	.9231652504
65+	.9288826944	.9342038051	.9391527328	.9437525585	.9480252870
70+	.9519918499	.9556721164	.9590849116	.9622480392	.9651783095
75+	.9678915707	.9704027424	.9727258514	.9748740687	.9768597473

DIRICHLET 1 PROBABILITY, B= 2

N	0	1	2	3	4
			P=1/10, R= 3		
80+	.9786944601	.9803890383	.9819536087	.9833976311	.9847299344
85+	.9859587519	.9870917557	.9881360894	.9890983999	.9899848680
90+	.9908012372	.9915528414	.9922446314	.9928811997	.9934668042
* 115+	.9989642735	.9990527696	.9991338149	.9992080254	.9992759674
			P=1/10, R= 4		
60+	.7397325211	.7547552335	.7690981327	.7827722661	.7957909236
65+	.8081693004	.8199241838	.8310736646	.8416368731	.8516337385
70+	.8610847726	.8700108747	.8784331588	.8863728010	.8938509049
75+	.9008883856	.9075058696	.9137236092	.9195614112	.9250385777
80+	.9301738580	.9349854109	.9394907753	.9437068495	.9476498773
85+	.9513354395	.9547784523	.9579931688	.9609931853	.9637914506
90+	.9664002787	.9688313628	.9710957924	.9732040709	.9751661350
95+	.9769913748	.9786886543	.9802663328	.9817322856	.9830939259
100+	.9843582256	.9855317359	.9866206085	.9876306150	.9885671670
105+	.9894353346	.9902398652	.9909852011	.9916754965	.9923146339
* 130+	.9986271250	.9987376118	.9988393829	.9989331090	.9990194107
			P=1/10, R= 5		
70+	.7021730248	.7174160648	.7320911956	.7461994768	.7597442947
75+	.7727311245	.7851673003	.7970617946	.8084250088	.8192685754
80+	.8296051720	.8394483486	.8488123665	.8577120512	.8661626569
85+	.8741797427	.8817790623	.8889764627	.8957877956	.9022288373
90+	.9083152196	.9140623683	.9194854508	.9245993311	.9294185315
95+	.9339572016	.9382290917	.9422475333	.9460254230	.9495752110
100+	.9529088942	.9560380114	.9589736427	.9617264108	.9643064848
105+	.9667235861	.9689869958	.9711055636	.9730877176	.9749414758
110+	.9766744579	.9782938972	.9798066541	.9812192291	.9825377763
115+	.9837681168	.9849157525	.9859858792	.9869834006	.9879129406
120+	.9887788571	.9895852540	.9903359937	.9910347092	.9916848156
* 150+	.9989050094	.9989895371	.9990676833	.9991399169	.9992066733
			P=1/10, R= 6		
85+	.7421411304	.7549198069	.7672208283	.7790484784	.7904083961
90+	.8013074300	.8117534998	.8217554629	.8313229887	.8404664398
95+	.8491967613	.8575253763	.8654640903	.8730250022	.8802204226
100+	.8870627999	.8935646525	.8997385075	.9055968467	.9111520573
105+	.9164163888	.9214019155	.9261205029	.9305837802	.9348031150
110+	.9387895941	.9425540064	.9461068293	.9494582188	.9526180013
115+	.9555956687	.9584003752	.9610409359	.9635258279	.9658631921
120+	.9680608364	.9701262410	.9720665632	.9738886444	.9755990173
125+	.9772039131	.9787092704	.9801207436	.9814437115	.9826832870
130+	.9838443260	.9849314369	.9859489900	.9869011266	.9877917688
135+	.9886246281	.9894032148	.9901308467	.9908106580	.9914456074
* 165+	.9987618052	.9988533034	.9989382108	.9990169873	.9990900621
			P=1/10, R= 7		
95+	.7155354346	.7284446910	.7409350440	.7530065989	.7646607903
100+	.7759002730	.7867288162	.7971511983	.8071731066	.8168010390
105+	.8260422110	.8349044664	.8433961925	.8515262402	.8593038492
110+	.8667385774	.8738402357	.8806188271	.8870844914	.8932474534
115+	.8991179762	.9047063188	.9100226976	.9150772518	.9198800127
120+	.9244408764	.9287695797	.9328756795	.9367685344	.9404572898
125+	.9439508648	.9472579415	.9503869569	.9533460961	.9561432876
130+	.9587862001	.9612822408	.9636385545	.9658620249	.9679592757
135+	.9699366735	.9718003310	.9735561113	.9752096321	.9767662713
140+	.9782311725	.9796092505	.9809051981	.9821234922	.9832684004
145+	.9843439882	.9853541249	.9863024915	.9871925868	.9880277346
150+	.9888110903	.9895456477	.9902342458	.9908795747	.9914841825
* 180+	.9986665902	.9987614022	.9988496603	.9989318019	.9990082365

DIRICHLET 1 PROBABILITY, B= 2

N	0	1	2	3	4
			P=1/10,	R= 8	
110+	.7530427631	.7640987817	.7747858716	.7851067163	.7950648418
115+	.8046645404	.8139107972	.8228092195	.8313659677	.8395876903
120+	.8474814613	.8550547209	.8623152198	.8692709664	.8759301777
125+	.8823012334	.8883926330	.8942129566	.8997708288	.9050748848
130+	.9101337410	.9149559667	.9195500597	.9239244242	.9280873506
135+	.9320469984	.9358113805	.9393883503	.9427855897	.9460106000
140+	.9490706934	.9519729869	.9547243968	.9573316348	.9598012055
145+	.9621394046	.9643523182	.9664458229	.9684255867	.9702970710
150+	.9720655320	.9737360243	.9753134036	.9768023306	.9782072749
155+	.9795325191	.9807821638	.9819601318	.9830701736	.9841158717
160+	.9851006465	.9860277608	.9869003256	.9877213047	.9884935207
165+	.9892196593	.9899022753	.9905437969	.9911465313	.9917126691
* 200+	.9990300809	.9990974471	.9991602651	.9992188315	.9992734243
			P=1/10,	R= 9	
120+	.7324085133	.7435824464	.7544223461	.7649286601	.7751026609
125+	.7849463867	.7944625816	.8036546375	.8125265375	.8210828004
130+	.8293284269	.8372688484	.8449098775	.8522576604	.8593186322
135+	.8660994741	.8726070734	.8788484852	.8848308975	.8905615980
140+	.8960479430	.9012973295	.9063171687	.9111148618	.9156977785
145+	.9200732366	.9242484844	.9282306839	.9320268968	.9356440716
150+	.9390890321	.9423684676	.9454889248	.9484567998	.9512783331
155+	.9539596038	.9565065262	.9589248466	.9612201413	.9633978153
160+	.9654631015	.9674210610	.9692765833	.9710343882	.9726990265
165+	.9742748826	.9757661770	.9771769685	.9785111579	.9797724904
170+	.9809645600	.9820908124	.9831545491	.9841589313	.9851069838
175+	.9860015988	.9868455407	.9876414492	.9883918442	.9890991296
180+	.9897655973	.9903934315	.9909847124	.9915414203	.9920654392
* 215+	.9990058916	.9990728674	.9991354648	.9991939596	.9992486111
			P=1/10,	R=10	
130+	.7137065982	.7249206146	.7358325366	.7464411621	.7567460715
135+	.7667475812	.7764466969	.7858450667	.7949449347	.8037490949
140+	.8122608471	.8204839517	.8284225884	.8360813139	.8434650222
145+	.8505789067	.8574284235	.8640192560	.8703572824	.8764485440
150+	.8822992157	.8879155781	.8933039919	.8984708732	.9034226709
155+	.9081658464	.9127068539	.9170521229	.9212080420	.9251809445
160+	.9289770946	.9326026760	.9360637811	.9393664013	.9425164188
165+	.9455195995	.9483815865	.9511078950	.9537039076	.9561748709
170+	.9585258926	.9607619391	.9628878342	.9649082581	.9668277469
175+	.9686506929	.9703813447	.9720238088	.9735820502	.9750598944
180+	.9764610294	.9777890074	.9790472479	.9802390395	.9813675430
185+	.9824357945	.9834467081	.9844030790	.9853075867	.9861627983
190+	.9869711719	.9877350594	.9884567105	.9891382753	.9897818081
195+	.9903892706	.9909625348	.9915033866	.9920135288	.9924945842
* 230+	.9990023111	.9990677014	.9991289392	.9991862782	.9992399573

DIRICHLET 1   PROBABILITY,   B= 3

N	0	1	2	3	4
			P=1/ 4, R= 1		
5◆	.3808593750	.5126953125	.6229248047	.7113647461	.7806015015
10◆	.8339881897	.8747591972	.9057033062	.9290944040	.9467292577
15◆	.9600011688	.9699779889	.9774720477	.9830983137	.9873208743
20◆	.9904892252	.9928662036	.9946492951	.9959867925	.9969900050
✱ 25◆	.9977424590	.9983068219	.9987301053	.9990475734	.9992856772
			P=1/ 4, R= 2		
15◆	.7657258920	.8130944995	.8516291594	.8827009130	.9075814339
20◆	.9273939968	.9430998579	.9555038944	.9652695919	.9729375821
25◆	.9789445188	.9836406511	.9873053777	.9901605789	.9923817995
✱ 35◆	.9983896512	.9987604481	.9990464948	.9992669909	.9994368335
			P=1/ 4, R= 3		
20◆	.7344619679	.7813973683	.8210210051	.8541514449	.8816341705
25◆	.9042819163	.9228415440	.9379786687	.9502735712	.9602238710
30◆	.9682509108	.9747078761	.9798884178	.9840350489	.9873469190
35◆	.9899867773	.9920870672	.9937551713	.9950778671	.9961250725
✱ 45◆	.9990979183	.9992947998	.9994491735	.9995700999	.9996647381
			P=1/ 4, R= 4		
25◆	.7204939672	.7652059001	.8038114867	.8368366564	.8648670482
30◆	.8884989164	.9083073193	.9248271057	.9385429847	.9498857814
35◆	.9592327183	.9669101612	.9731977445	.9783331368	.9825169654
40◆	.9859176013	.9886756254	.9909078890	.9927111292	.9941651418
✱ 50◆	.9985054316	.9988137547	.9990594514	.9992550069	.9994104742
			P=1/ 4, R= 5		
30◆	.7157268846	.7578037975	.7946770429	.8267126075	.8543388374
35◆	.8780088824	.8981743822	.9152680274	.9296928503	.9418164373
40◆	.9519686191	.9604415207	.9674911362	.9733398207	.9781792670
45◆	.9821736696	.9854628788	.9881654172	.9903812858	.9921945191
✱ 55◆	.9978363923	.9982607707	.9986035417	.9988800242	.9991027474
			P=1/ 4, R= 6		
35◆	.7162592260	.7557119960	.7906455126	.8213336201	.8481061105
40◆	.8713202950	.8913400997	.9085213644	.9232020587	.9356962683
45◆	.9462909860	.9552449219	.9627887193	.9691261030	.9744356098
50◆	.9788726428	.9825716672	.9856484198	.9882020493	.9903171316
✱ 65◆	.9989960006	.9991889336	.9993454944	.9994723847	.9995751056
			P=1/ 4, R= 7		
40◆	.7199946387	.7569675615	.7899535101	.8191703024	.8448833037
45◆	.8673837493	.8869722779	.9039468918	.9185945366	.9311855528
50◆	.9419703385	.9511776719	.9590142428	.9656650339	.9712942766
55◆	.9760467688	.9800493994	.9834127628	.9862327854	.9885923058
60◆	.9905625741	.9922046469	.9935706668	.9947050219	.9956453864
✱ 70◆	.9986891429	.9989313970	.9991298598	.9992922389	.9994249296
			P=1/ 4, R= 8		
45◆	.7257159440	.7603887204	.7914994849	.8192301497	.8438021230
50◆	.8654595719	.8844564147	.9010465453	.9154767633	.9279819016
55◆	.9387816927	.9480789782	.9560589277	.9628889983	.9687194165
60◆	.9736840146	.9779012908	.9814755959	.9844983730	.9870493984
65◆	.9891979879	.9910041427	.9925196213	.9937889266	.9948502072
✱ 75◆	.9983728893	.9986629719	.9989026066	.9991003060	.9992631997
			P=1/ 4, R= 9		
50◆	.7326749356	.7652277167	.7945659546	.8208473375	.8442620245
55◆	.8650195525	.8833384861	.8994384846	.9135344219	.9258322075
60◆	.9365259786	.9457963753	.9538096503	.9607174069	.9666567976
65◆	.9717510472	.9761101953	.9798319746	.9830027647	.9856985718
70◆	.9879860033	.9899232107	.9915607866	.9929426035	.9941065907
✱ 80◆	.9980567790	.9983918930	.9986707738	.9989025476	.9990949205
			P=1/ 4, R=10		
55◆	.7403915827	.7709928455	.7986696278	.8235624304	.8458383230

DIRICHLET 1 PROBABILITY, B= 3

N	0	1	2	3	4
			P=1/ 4, R=10		
60+	.8656805694	.8832803259	.8988301801	.9125192757	.9245297676
65+	.9350343659	.9441947559	.9521607042	.9590696945	.9650469595
70+	.9702058020	.9746481192	.9784650614	.9817377723	.9845381702
75+	.9869297402	.9889683141	.9907028229	.9921760107	.9934251027
* 90+	.9991084068	.9992616709	.9993892355	.9994952833	.9995833410
			P=1/ 5, R= 1		
10+	.6959124480	.7531439309	.8003551396	.8389847532	.8704068452
15+	.8958563649	.9164034105	.9329536336	.9462614156	.9569482240
20+	.9655220286	.9723956903	.9779033919	.9823148169	.9858471154
25+	.9886748494	.9909381738	.9927495155	.9941989983	.9953588302
* 35+	.9987831070	.9990264753	.9992211740	.9993769355	.9995015462
			P=1/ 5, R= 2		
15+	.5464909895	.6099618634	.6668240143	.7170226269	.7608185383
20+	.7986645360	.8311141948	.8587582912	.8821828512	.9019432692
25+	.9185498603	.9324612248	.9440827353	.9537682097	.9618234250
30+	.9685105638	.9740529989	.9786400442	.9824314511	.9855615328
35+	.9881428642	.9902695490	.9920200679	.9934597395	.9946428307
* 45+	.9983993900	.9986933647	.9989337762	.9991302900	.9992908482
			P=1/ 5, R= 3		
25+	.7189130758	.7572569201	.7911720104	.8209597216	.8469645521
30+	.8695483620	.8890720189	.9058829575	.9203072942	.9326453372
35+	.9431695490	.9521242204	.9597262892	.9661668821	.9716132733
40+	.9762110410	.9800862743	.9833477279	.9860888649	.9883897500
45+	.9903187745	.9919342093	.9932855881	.9944149290	.9953578071
* 55+	.9984909134	.9987513316	.9989673862	.9991465177	.9992949425
			P=1/ 5, R= 4		
30+	.6544074757	.6953070352	.7324702390	.7659753863	.7959764763
35+	.8226788058	.8463191558	.8671502620	.8854290574	.9014080994
40+	.9153295940	.9274214805	.9378951049	.9469440898	.9547440764
45+	.9614530843	.9672122877	.9721470567	.9763681493	.9799729701
50+	.9830468367	.9856642128	.9878898811	.9897800383	.9913833050
* 65+	.9987500130	.9989556388	.9991279750	.9992723080	.9993931045
			P=1/ 5, R= 5		
40+	.7793935558	.8059261245	.8297365459	.8510057201	.8699259513
45+	.8866933411	.9015020309	.9145399985	.9259861321	.9360083260
50+	.9447623776	.9523914974	.9590262758	.9647849812	.9697740878
55+	.9740889547	.9778145943	.9810264863	.9837914005	.9861682069
60+	.9882086537	.9899581037	.9914562210	.9927376060	.9938323755
* 75+	.9990300879	.9991836841	.9993133887	.9994228387	.9995151321
			P=1/ 5, R= 6		
45+	.7400100859	.7685040581	.7944564285	.8179791651	.8392057755
50+	.8582835263	.8753670953	.8906135264	.9041783291	.9162125580
55+	.9268607125	.9362593095	.9445359945	.9518090797	.9581874081
60+	.9637704652	.9686486686	.9729037849	.9766094284	.9798316093
65+	.9826293059	.9850550408	.9871554472	.9889718150	.9905406091
* 80+	.9983485759	.9985974342	.9988096219	.9989903969	.9991442890
			P=1/ 5, R= 7		
50+	.7043830621	.7341499915	.7615966171	.7867773662	.8097739526
55+	.8306881945	.8496358297	.8667413183	.8821335716	.8959425258
60+	.9082964609	.9193199646	.9291324417	.9378470770	.9455701691
65+	.9524007609	.9584305038	.9637437023	.9684174929	.9725221217
70+	.9761212901	.9792725439	.9820276879	.9844332105	.9865307077
75+	.9883572969	.9899460166	.9913262051	.9925238579	.9935619612
* 90+	.9988218152	.9989948325	.9991430181	.9992698383	.9993782921
			P=1/ 5, R= 8		
55+	.6721953908	.7027314335	.7311796369	.7575484850	.7818764139
60+	.8042255637	.8246761520	.8433215312	.8602639422	.8756109420

B 30

DIRICHLET 1 PROBABILITY, B= 3

N	0	1	2	3	4
			P=1/ 5, R= 8		
65+	.8894724605	.9019584274	.9131769040	.9232326542	.9322260885
70+	.9402525228	.9474016949	.9537574923	.9593978464	.9643947595
75+	.9688144308	.9727174575	.9761590899	.9791895213	.9818542010
80+	.9841941578	.9862463260	.9880438678	.9896164857	.9909907232
* 95+	.9982132202	.9984647630	.9986818279	.9988689883	.9990302357
			P=1/ 5, R= 9		
60+	.6430850023	.6740229096	.7031006254	.7302909772	.7555973697
65+	.7790485613	.8006937637	.8205981674	.8388389528	.8555018059
70+	.8706779354	.8844615655	.8969478701	.9082313038	.9184042864
75+	.9275561935	.9357726090	.9431348000	.9497193750	.9555980931
80+	.9608377943	.9655004265	.9696431458	.9733184740	.9765744965
85+	.9794550886	.9820001596	.9842459076	.9862250769	.9879672136
90+	.9894989158	.9908440753	.9920241086	.9930581752	.9939633839
* 105+	.9987679316	.9989383132	.9990857466	.9992132253	.9993233675
			P=1/ 5, R=10		
70+	.7552099557	.7777906968	.7987120952	.8180286739	.8358057149
75+	.8521163810	.8670392165	.8806560158	.8930500367	.9043045295
80+	.9145015514	.9237210317	.9320400563	.9395323397	.9462678563
85+	.9523126047	.9577284805	.9625732390	.9669005270	.9707599716
90+	.9741973093	.9772545476	.9799701471	.9823792193	.9845137307
95+	.9864027116	.9880724632	.9895467611	.9908470528	.9919926467
* 110+	.9982562369	.9984886757	.9986910369	.9988670733	.9990200909
			P=1/ 6, R= 1		
10+	.5665312786	.6304308214	.6864079349	.7349009860	.7765557295
15+	.8121040426	.8422883789	.8678175160	.8893429935	.9074487797
20+	.9226490724	.9353908472	.9460589554	.9549823804	.9624408043
25+	.9686709886	.9738726993	.9782140537	.9818362468	.9848576732
30+	.9873774837	.9894786291	.9912304527	.9926908851	.9939082984
* 40+	.9979591378	.9982992363	.9985826668	.9988188689	.9990157107
			P=1/ 6, R= 2		
20+	.6397077398	.6833941414	.7228203201	.7581304005	.7895490840
25+	.8173504298	.8418340539	.8633076239	.8820744796	.8984252864
30+	.9126327728	.9249487643	.9356028784	.9448023868	.9527328650
35+	.9595593473	.9654277777	.9704666091	.9747884467	.9784916639
40+	.9816619465	.9843737370	.9866915640	.9886712517	.9903610090
* 55+	.9984103639	.9986532040	.9988592567	.9990340381	.9991822476
			P=1/ 6, R= 3		
30+	.7094464183	.7416678656	.7709522418	.7974291577	.8212575725
35+	.8426145925	.8616865761	.8786622688	.8937276666	.9070623142
40+	.9188367637	.9292109545	.9383333067	.9463403558	.9533567877
45+	.9594957595	.9648594176	.9695395422	.9736182667	.9771688300
50+	.9802563345	.9829384874	.9852663127	.9872848225	.9890336443
55+	.9905475998	.9918572350	.9929893039	.9939672045	.9948113725
* 70+	.9990404209	.9991788490	.9992975511	.9993992970	.9994864745
			P=1/ 6, R= 4		
40+	.7668863995	.7913746153	.8136922268	.8339552072	.8522898847
45+	.8688281476	.8837036137	.8970486450	.9089920897	.9196576348
50+	.9291626627	.9376175170	.9451250931	.9517806827	.9576720129
55+	.9628794288	.9674761801	.9715287790	.9750974034	.9782363246
60+	.9809943457	.9834152367	.9855381598	.9873980775	.9890261404
65+	.9904500509	.9916944023	.9927809930	.9937291150	.9945558182
* 80+	.9988881120	.9990402299	.9991718725	.9992857420	.9993841928
			P=1/ 6, R= 5		
45+	.6908718102	.7191902289	.7455227226	.7699013783	.7923807490
50+	.8130326707	.8319416982	.8492011718	.8649098954	.8791693859
55+	.8920816422	.9037473753	.9142646415	.9237278182	.9322268714
60+	.9398468621	.9466676504	.9527637561	.9582043445	.9630533079

DIRICHLET 1 PROBABILITY, B= 3

N	0	1	2	3	4

P=1/ 6, R= 5

N	0	1	2	3	4
65+	.9673694196	.9712065404	.9746138621	.9776361738	.9803141420
70+	.9826845960	.9847808109	.9866327857	.9882675116	.9897092279
75+	.9909796649	.9920982713	.9930824275	.9939476429	.9947077384
* 90+	.9988320996	.9989849126	.9991181137	.9992341616	.9993352156

P=1/ 6, R= 6

55+	.7543587175	.7765967924	.7972062674	.8162442652	.8337775485
60+	.8498797932	.8646292402	.8781067097	.8903939507	.9015722961
65+	.9117215875	.9209193389	.9292401055	.9367550266	.9435315158
70+	.9496330726	.9551191932	.9600453607	.9644630982	.9684200710
75+	.9719602239	.9751239457	.9779482517	.9804669762	.9827109718
80+	.9847083093	.9864844754	.9880625661	.9894634732	.9907060619
* 100+	.9988452791	.9989907178	.9991182308	.9992299704	.9993278386

P=1/ 6, R= 7

60+	.6950452541	.7198256132	.7431055698	.7648977112	.7852291903
65+	.8041390523	.8216757842	.8378951097	.8528580388	.8666291689
70+	.8792752290	.8908638503	.9014625461	.9111378808	.9199548062
75+	.9279761455	.9352622051	.9418704958	.9478555458	.9532687914
80+	.9581585312	.9625699317	.9665450739	.9701230321	.9733399774
85+	.9762292990	.9788217377	.9811455288	.9832265474	.9850884573
90+	.9867528571	.9882394249	.9895660584	.9907490091	.9918030114
* 110+	.9989032694	.9990369189	.9991546592	.9992583310	.9993495695

P=1/ 6, R= 8

70+	.7562748889	.7762003114	.7948326766	.8122078140	.8283685800
75+	.8433632491	.8572440817	.8700660690	.8818858465	.8927607686
80+	.9027481307	.9119045272	.9202853302	.9279442770	.9349331512
85+	.9413015468	.9470967016	.9523633913	.9571438731	.9614778704
90+	.9654025917	.9689527767	.9721607630	.9750565695	.9776679921
95+	.9800207078	.9821383838	.9840427900	.9857539132	.9872900693
100+	.9886680145	.9899030537	.9910091444	.9919989969	.9928841698
* 120+	.9989872262	.9991071281	.9992131829	.9993069422	.9993897906

P=1/ 6, R= 9

75+	.7074893350	.7293552808	.7500096814	.7694618960	.7877308689
80+	.8048435757	.8208335735	.8357396716	.8496047292	.8624745811
85+	.8743970932	.8854213398	.8955968976	.9049732489	.9135992829
90+	.9215228879	.9287906231	.9354474621	.9415365979	.9470993028
95+	.9521748340	.9568003782	.9610110301	.9648397969	.9683176266
100+	.9714734528	.9743342545	.9769251262	.9792693564	.9813885115
105+	.9833025229	.9850297766	.9865872026	.9879903646	.9892535470
110+	.9903898407	.9914112252	.9923286483	.9931521014	.9938906921
* 130+	.9990836970	.9991894409	.9992832931	.9993665497	.9994403710

P=1/ 6, R=10

85+	.7645266518	.7823846320	.7991741006	.8149218274	.8296595281
90+	.8434228520	.8562504658	.8681832354	.8792635024	.8895344540
95+	.8990395787	.9078222033	.9159251053	.9233901918	.9302582400
100+	.9365686920	.9423594973	.9476669962	.9525258409	.9569689463
105+	.9610274672	.9647307980	.9681065900	.9711807838	.9739776538
110+	.9765198613	.9788285156	.9809232396	.9828222392	.9845423747
115+	.9860992336	.9875072029	.9887795411	.9899284487	.9909651367
* 135+	.9985244729	.9986880202	.9988339703	.9989641488	.9990802007

P=1/ 7, R= 1

15+	.7219465280	.7589781918	.7914790325	.8198832913	.8446223419
20+	.8661089222	.8847279290	.9008317410	.9147385804	.9267328461
25+	.9370666653	.9459621409	.9536139431	.9601920088	.9658442008
30+	.9706988343	.9748670212	.9784448062	.9815150894	.9841493401
35+	.9864091120	.9883473757	.9900096849	.9914351932	.9926575372
* 50+	.9986518629	.9988444328	.9990094987	.9991509881	.9992722678

DIRICHLET 1   PROBABILITY,   B= 3

N	0	1	2	3	4
			P=1/ 7, R= 2		
25+	.6944461362	.7268856517	.7564553017	.7832829871	.8075230321
30+	.8293458086	.8489295964	.8664543639	.8820971543	.8960287837
35+	.9084115907	.9193980106	.9291297861	.9377376571	.9453414026
40+	.9520501362	.9579627734	.9631686109	.9677479711	.9717728777
45+	.9753077347	.9784099938	.9811307946	.9835155718	.9856046227
50+	.9874336336	.9890341640	.9904340892	.9916580023	.9927275786
* 65+	.9984200753	.9986266805	.9988065018	.9989629745	.9990990993
			P=1/ 7, R= 3		
35+	.7030863203	.7308095446	.7564520577	.7800762850	.8017632093
40+	.8216070908	.8397110097	.8561831789	.8711339538	.8846734588
45+	.8969097493	.9079474291	.9178866479	.9268224143	.9348441635
50+	.9420355280	.9484742687	.9542323293	.9593759818	.9639660402
55+	.9680581201	.9717029285	.9749465718	.9778308703	.9803936725
60+	.9826691632	.9846881620	.9864784072	.9880648257	.9894697860
65+	.9907133341	.9918134130	.9927860646	.9936456163	.9944048513
* 80+	.9986497557	.9988155560	.9989612723	.9990892974	.9992017450
			P=1/ 7, R= 4		
45+	.7215283033	.7454260510	.7676896237	.7883628537	.8075014334
50+	.8251699759	.8414394542	.8563850098	.8700841073	.8826150057
55+	.8940555170	.9044820159	.9139686720	.9225868698	.9304047904
60+	.9374871272	.9438949136	.9496854405	.9549122467	.9596251671
65+	.9638704244	.9676907543	.9711255549	.9742110521	.9769804757
70+	.9794642404	.9816901284	.9836834706	.9854673227	.9870626369
75+	.9884884251	.9897619148	.9908986967	.9919128627	.9928171356
* 95+	.9989762963	.9990961782	.9992022508	.9992960731	.9993790330
			P=1/ 7, R= 5		
55+	.7426641599	.7634519595	.7828973063	.8010370728	.8179157135
60+	.8335834867	.8480948854	.8615072701	.8738796966	.8852719241
65+	.8957435886	.9053535267	.9141592321	.9222164293	.9295787493
70+	.9362974930	.9424214682	.9479968903	.9530673335	.9576737257
75+	.9618543766	.9656450339	.9690789594	.9721870220	.9749978012
80+	.9775376991	.9798310567	.9819002726	.9837659217	.9854468731
85+	.9869604050	.9883223160	.9895470324	.9906477107	.9916363348
* 105+	.9986750100	.9988227110	.9989542795	.9990714342	.9991757165
			P=1/ 7, R= 6		
65+	.7639499170	.7821612032	.7992398737	.8152191523	.8301371485
70+	.8440357165	.8569594423	.8689547539	.8800691507	.8903505439
75+	.8998466997	.9086047769	.9166709483	.9240900983	.9309055857
80+	.9371590656	.9428903612	.9481373777	.9529360538	.9573203421
85+	.9613222161	.9649716966	.9682968950	.9713240697	.9740776916
90+	.9765805177	.9788536693	.9809167136	.9827877466	.9844834773
95+	.9860193100	.9874094266	.9886668655	.9898035986	.9908306054
* 120+	.9991013706	.9991985743	.9992854840	.9993631625	.9994325660
			P=1/ 7, R= 7		
70+	.6888445303	.7100286230	.7301677561	.7492624782	.7673214868
75+	.7843604939	.8004011492	.8154700329	.8295977258	.8428179621
80+	.8551668667	.8666822756	.8774031370	.8873689905	.8966195185
85+	.9051941647	.9131318146	.9204705320	.9272473452	.9334980787
90+	.9392572232	.9445578409	.9494315001	.9539082353	.9580165290
95+	.9617833121	.9652339787	.9683924139	.9712810310	.9739208171
100+	.9763313841	.9785310246	.9805367706	.9823644542	.9840287693
105+	.9855433335	.9869207493	.9881726646	.9893098315	.9903421635
* 130+	.9989581449	.9990666571	.9991641334	.9992516642	.9993302361
			P=1/ 7, R= 8		
80+	.7191228139	.7378357871	.7556170354	.7724748285	.7884229362
85+	.8034798279	.8176679181	.8310128671	.8435429370	.8552884077
90+	.8662810505	.8765536598	.8861396405	.8950726473	.9033862744

B 33

DIRICHLET 1 PROBABILITY, B= 3

N	0	1	2	3	4
			$P=1/7$, $R=8$		
95+	.9111137901	.9182879142	.9249406331	.9311030496	.9368052636
100+	.9420762803	.9469439414	.9514348789	.9555744856	.9593869017
105+	.9628950149	.9661204709	.9690836936	.9718039124	.9742991958
110+	.9765864895	.9786816582	.9805995300	.9823539422	.9839577888
115+	.9854230677	.9867609284	.9879817187	.9890950309	.9901097466
* 140+	.9988417543	.9989583408	.9990634940	.9991583001	.9992437461
			$P=1/7$, $R=9$		
90+	.7463804087	.7629738160	.7787344049	.7936749154	.8078118310
95+	.8211648009	.8337561005	.8456101332	.8567529745	.8672119599
100+	.8770153143	.8861918224	.8947705389	.9027805351	.9102506810
105+	.9172094589	.9236848063	.9297039867	.9352934834	.9404789158
110+	.9452849749	.9497353758	.9538528253	.9576590024	.9611745498
115+	.9644190758	.9674111633	.9701683861	.9727073310	.9750436239
120+	.9771919599	.9791661358	.9809790852	.9826429143	.9841689395
125+	.9855677244	.9868491178	.9880222910	.9890957744	.9900774937
* 150+	.9987523320	.9988740882	.9989842908	.9990839995	.9991741804
			$P=1/7$, $R=10$		
95+	.6856450653	.7042351955	.7220589074	.7391113590	.7553927392
100+	.7709077385	.7856650301	.7996767685	.8129581101	.8255267600
105+	.8374025487	.8486070397	.8591631689	.8690949170	.8784270130
110+	.8871846697	.8953933483	.9030785523	.9102656472	.9169797059
115+	.9232453768	.9290867728	.9345273802	.9395899843	.9442966113
120+	.9486684832	.9527259860	.9564886481	.9599751285	.9632032130
125+	.9661898174	.9689509968	.9715019598	.9738570870	.9760299526
130+	.9780333491	.9798793135	.9815791559	.9831434888	.9845822566
135+	.9859047666	.9871197188	.9882352370	.9892588980	.9901977615
* 160+	.9986883300	.9988127633	.9989257375	.9990282702	.9991212928
			$P=1/8$, $R=1$		
15+	.6344215801	.6753244251	.7122859238	.7455160316	.7752653259
20+	.8018046414	.8254105363	.8463552406	.8648999859	.8812908408
25+	.8957563694	.9085065916	.9197328474	.9296082722	.9382886649
30+	.9459135917	.9526076125	.9584815523	.9636337621	.9681513355
35+	.9721112581	.9755814769	.9786218845	.9812852168	.9836178647
40+	.9856606060	.9874492608	.9890152758	.9903862457	.9915863744
* 60+	.9990055965	.9991298850	.9992386404	.9993338036	.9994170731
			$P=1/8$, $R=2$		
30+	.7298878370	.7553191267	.7787145987	.8001693237	.8197885806
35+	.8376837039	.8539687486	.8687578671	.8821632925	.8942938320
40+	.9052537820	.9151421868	.9240523733	.9320717040	.9392815000
45+	.9457570936	.9515679769	.9567780204	.9614457390	.9656245895
50+	.9693632847	.9727061155	.9756932713	.9783611546	.9807426844
55+	.9828675859	.9847626651	.9864520659	.9879575109	.9892985242
60+	.9904926380	.9915555821	.9925014582	.9933428994	.9940912150
* 75+	.9984282720	.9986079395	.9987672510	.9989084876	.9990336782
			$P=1/8$, $R=3$		
40+	.6985207101	.7228230126	.7455776133	.7668171587	.7865861763
45+	.8049383745	.8219342740	.8376391606	.8521213435	.8654506941
50+	.8776974409	.8889311913	.8992201536	.9086305330	.9172260761
55+	.9250677418	.9322134776	.9387180840	.9446331496	.9500070441
60+	.9548849569	.9593089709	.9633181636	.9669487270	.9702341023
65+	.9732051224	.9758901601	.9783152776	.9805043751	.9824793375
70+	.9842601757	.9858651637	.9873109695	.9886127786	.9897844121
75+	.9908384363	.9917862666	.9926382639	.9934038241	.9940914619
* 95+	.9990184739	.9991242413	.9992187588	.9993032042	.9993786344
			$P=1/8$, $R=4$		
50+	.6848430864	.7075593830	.7290433380	.7493044061	.7683619225
55+	.7862434183	.8029830727	.8186203117	.8331985570	.8467641222

DIRICHLET 1 PROBABILITY, B= 3

N	0	1	2	3	4
			P=1/ 8, R= 4		
60+	.8593652522	.8710512961	.8818720054	.8918769462	.9011150141
65+	.9096340411	.9174804841	.9246991836	.9313331844	.9374236088
70+	.9430095744	.9481281494	.9528143390	.9571010978	.9610193616
75+	.9645980971	.9678643630	.9708433809	.9735586137	.9760318478
80+	.9782832788	.9803315976	.9821940776	.9838866598	.9854240372
85+	.9868197355	.9880861926	.9892348330	.9902761404	.9912197261
110+	.9990372991	.9991358474	.9992244741	.9993041581	.9993757841
			P=1/ 8, R= 5		
60+	.6802502212	.7014159111	.7215591687	.7406807262	.7587890954
65+	.7758994711	.7920326968	.8072143012	.8214736129	.8348429565
70+	.8473569300	.8590517630	.8699647521	.8801337702	.8895968436
75+	.8983917937	.9065559357	.9141258300	.9211370809	.9276241765
80+	.9336203656	.9391575665	.9442663040	.9489756693	.9533133014
85+	.9573053841	.9609766586	.9643504464	.9674486824	.9702919554
90+	.9728995534	.9752895148	.9774786810	.9794827519	.9813163420
95+	.9829930374	.9845254517	.9859252821	.9872033628	.9883697190
100+	.9894336168	.9904036132	.9912876026	.9920928621	.9928260937
120+	.9985482910	.9986892482	.9988168060	.9989322053	.9990365759
			P=1/ 8, R= 6		
70+	.6807970277	.7005490615	.7194255304	.7374241569	.7545487674
75+	.7708085414	.7862172924	.8007927851	.8145560967	.8275310262
80+	.8397435518	.8512213401	.8619933035	.8720892069	.8815393204
85+	.8903741163	.8986240063	.9063191177	.9134891037	.9201629863
90+	.9263690269	.9321346233	.9374862292	.9424492937	.9470482187
95+	.9513063315	.9552458703	.9588879814	.9622527257	.9653590927
100+	.9682250219	.9708674291	.9733022365	.9755444066	.9776079786
105+	.9795061061	.9812510965	.9828544502	.9843269007	.9856784542
110+	.9869184287	.9880554920	.9890976995	.9900525300	.9909269213
135+	.9987761803	.9988908637	.9989950440	.9990896564	.9991755559
			P=1/ 8, R= 7		
80+	.6844294322	.7029131000	.7206289035	.7375739142	.7537500381
85+	.7691634783	.7838242144	.7977455027	.8109434031	.8234363357
90+	.8352446686	.8463903390	.8568965068	.8667872418	.8760872429
95+	.8848215876	.8930155124	.9006942194	.9078827101	.9146056425
100+	.9208872107	.9267510442	.9322201261	.9373167266	.9420623525
105+	.9464777084	.9505826704	.9543962690	.9579366811	.9612212294
110+	.9642663881	.9670877941	.9697002626	.9721178066	.9743536587
115+	.9764202964	.9783294681	.9800922210	.9817189299	.9832193262
120+	.9846025279	.9858770681	.9870509251	.9881315506	.9891258980
125+	.9900404501	.9908812453	.9916539034	.9923636508	.9930153441
150+	.9989975503	.9990887163	.9991717887	.9992474653	.9993163859
			P=1/ 8, R= 8		
90+	.6899790738	.7073253642	.7239860439	.7399585078	.7552440274
95+	.7698473514	.7837763150	.7970414634	.8096556920	.8216339059
100+	.8329927015	.8437500700	.8539251236	.8635378460	.8726088649
105+	.8811592475	.8892103170	.8967834910	.9039001377	.9105814520
110+	.9168483476	.9227213654	.9282205963	.9333656169	.9381754376
115+	.9426684610	.9468624502	.9507745061	.9544210511	.9578178210
120+	.9609798612	.9639215291	.9666565001	.9691977778	.9715577063
125+	.9737479863	.9757796925	.9776632925	.9794086676	.9810251340
130+	.9825214648	.9839059128	.9851862326	.9863697038	.9874631530
135+	.9884729766	.9894051620	.9902653092	.9910586516	.9917900760
160+	.9987237405	.9988354709	.9989376808	.9990311551	.9991166166
			P=1/ 8, R= 9		
100+	.6967371483	.7130590511	.7287603613	.7438391692	.7582967117
105+	.7721370648	.7853668433	.7979949085	.8100320887	.8214909126
110+	.8323853582	.8427306175	.8525428783	.8618391223	.8706369407

DIRICHLET 1 PROBABILITY, B= 3

N	0	1	2	3	4
			P=1/ 8, R= 9		
115+	.8789543662	.8868097211	.8942214811	.9012081543	.9077881741
120+	.9139798054	.9198010631	.9252696426	.9304028603	.9352176043
125+	.9397302936	.9439568455	.9479126508	.9516125540	.9550708412
130+	.9583012312	.9613168729	.9641303450	.9667536603	.9691982725
135+	.9714750850	.9735944626	.9755662446	.9773997589	.9791038382
140+	.9806868357	.9821566430	.9835207070	.9847860481	.9859592779
145+	.9870466175	.9880539148	.9889866624	.9898500149	.9906488056
* 175+	.9989994314	.9990851111	.9991636520	.9992356293	.9993015740
			P=1/ 8, R=10		
110+	.7042523354	.7196462772	.7344725320	.7487299875	.7624201131
115+	.7755467179	.7881157129	.8001348805	.8116136516	.8225628938
120+	.8329947099	.8429222487	.8523595285	.8613212728	.8698227591
125+	.8778796800	.8855080177	.8927239291	.8995436438	.9059833727
130+	.9120592267	.9177871460	.9231828378	.9282617237	.9330388938
135+	.9375290690	.9417465694	.9457052893	.9494186773	.9528997212
140+	.9561609380	.9592143673	.9620715680	.9647436188	.9672411207
145+	.9695742023	.9717525265	.9737852997	.9756812815	.9774487966
150+	.9790957464	.9806296228	.9820575212	.9833861551	.9846218698
155+	.9857706574	.9868381713	.9878297405	.9887503839	.9896048248
160+	.9903975047	.9911325969	.9918140199	.9924454505	.9930303358
* 185+	.9987981493	.9988978070	.9989894430	.9990736798	.9991510941
			P=1/ 9, R= 1		
20+	.7348961539	.7622312573	.7869930082	.8093667251	.8295386680
25+	.8476914243	.8640007168	.8786333063	.8917457210	.9034836047
30+	.9139815160	.9233630512	.9317411924	.9392188044	.9458892245
35+	.9518369033	.9571380656	.9618613691	.9660685462	.9698150180
40+	.9731504745	.9761194170	.9787616610	.9811127996	.9832046277
45+	.9850655298	.9867208312	.9881931176	.9895025224	.9906669867
* 65+	.9985805064	.9987382011	.9988783801	.9990029883	.9991137548
			P=1/ 9, R= 2		
35+	.7545011255	.7752375314	.7944653126	.8122537305	.8286764623
40+	.8438097012	.8577306020	.8705160286	.8822415657	.8929807541
45+	.9028045187	.9117807547	.9199740472	.9274454989	.9342526448
50+	.9404494376	.9460862868	.9512101404	.9558645981	.9600900470
55+	.9639238127	.9674003212	.9705512648	.9734057708	.9759905681
60+	.9783301510	.9804469373	.9823614208	.9840923168	.9856566991
65+	.9870701307	.9883467854	.9894995614	.9905401887	.9914793267
* 90+	.9990848556	.9991782288	.9992621512	.9993375701	.9994053390
			P=1/ 9, R= 3		
45+	.6950842902	.7167057208	.7371323986	.7563824606	.7744818793
50+	.7914629808	.8073631106	.8222234464	.8360879539	.8490024803
55+	.8610139746	.8721698266	.8825173128	.8921031391	.9009730687
60+	.9091716268	.9167418700	.9237252139	.9301613092	.9360879592
65+	.9415410727	.9465546458	.9511607675	.9553896448	.9592696433
70+	.9628273397	.9660875839	.9690735675	.9718068971	.9743076707
75+	.9765945552	.9786848653	.9805946404	.9823387218	.9839308262
80+	.9853836187	.9867087819	.9879170820	.9890184329	.9900219560
* 105+	.9987305934	.9988515872	.9989612086	.9990605084	.9991504423
			P=1/ 9, R= 4		
60+	.7522643647	.7689210745	.7846728222	.7995420690	.8135545440
65+	.8267385768	.8391244974	.8507440997	.8616301698	.8718160715
70+	.8813353896	.8902216227	.8985079241	.9062268857	.9134103595
75+	.9200893141	.9262937218	.9320524716	.9373933069	.9423427819
80+	.9469262367	.9511677862	.9550903222	.9587155251	.9620638851
85+	.9651547290	.9680062539	.9706355643	.9730587126	.9752907418
90+	.9773457298	.9792368347	.9809763401	.9825757003	.9840455852
95+	.9853959236	.9866359465	.9877742284	.9888187267	.9897768210

DIRICHLET 1 PROBABILITY, B= 3

N	0	1	2	3	4
			P=1/ 9, R= 4		
100+	.9906553491	.9914606426	.9921985601	.9928745190	.9934935256
* 125+	.9990819337	.9991654082	.9992414189	.9993106197	.9993736087
			P=1/ 9, R= 5		
70+	.7224105112	.7392262403	.7552622633	.7705277254	.7850352998
75+	.7988006773	.8118420886	.8241798627	.8358360216	.8468339111
80+	.8571978678	.8669529218	.8761245318	.8847383526	.8928200326
85+	.9003950384	.9074885056	.9141251120	.9203289736	.9261235579
90+	.9315316160	.9365751289	.9412752678	.9456523660	.9497259018
95+	.9535144891	.9570358774	.9603069564	.9633437679	.9661615212
100+	.9687746124	.9711966470	.9734404650	.9755181669	.9774411424
105+	.9792200990	.9808650912	.9823855506	.9837903152	.9850876587
110+	.9862853197	.9873905299	.9884100414	.9893501539	.9902167402
* 140+	.9990095804	.9990956039	.9991743033	.9992462871	.9993121151
			P=1/ 9, R= 6		
80+	.6992880378	.7159640052	.7319569419	.7472949521	.7619554885
85+	.7759549748	.7893034420	.8020131829	.8140984258	.8255750292
90+	.8364601985	.8467722256	.8565302497	.8657540417	.8744638089
95+	.8826800206	.8904232534	.8977140552	.9045728260	.9110197155
100+	.9170745349	.9227566831	.9280850846	.9330781388	.9377536805
105+	.9421289478	.9462205594	.9500444987	.9536161040	.9569500644
110+	.9600604204	.9629605689	.9656632709	.9681806633	.9705242718
115+	.9727050272	.9747332822	.9766188301	.9783709240	.9799982973
120+	.9815091838	.9829113388	.9842120601	.9854182085	.9865362283
125+	.9875721680	.9885316997	.9894201388	.9902424626	.9910033284
* 155+	.9989973519	.9990810840	.9991579784	.9992285785	.9992933859
			P=1/ 9, R= 7		
95+	.7571642932	.7706061260	.7834696033	.7957634354	.8074980378
100+	.8186852986	.8293383601	.8394714141	.8490995121	.8582383914
105+	.8669043141	.8751139223	.8828841063	.8902318866	.8971743085
110+	.9037283493	.9099108365	.9157383766	.9212272943	.9263935798
115+	.9312528456	.9358202903	.9401106690	.9441382711	.9479169020
120+	.9514598717	.9547799862	.9578895437	.9608003344	.9635236424
125+	.9660702513	.9684504508	.9706740466	.9727503703	.9746882922
130+	.9764962334	.9781821805	.9797536997	.9812179518	.9825817071
135+	.9838513617	.9850329519	.9861321703	.9871543809	.9881046342
140+	.9889876818	.9898079913	.9905697599	.9912769286	.9919331951
* 170+	.9990271882	.9991057038	.9991780345	.9992446534	.9993059985
			P=1/ 9, R= 8		
105+	.7416154961	.7550184324	.7678968036	.7802554417	.7921008432
110+	.8034409808	.8142851229	.8246436628	.8345279565	.8439501714
115+	.8529231452	.8614602547	.8695752956	.8772823721	.8845957966
120+	.8915299989	.8980994445	.9043185619	.9102016779	.9157629611
125+	.9210163719	.9259756202	.9306541287	.9350650021	.9392210017
130+	.9431345248	.9468175882	.9502818160	.9535384309	.9565982482
135+	.9594716730	.9621687004	.9646989164	.9670715023	.9692952396
140+	.9713785172	.9733293389	.9751553326	.9768637599	.9784615267
145+	.9799551938	.9813509887	.9826548168	.9838722734	.9850086555
150+	.9860689736	.9870579638	.9879800993	.9888396022	.9896404544
155+	.9903864094	.9910810027	.9917275623	.9923292196	.9928889186
* 180+	.9986265138	.9987328969	.9988312662	.9989222050	.9990062562
			P=1/ 9, R= 9		
115+	.7286805743	.7419625488	.7547670170	.7670960913	.7789534544
120+	.7903442069	.8012747196	.8117524913	.8217860123	.8313846352
125+	.8405584524	.8493181812	.8576750562	.8656407299	.8732271798
130+	.8804466242	.8873114444	.8938341140	.9000271356	.9059029832
135+	.9114740515	.9167526102	.9217507643	.9264804197	.9309532527
140+	.9351806850	.9391738617	.9429436339	.9465005442	.9498548153

SOBEL, UPPULURI AND FRANKOWSKI

DIRICHLET 1 PROBABILITY, B= 3

N	0	1	2	3	4
		P=1/ 9,	R= 9		
145+	.9530163418	.9559946841	.9587990649	.9614383674	.9639211354
150+	.9662555757	.9684495602	.9705106314	.9724460069	.9742625862
155+	.9759669576	.9775654059	.9790639206	.9804682046	.9817836827
160+	.9830155114	.9841685876	.9852475584	.9862568300	.9872005780
165+	.9880827556	.9889071040	.9896771604	.9903962679	.9910675836
* 195+	.9987484971	.9988428816	.9989303507	.9990113935	.9990864660
		P=1/ 9,	R=10		
125+	.7178740304	.7309845296	.7436584018	.7558958564	.7676985574
130+	.7790695001	.7900128903	.8005340259	.8106391825	.8203355039
135+	.8296308965	.8385339299	.8470537421	.8551999506	.8629825696
140+	.8704119327	.8774986206	.8842533958	.8906871414	.8968108063
145+	.9026353548	.9081717213	.9134307698	.9184232577	.9231598041
150+	.9276508612	.9319066904	.9359373410	.9397526325	.9433621395
155+	.9467751799	.9500008045	.9530477899	.9559246330	.9586395470
160+	.9612004593	.9636150116	.9658905599	.9680341768	.9700526542
165+	.9719525076	.9737399802	.9754210481	.9770014268	.9784865768
170+	.9798817105	.9811917992	.9824215806	.9835755656	.9846580464
175+	.9856731038	.9866246149	.9875162608	.9883515341	.9891337464
180+	.9898660358	.9905513741	.9911925741	.9917922963	.9923530561
* 210+	.9988766147	.9989593497	.9990361699	.9991074829	.9991736698
		P=1/10,	R= 1		
20+	.6690597587	.6988546029	.7263138401	.7515470432	.7746761890
25+	.7958301764	.8151406122	.8327386315	.8487525554	.8633062152
30+	.8765178060	.8884991512	.8993552860	.9091842818	.9180772522
35+	.9261184912	.9333857053	.9399503095	.9458777654	.9512279411
40+	.9560554829	.9604101851	.9643373543	.9678781600	.9710699708
45+	.9739466707	.9765389582	.9788746239	.9809788092	.9828742454
50+	.9845814738	.9861190483	.9875037206	.9887506100	.9898733580
55+	.9908842692	.9917944399	.9926138739	.9933515888	.9940157118
* 75+	.9988901963	.9990011605	.9991010315	.9991909180	.9992718179
		P=1/10,	R= 2		
35+	.6665929826	.6903971185	.7128659848	.7340190434	.7538856299
40+	.7725029471	.7899142909	.8061675014	.8213136264	.8354057826
45+	.8484981946	.8606453962	.8719015752	.8823200434	.8919528175
50+	.9008502938	.9090610043	.9166314419	.9236059424	.9300266149
55+	.9359333118	.9413636300	.9463529390	.9509344280	.9551391693
60+	.9589961936	.9625325728	.9657735098	.9687424305	.9714610788
65+	.9739496108	.9762266883	.9783095710	.9802142054	.9819553108
70+	.9835446423	.9850001695	.9863279513	.9875404072	.9886472845
75+	.9896575410	.9905794044	.9914204280	.9921875421	.9928871027
* 100+	.9990350463	.9991235638	.9992040282	.9992771647	.9993436341
		P=1/10,	R= 3		
50+	.6924042730	.7118716491	.7303895525	.7479685172	.7646244096
55+	.7803775663	.7952520023	.8092746922	.8224749226	.8348837158
60+	.8465333211	.8574567684	.8676874832	.8772589546	.8862044552
65+	.8945568049	.9023481767	.9096099386	.9163725274	.9226653514
70+	.9285167175	.9339537801	.9390025076	.9436876660	.9480328140
75+	.9520603109	.9557913323	.9592458940	.9624428816	.9654000850
80+	.9681342364	.9706610506	.9729952677	.9751506963	.9771402578
85+	.9789760302	.9806692919	.9822305648	.9836696562	.9849956998
90+	.9862171952	.9873420461	.9883775970	.9893306680	.9902075883
* 120+	.9990072976	.9990926738	.9991708055	.9992422974	.9993077048
		P=1/10,	R= 4		
65+	.7228097734	.7390737851	.7545968555	.7693889847	.7834631671
70+	.7968349456	.8095219977	.8215437541	.8329210503	.8436758108
75+	.8538307650	.8634091930	.8724347014	.8809310254	.8889218569
80+	.8964306963	.9034807253	.9100947002	.9162948627	.9221028672

B 38

DIRICHLET 1 PROBABILITY, B= 3

N	0	1	2	3	4
			P=1/10, R= 4		
85+	.9275397223	.9326257456	.9373805293	.9418229163	.9459709851
90+	.9498420419	.9534526200	.9568184835	.9599546371	.9628753391
95+	.9655941174	.9681237892	.9704764814	.9726636538	.9746961228
100+	.9765840863	.9783371489	.9799643474	.9814741768	.9828746152
105+	.9841731492	.9853767987	.9864921408	.9875253333	.9884821375
110+	.9893679404	.9901877754	.9909463430	.9916480303	.9922969296
* 135+	.9986483973	.9987580793	.9988590227	.9989519081	.9990373647
			P=1/10, R= 5		
80+	.7525019472	.7662761319	.7794457657	.7920210237	.8040138254
85+	.8154375774	.8263069331	.8366375710	.8464459904	.8557493257
90+	.8645651766	.8729114568	.8808062574	.8882677262	.8953139608
95+	.9019629155	.9082323207	.9141396128	.9197018762	.9249357936
100+	.9298576056	.9344830780	.9388274768	.9429055481	.9467315057
105+	.9503190214	.9536812217	.9568306869	.9597794542	.9625390233
110+	.9651203648	.9675339301	.9697896634	.9718970149	.9738649552
115+	.9757019903	.9774161776	.9790151423	.9805060936	.9818958419
120+	.9831908149	.9843970750	.9855203351	.9865659752	.9875390584
125+	.9884443457	.9892863122	.9900691609	.9907968370	.9914730421
* 155+	.9989026001	.9989880698	.9990670184	.9991399312	.9992072584
			P=1/10, R= 6		
90+	.7132665903	.7276661345	.7415319748	.7548662576	.7676731098
95+	.7799584322	.7917297007	.8029957759	.8137667225	.8240536394
100+	.8338684999	.8432240032	.8521334376	.8606105532	.8686694472
105+	.8763244578	.8835900698	.8904808288	.8970112651	.9031958259
110+	.9090488159	.9145843456	.9198162862	.9247582310	.9294234638
115+	.9338249312	.9379752210	.9418865448	.9455707248	.9490391838
120+	.9523029389	.9553725983	.9582583598	.9609700126	.9635169403
125+	.9659081256	.9681521575	.9702572378	.9722311909	.9740814723
130+	.9758151795	.9774390625	.9789595349	.9803826853	.9817142891
135+	.9829598202	.9841244630	.9852131234	.9862304415	.9871808023
140+	.9880683474	.9888969860	.9896704054	.9903920824	.9910652924
* 170+	.9987368385	.9988310932	.9989184796	.9989994838	.9990745589
			P=1/10, R= 7		
105+	.7474690183	.7598484264	.7717647097	.7832226694	.7942283385
110+	.8047888439	.8149122760	.8246075636	.8338843566	.8427529156
115+	.8512240087	.8593088157	.8670188397	.8743658259	.8813616867
120+	.8880184346	.8943481206	.9003627793	.9060743798	.9114947817
125+	.9166356969	.9215086560	.9261249789	.9304957502	.9346317980
130+	.9385436761	.9422416504	.9457356868	.9490354431	.9521502624
135+	.9550891695	.9578608682	.9604737414	.9629358520	.9652549454
140+	.9674384533	.9694934981	.9714268989	.9732451771	.9749545639
145+	.9765610074	.9780701806	.9794874894	.9808180809	.9820668523
150+	.9832384592	.9843373247	.9853676475	.9863334115	.9872383936
155+	.9880861726	.9888801376	.9896234961	.9903192819	.9909703633
* 190+	.9990502756	.9991191764	.9991831994	.9992426797	.9992979304
			P=1/10, R= 8		
115+	.7178232638	.7305923938	.7429429357	.7548755745	.7663922924
120+	.7774962569	.7881917116	.7984838702	.8083788139	.8178833930
125+	.8270051340	.8357521497	.8441330561	.8521568931	.8598330512
130+	.8671712024	.8741812372	.8808732058	.8872572645	.8933436267
135+	.8991425183	.9046641376	.9099186192	.9149160015	.9196661989
140+	.9241789757	.9284639250	.9325304493	.9363877447	.9400447871
145+	.9435103212	.9467928513	.9499006342	.9528416741	.9556237189
150+	.9582542579	.9607405212	.9630894798	.9653078470	.9674020808
155+	.9693783868	.9712427223	.9730008003	.9746580948	.9762198459
160+	.9776910658	.9790765445	.9803808562	.9816083660	.9827632359
165+	.9838494321	.9848707312	.9858307273	.9867328382	.9875803128

DIRICHLET 1 PROBABILITY,   B= 3

N	0	1	2	3	4
		P=1/10,	R= 8		
170+	.9883762371	.9891235407	.9898250035	.9904832619	.9911008149
* 205+	.9989865076	.9990576171	.9991238680	.9991855816	.9992430590
		P=1/10,	R= 9		
130+	.7520093107	.7631204685	.7738564971	.7842204920	.7942164000
135+	.8038489377	.8131235133	.8220461510	.8306234192	.8388623612
140+	.8467704310	.8543554311	.8616254557	.8685888361	.8752540912
145+	.8816298806	.8877249622	.8935481522	.8991082896	.9044142032
150+	.9094746818	.9142984479	.9188941334	.9232702585	.9274352130
155+	.9313972397	.9351644202	.9387446629	.9421456924	.9453750411
160+	.9484400421	.9513478240	.9541053061	.9567191963	.9591959884
165+	.9615419622	.9637631828	.9658655020	.9678545595	.9697357856
170+	.9715144038	.9731954337	.9747836956	.9762838135	.9777002206
175+	.9790371628	.9802987046	.9814887332	.9826109643	.9836689469
180+	.9846660688	.9856055620	.9864905076	.9873238416	.9881083600
185+	.9888467237	.9895414643	.9901949887	.9908095842	.9913874233
* 220+	.9989495487	.9990210622	.9990878429	.9991501932	.9992083974
		P=1/10,	R=10		
140+	.7282668627	.7396700915	.7507274184	.7614395039	.7718078879
145+	.7818349219	.7915237020	.8008780035	.8099022175	.8186012889
150+	.8269806579	.8350462030	.8428041872	.8502612065	.8574241417
155+	.8643001122	.8708964341	.8772205791	.8832801381	.8890827863
160+	.8946362515	.8999482848	.9050266335	.9098790174	.9145131060
165+	.9189364992	.9231567088	.9271811432	.9310170931	.9346717192
170+	.9381520413	.9414649296	.9446170966	.9476150908	.9504652911
175+	.9531739033	.9557469562	.9581902999	.9605096045	.9627103589
180+	.9647978717	.9667772714	.9686535078	.9704313539	.9721154079
185+	.9737100961	.9752196754	.9766482372	.9779997101	.9792778642
190+	.9804863146	.9816285257	.9827078149	.9837273575	.9846901899
195+	.9855992150	.9864572059	.9872668102	.9880305545	.9887508488
200+	.9894299902	.9900701677	.9906734657	.9912418687	.9917772647
* 235+	.9989353575	.9990058812	.9990718701	.9991336055	.9991913517

DIRICHLET 1   PROBABILITY,   B= 4

N	0	1	2	3	4
			P=1/ 5, R= 1		
10+	.6063636480	.6780026880	.7381156823	.7879072481	.8287692521
15+	.8620793266	.8891009573	.9109429198	.9285514908	.9427194307
20+	.9541024653	.9632381160	.9705641755	.9764355486	.9811389632
25+	.9849054850	.9879209765	.9903347343	.9922665593	.9938125105
* 35+	.9983775104	.9987019877	.9989615778	.9991692548	.9993353994
			P=1/ 5, R= 2		
20+	.7356539036	.7775087097	.8134342491	.8440531833	.8699985608
25+	.8918786733	.9102569446	.9256420716	.9384848762	.9491793437
30+	.9580660996	.9654371546	.9715411560	.9765886696	.9807572098
35+	.9841958680	.9870294690	.9893622400	.9912810084	.9928579616
* 45+	.9978659055	.9982578522	.9985783887	.9988403995	.9990544722
			P=1/ 5, R= 3		
25+	.6341755547	.6825789486	.7258783050	.7642499675	.7979871318
30+	.8274516011	.8530379031	.8751477152	.8941724749	.9104822500
35+	.9244192419	.9362946101	.9463876005	.9549462087	.9621888107
40+	.9683063548	.9734648277	.9778078014	.9814589312	.9845243267
45+	.9870947472	.9892476023	.9910487476	.9925540825	.9938109576
* 55+	.9979879244	.9983351345	.9986231981	.9988620341	.9990599302
			P=1/ 5, R= 4		
35+	.7663947855	.7970871084	.8242703925	.8482218495	.8692301090
40+	.8875829925	.9035589733	.9174215336	.9294157371	.9397664539
45+	.9486777822	.9563333094	.9628969392	.9685140745	.9733130063
50+	.9774063951	.9808927691	.9838579841	.9863766115	.9885132345
55+	.9903236387	.9918558960	.9931513406	.9942454418	.9951685793
* 65+	.9983333741	.9986075338	.9988373101	.9990297506	.9991908104
			P=1/ 5, R= 5		
40+	.7105889399	.7446853646	.7754866322	.8031506637	.8278699587
45+	.8498574216	.8693353183	.8865269324	.9016504752	.9149148275
50+	.9265167360	.9366391380	.9454503422	.9531038430	.9597385891
55+	.9654795660	.9704385839	.9747151885	.9783976323	.9815638627
60+	.9842824936	.9866137396	.9886102988	.9903181738	.9917774296
* 75+	.9987067958	.9989115868	.9990845235	.9992304551	.9993535118
			P=1/ 5, R= 6		
45+	.6603722179	.6965967531	.7298575307	.7602078073	.7877492725
50+	.8126182864	.8349743764	.8549908797	.8728475279	.8887247259
55+	.9027992715	.9152412687	.9262120102	.9358626315	.9443333667
60+	.9517532639	.9582402430	.9639014020	.9688334944	.9731235213
65+	.9768493898	.9800806048	.9828789676	.9852992622	.9873899176
70+	.9891936350	.9907479764	.9920859103	.9932363144	.9942244339
* 85+	.9990335998	.9991819598	.9993079751	.9994149352	.9995056583
			P=1/ 5, R= 7		
55+	.7766482359	.8013062956	.8236558005	.8438342129	.8619879082
60+	.8782670917	.8928218306	.9057990411	.9173402776	.9275801817
65+	.9366454666	.9446543243	.9517161654	.9579316105	.9633926705
70+	.9681830623	.9723786176	.9760477533	.9792519747	.9820463941
75+	.9844802486	.9865974062	.9884368527	.9900331540	.9914168899
* 90+	.9984291037	.9986597883	.9988573654	.9990264565	.9991710599
			P=1/ 5, R= 8		
60+	.7423613036	.7688180628	.7930537248	.8151642459	.8352607907
65+	.8534643638	.8699014017	.8847002272	.8979882640	.9098899056
70+	.9205249361	.9300074113	.9384449147	.9459381165	.9525805709
75+	.9584587000	.9636519174	.9682328567	.9722676729	.9758163940
80+	.9789333027	.9816673335	.9840624737	.9861581591	.9879896584
85+	.9895884413	.9909825279	.9921968165	.9932533892	.9941717957
* 100+	.9988920734	.9990512856	.9991881414	.9993056940	.9994065939
			P=1/ 5, R= 9		
65+	.7099460768	.7377750264	.7635059984	.7871966137	.8089242024

DIRICHLET 1 PROBABILITY, B= 4

N	0	1	2	3	4
		P=1/ 5,	R= 9		
70+	.8287805146	.8468672005	.8632920247	.8781657545	.8915996538
75+	.9037035082	.9145841095	.9243441292	.9330813177	.9408879715
80+	.9478506183	.9540498758	.9595604471	.9644512206	.9687854489
85+	.9726209844	.9760105535	.9790020571	.9816388833	.9839602256
90+	.9860013993	.9877941500	.9893669517	.9907452909	.9919519357
* 105+	.9983572593	.9985844296	.9987810039	.9989509729	.9990978274
		P=1/ 5,	R=10		
70+	.6794552345	.7083011967	.7351930520	.7601543463	.7832313184
75+	.8044879208	.8240014209	.8418585904	.8581524637	.8729796285
80+	.8864380018	.8986250402	.9096363292	.9195645018	.9284984357
85+	.9365226852	.9437171069	.9501566435	.9559112361	.9610458354
90+	.9656204925	.9696905080	.9733066246	.9765152498	.9793586985
95+	.9818754465	.9841003886	.9860650961	.9877980687	.9893249793
100+	.9906689080	.9918505650	.9928885006	.9937993020	.9945977770
* 115+	.9988706714	.9990244754	.9991578689	.9992734773	.9993736011
		P=1/ 6,	R= 1		
15+	.7539580629	.7927220494	.8257662691	.8538004337	.8774954735
20+	.8974643443	.9142541630	.9283452858	.9401543989	.9500397199
25+	.9583071095	.9652163622	.9709872536	.9758051203	.9798258713
30+	.9831804059	.9859784577	.9883119053	.9902576036	.9918797913
* 45+	.9989063571	.9990886190	.9992405079	.9993670846	.9994725670
		P=1/ 6,	R= 2		
25+	.7604481081	.7919637978	.8197802939	.8442168297	.8655980227
30+	.8842413508	.9004489228	.9145024767	.9266607368	.9371584460
35+	.9462065492	.9539931322	.9606848269	.9664284744	.9713528984
40+	.9755706891	.9791799314	.9822658371	.9849022563	.9871530581
45+	.9890733779	.9907107341	.9921060211	.9932943870	.9943060058
* 60+	.9990771966	.9992191598	.9993394362	.9994413112	.9995275772
		P=1/ 6,	R= 3		
30+	.6241976649	.6643159590	.7011855476	.7348323872	.7653505564
35+	.7928827819	.8176042894	.8397098202	.8594034889	.8768910933
40+	.8923744720	.9060475295	.9180935871	.9286837652	.9379761505
45+	.9461155462	.9532336433	.9594494858	.9648701296	.9695914219
50+	.9736988421	.9772683649	.9803673151	.9830551940	.9853844640
55+	.9874012821	.9891461800	.9906546870	.9919578979	.9930829864
* 70+	.9987205869	.9989051501	.9990634142	.9991990716	.9993153056
		P=1/ 6,	R= 4		
40+	.6959287487	.7270172789	.7555588986	.7816354249	.8053564255
45+	.8268505730	.8462585076	.8637270842	.8794048470	.8934385633
50+	.9059706503	.9171373404	.9270674443	.9358815916	.9436918434
55+	.9506015905	.9567056647	.9620906046	.9668350301	.9710100863
60+	.9746799308	.9779022404	.9807287208	.9832056087	.9853741553
65+	.9872710870	.9889290399	.9903769634	.9916404955	.9927423067
* 80+	.9985175159	.9987203304	.9988958471	.9990476683	.9991789325
		P=1/ 6,	R= 5		
50+	.7546548764	.7789896365	.8013127175	.8217177257	.8403090595
55+	.8571977483	.8724980545	.8863247550	.8987910191	.9100067957
60+	.9200776305	.9291038387	.9371799686	.9443944985	.9508297171
65+	.9565617468	.9616606747	.9661907623	.9702107113	.9737739663
70+	.9769290398	.9797198477	.9821860466	.9843633652	.9862839267
75+	.9879765568	.9894670762	.9907785749	.9919316695	.9929447408
* 90+	.9984428346	.9986465757	.9988241702	.9989788956	.9991136306
		P=1/ 6,	R= 6		
55+	.6799594403	.7080878744	.7343386175	.7587353091	.7813227288
60+	.8021621558	.8213272661	.8389005757	.8549704141	.8696283950
65+	.8829673401	.8950796088	.9060557839	.9159836634	.9249475146
70+	.9330275458	.9402995605	.9468347577	.9526996520	.9579560866

DIRICHLET 1   PROBABILITY,   B= 4

N	0	1	2	3	4
			P=1/ 6,	R= 6	
75+	.9626613183	.9668681579	.9706251496	.9739767788	.9769636969
80+	.9796229574	.9819882543	.9840901610	.9859563630	.9876118845
85+	.9890793043	.9903789618	.9915291512	.9925463031	.9934451541
* 100+	.9984604048	.9986543144	.9988243255	.9989733069	.9991037943
			P=1/ 6,	R= 7	
65+	.7431676558	.7656873263	.7866149993	.8060025851	.8239115565
70+	.8404103559	.8555721605	.8694729867	.8821901083	.8938007589
75+	.9043810884	.9140053403	.9227452226	.9306694424	.9378433788
80+	.9443288723	.9501841083	.9554635777	.9602180994	.9644948904
85+	.9683376732	.9717868100	.9748794570	.9776497310	.9801288841
90+	.9823454814	.9843255793	.9860929007	.9876690055	.9890734558
95+	.9903239726	.9914365858	.9924257753	.9933046031	.9940848377
* 110+	.9985377204	.9987159126	.9988728943	.9990111194	.9991327678
			P=1/ 6,	R= 8	
70+	.6822715285	.7074947633	.7312291150	.7534845758	.7742857708
75+	.7936692777	.8116811847	.8283749018	.8438092304	.8580466865
80+	.8711520646	.8831912253	.8942300888	.9043338127	.9135661323
85+	.9219888460	.9296614233	.9366407195	.9429807804	.9487327218
90+	.9539446712	.9586617600	.9629261566	.9667771308	.9702511435
95+	.9733819548	.9762007446	.9787362426	.9810148637	.9830608454
100+	.9848963851	.9865417763	.9880155409	.9893345573	.9905141833
* 120+	.9986496574	.9988095210	.9989509231	.9990759324	.9991863945
			P=1/ 6,	R= 9	
80+	.7440103173	.7645515993	.7837841869	.8017429343	.8184699303
85+	.8340128463	.8484234739	.8617564477	.8740681446	.8854157483
90+	.8958564646	.9054468744	.9142424093	.9222969348	.9296624289
95+	.9363887420	.9425234266	.9481116271	.9531960182	.9578167852
100+	.9620116369	.9658158466	.9692623133	.9723816400	.9752022236
105+	.9777503540	.9800503190	.9821245121	.9839935421	.9856763423
110+	.9871902783	.9885512528	.9897738076	.9908712217	.9918556045
* 130+	.9987782800	.9989192676	.9990444006	.9991554070	.9992538335
			P=1/ 6,	R=10	
85+	.6926267091	.7152953811	.7367252800	.7569244263	.7759109852
90+	.7937115907	.8103597953	.8258946589	.8403594808	.8538006751
95+	.8662667861	.8778076352	.8884735935	.8983149669	.9073814863
100+	.9157218906	.9233835922	.9304124164	.9368524033	.9427456649
105+	.9481322891	.9530502837	.9575355523	.9616218990	.9653410546
110+	.9687227213	.9717946306	.9745826130	.9771106755	.9794010838
115+	.9814744494	.9833498180	.9850447582	.9865754504	.9879567734
120+	.9892023900	.9903248284	.9913355618	.9922450840	.9930629815
* 140+	.9989114889	.9990342763	.9991435769	.9992408252	.9993273092
			P=1/ 7,	R= 1	
15+	.6415180542	.6874771821	.7283349019	.7644172623	.7961121360
20+	.8238322536	.8479903499	.8689832523	.8871823887	.9029287840
25+	.9165311052	.9282657110	.9383779566	.9470842346	.9545743913
30+	.9610142824	.9665483121	.9713018602	.9753835436	.9788872841
35+	.9818941732	.9844741382	.9866874168	.9885858557	.9902140485
* 50+	.9982025827	.9984593143	.9986793820	.9988680201	.9990297161
			P=1/ 7,	R= 2	
30+	.7763339951	.8015113577	.8241639535	.8444772818	.8626396357
35+	.8788368285	.8932484303	.9060452077	.9173875043	.9274243417
40+	.9362930605	.9441193559	.9510175934	.9570913145	.9624338647
45+	.9671290911	.9712520710	.9748698445	.9780421306	.9808220141
50+	.9832565941	.9853875886	.9872518949	.9888821027	.9903069620
* 70+	.9989566878	.9990939551	.9992132964	.9993170318	.9994071844
			P=1/ 7,	R= 3	
40+	.7663256217	.7895576790	.8108044067	.8301746824	.8477842855

DIRICHLET 1   PROBABILITY,   B= 4

N	0	1	2	3	4
			P=1/ 7, R= 3		
45+	.8637522285	.8781978580	.8912386192	.9029883818	.9135562329
50+	.9230456516	.9315539903	.9391722008	.9459847488	.9520696727
55+	.9574987507	.9623377431	.9666466887	.9704802347	.9738879843
60+	.9769148520	.9796014173	.9819842695	.9840963408	.9859672225
65+	.9876234639	.9890888518	.9903846710	.9915299445	.9925416556
* 85+	.9990673269	.9991828840	.9992842932	.9993732619	.9994512954
			P=1/ 7, R= 4		
50+	.7708238276	.7917283758	.8110199222	.8287738150	.8450712118
55+	.8599967154	.8736364253	.8860763598	.8974012059	.9076933518
60+	.9170321611	.9254934495	.9331491313	.9400670038	.9463106428
65+	.9519393876	.9570083953	.9615687474	.9656675963	.9693483392
70+	.9726508111	.9756114891	.9782637020	.9806378410	.9827615676
75+	.9846600160	.9863559882	.9878701403	.9892211590	.9904259284
* 95+	.9986351027	.9987949354	.9989363580	.9990614488	.9991720575
			P=1/ 7, R= 5		
55+	.6664135832	.6924272057	.7169379455	.7399507155	.7614863679
60+	.7815788046	.8002723420	.8176193491	.8336781665	.8485112999
65+	.8621838749	.8747623354	.8863133633	.8969029987	.9065959348
70+	.9154549678	.9235405795	.9309106326	.9376201625	.9437212476
75+	.9492629455	.9542912828	.9588492865	.9629770492	.9667118197
80+	.9700881124	.9731378300	.9758903961	.9783728913	.9806101933
85+	.9826251159	.9844385460	.9860695775	.9875356406	.9888526252
90+	.9900349996	.9910959215	.9920473438	.9929001135	.9936640640
* 110+	.9990246957	.9991347417	.9992325946	.9993195761	.9993968686
			P=1/ 7, R= 6		
65+	.6929940530	.7159553299	.7376191859	.7579984953	.7771166486
70+	.7950056717	.8117045077	.8272574721	.8417128828	.8551218608
75+	.8675372945	.8790129566	.8896027612	.8993601489	.9083375863
80+	.9165861651	.9241552918	.9310924522	.9374430430	.9432502583
85+	.9485550235	.9533959681	.9578094303	.9618294873	.9654880063
90+	.9688147114	.9718372621	.9745813410	.9770707472	.9793274932
95+	.9813719039	.9832227158	.9848971746	.9864111316	.9877791370
100+	.9890145295	.9901295232	.9911352892	.9920420341	.9928590736
* 120+	.9988018557	.9989314542	.9990473290	.9991508965	.9992434315
			P=1/ 7, R= 7		
75+	.7186970002	.7390604222	.7582874016	.7763959294	.7934110414
80+	.8093635383	.8242888200	.8382258367	.8512161572	.8633031496
85+	.8745312712	.8849454582	.8945906095	.9035111548	.9117507003
90+	.9193517415	.9263554372	.9328014359	.9387277488	.9441706606
95+	.9491646758	.9537424916	.9579349947	.9617712786	.9652786747
100+	.9684827972	.9714075978	.9740754265	.9765070995	.9787219690
105+	.9807379968	.9825718276	.9842388629	.9857533344	.9871283753
110+	.9883760905	.9895076233	.9905332212	.9914622971	.9923034890
* 130+	.9986108979	.9987555725	.9988855344	.9990022371	.9991069957
			P=1/ 7, R= 8		
85+	.7429516762	.7610734276	.7781912086	.7943259289	.8095032485
90+	.8237526811	.8371067803	.8496004084	.8612700871	.8721534262
95+	.8822886266	.8917140532	.9004678713	.9085877422	.9161105708
100+	.9230723015	.9295077558	.9354505067	.9409327859	.9459854193
105+	.9506377857	.9549177965	.9588518929	.9624650572	.9657808355
110+	.9688213705	.9716074420	.9741585125	.9764927779	.9786272212
115+	.9805776680	.9823588433	.9839844287	.9854671197	.9868186814
120+	.9880500039	.9891711560	.9901914367	.9911194252	.9919630291
* 145+	.9990943189	.9991867594	.9992699853	.9993448895	.9994122811
			P=1/ 7, R= 9		
95+	.7655307840	.7816995846	.7969767485	.8113845489	.8249484570
100+	.8376965008	.8496586856	.8608664746	.8713523282	.8811492986

DIRICHLET 1 PROBABILITY,   B= 4

N	0	1	2	3	4
			P=1/ 7,	R= 9	
105+	.8902906785	.8988096971	.9067392628	.9141117465	.9209588026
110+	.9273112234	.9331988233	.9386503500	.9436934181	.9483544640
115+	.9526587169	.9566301858	.9602916589	.9636647131	.9667697333
120+	.9696259376	.9722514096	.9746631346	.9768770387	.9789080320
125+	.9807700516	.9824761073	.9840383267	.9854680007	.9867756287
130+	.9879709627	.9890630505	.9900602781	.9909704098	.9918006280
* 155+	.9990076367	.9991058875	.9991946557	.9992748289	.9993472146
			P=1/ 7,	R=10	
100+	.7016069246	.7202615270	.7380600238	.7550057979	.7711073492
105+	.7863776187	.8008333467	.8144944718	.8273835744	.8395253644
110+	.8509462172	.8616737545	.8717364701	.8811633986	.8899838255
115+	.8982270348	.9059220920	.9130976604	.9197818465	.9260020724
120+	.9317849731	.9371563144	.9421409305	.9467626785	.9510444075
125+	.9550079401	.9586740655	.9620625415	.9651921043	.9680804852
130+	.9707444323	.9731997371	.9754612639	.9775429822	.9794580012
135+	.9812186055	.9828362914	.9843218042	.9856851748	.9869357569
140+	.9880822627	.9891327990	.9900949008	.9909755660	.9917812865
* 165+	.9989408953	.9990428712	.9991352910	.9992190214	.9992948535
			P=1/ 8,	R= 1	
20+	.7418623180	.7718334797	.7986220080	.8224886243	.8436943051
25+	.8624925729	.8791244712	.8938155314	.9067741894	.9181912308
30+	.9282399478	.9370767698	.9448421911	.9516618717	.9576478169
35+	.9628995762	.9675054171	.9715434470	.9750826658	.9781839413
40+	.9809009030	.9832807545	.9853650076	.9871901417	.9887881937
45+	.9901872841	.9914120844	.9924842318	.9934226963	.9942441043
* 60+	.9986741924	.9988398944	.9989848897	.9991117650	.9992227843
			P=1/ 8,	R= 2	
30+	.6517763404	.6832820835	.7125383535	.7395873598	.7644987146
35+	.7873621299	.8082812698	.8273686815	.8447416971	.8605191888
40+	.8748190557	.8877563291	.8994417891	.9099810005	.9194736844
45+	.9280133574	.9356871780	.9425759514	.9487542543	.9542906440
50+	.9592479290	.9636834777	.9676495512	.9711936462	.9743588394
55+	.9771841260	.9797047476	.9819525059	.9839560593	.9857412023
60+	.9873311256	.9887466580	.9900064894	.9911273765	.9921243305
* 80+	.9988595441	.9989906165	.9991067418	.9992096085	.9993007166
			P=1/ 8,	R= 3	
45+	.7455227677	.7671428054	.7872276146	.8058367783	.8230368506
50+	.8388991045	.8534976367	.8669077990	.8792049211	.8904632908
55+	.9007553565	.9101511213	.9187176988	.9265190039	.9336155555
60+	.9400643699	.9459189280	.9512291997	.9560417144	.9603996642
65+	.9643430333	.9679087444	.9711308165	.9740405300	.9766665935
70+	.9790353110	.9811707457	.9830948797	.9848277676	.9863876834
75+	.9877912594	.9890536186	.9901884975	.9912083624	.9921245168
* 95+	.9986913502	.9988323621	.9989583765	.9990709634	.9991715317
			P=1/ 8,	R= 4	
55+	.7218060208	.7429734895	.7628564621	.7814850758	.7988970122
60+	.8151358356	.8302495211	.8442891696	.8573079000	.8693599066
65+	.8804996679	.8907812925	.9002579866	.9089816301	.9170024460
70+	.9243687508	.9311267755	.9373205434	.9429917990	.9481799762
75+	.9529222001	.9572533162	.9612059399	.9648105227	.9680954312
80+	.9710870344	.9738097972	.9762863783	.9785377287	.9805831917
85+	.9824406008	.9841263761	.9856556181	.9870421976	.9882988424
90+	.9894372197	.9904680141	.9914010020	.9922451207	.9930085345
* 110+	.9987164459	.9988478338	.9989659949	.9990722340	.9991677305
			P=1/ 8,	R= 5	
65+	.7087261024	.7290571643	.7482940544	.7664521323	.7835535418
70+	.7996260150	.8147017800	.8288165761	.8420087768	.8543186167

DIRICHLET 1 PROBABILITY, B= 4

N	0	1	2	3	4
			P=1/ 8, R= 5		
75+	.8657875179	.8764575098	.8863707352	.8955690352	.9040936055
80+	.9119847160	.9192814878	.9260217185	.9322417519	.9379763844
85+	.9432588031	.9481205514	.9525915167	.9566999373	.9604724246
90+	.9639339971	.9671081250	.9700167816	.9726805009	.9751184388
95+	.9773484372	.9793870890	.9812498048	.9829508776	.9845035485
100+	.9859200692	.9872117643	.9883890904	.9894616933	.9904384627
* 125+	.9988412979	.9989550307	.9990578121	.9991506723	.9992345476
			P=1/ 8, R= 6		
75+	.7022850666	.7216702560	.7401048165	.7575966754	.7741595378
80+	.7898120140	.8045768057	.8184799585	.8315501793	.8438182231
85+	.8553163444	.8660778132	.8761364916	.8855264668	.8942817374
90+	.9024359470	.9100221630	.9170726934	.9236189403	.9296912837
95+	.9353189932	.9405301637	.9453516717	.9498091497	.9539269757
100+	.9577282751	.9612349340	.9644676203	.9674458127	.9701878352
105+	.9727108953	.9750311263	.9771636314	.9791225296	.9809210024
110+	.9825713401	.9840849895	.9854725990	.9867440644	.9879085725
115+	.9889746441	.9899501749	.9908424751	.9916583073	.9924039223
* 140+	.9990047235	.9990990512	.9991846239	.9992622339	.9993326041
			P=1/ 8, R= 7		
85+	.7001811337	.7186156913	.7362097615	.7529675867	.7688982478
90+	.7840150116	.7983347158	.8118771933	.8246647414	.8367216355
95+	.8480736879	.8587478519	.8687718681	.8781739531	.8869825260
100+	.8952259722	.9029324409	.9101296724	.9168448550	.9231045064
105+	.9289343780	.9343593799	.9394035232	.9440898785	.9484405485
110+	.9524766513	.9562183152	.9596846814	.9628939141	.9658632168
115+	.9686088538	.9711461751	.9734896448	.9756528725	.9776486453
120+	.9794889618	.9811850670	.9827474871	.9841860643	.9855099918
125+	.9867278477	.9878476288	.9888767828	.9898222402	.9906904448
* 150+	.9986634470	.9987849929	.9988957490	.9989966453	.9990885347
			P=1/ 8, R= 8		
95+	.7010051317	.7185246258	.7352901151	.7513041120	.7665731674
100+	.7811073712	.7949198770	.8080264529	.8204450632	.8321954805
105+	.8432989305	.8537777681	.8636551854	.8729549506	.8817011750
110+	.8899181089	.8976299627	.9048607522	.9116341666	.9179734567
115+	.9239013414	.9294399318	.9346106694	.9394342783	.9439307298
120+	.9481192164	.9520181362	.9556450843	.9590168520	.9621494311
125+	.9650580236	.9677570558	.9702601950	.9725803702	.9747297941
130+	.9767199869	.9785618021	.9802654522	.9818405357	.9832960644
135+	.9846404899	.9858817311	.9870272004	.9880838300	.9890580974
140+	.9899560499	.9907833291	.9915451939	.9922465426	.9928919348
* 165+	.9989263355	.9990215021	.9991084403	.9991878405	.9992603374
			P=1/ 8, R= 9		
105+	.7038507516	.7205049443	.7364746484	.7517616794	.7663712305
110+	.7803114820	.7935932262	.8062295118	.8182353087	.8296271967
115+	.8404230759	.8506419024	.8603034463	.8694280729	.8780365468
120+	.8861498566	.8937890604	.9009751495	.9077289309	.9140709253
125+	.9200212807	.9255996997	.9308253795	.9357169629	.9402925000
130+	.9445694185	.9485645024	.9522938780	.9557730056	.9590166770
135+	.9620390178	.9648534937	.9674729196	.9699094727	.9721747068
140+	.9742795693	.9762344197	.9780490485	.9797326982	.9812940838
145+	.9827414146	.9840824153	.9853243479	.9864740330	.9875378709
150+	.9885218625	.9894316294	.9902724343	.9910491994	.9917665258
* 175+	.9986659530	.9987801847	.9988848993	.9989808637	.9990687855
			P=1/ 8, R=10		
115+	.7081114841	.7239533449	.7391678376	.7537566250	.7677242071
120+	.7810776085	.7938260784	.8059808031	.8175546355	.8285618411
125+	.8390178607	.8489390927	.8583426915	.8672463851	.8756683092

DIRICHLET 1   PROBABILITY,   B= 4

N	0	1	2	3	4
			P=1/ 8,	R=10	
130+	.8836268579	.8911405511	.8982279158	.9049073829	.9111971963
135+	.9171153350	.9226794458	.9279067874	.9328141831	.9374179828
140+	.9417340325	.9457776509	.9495636119	.9531061332	.9564188694
145+	.9595149097	.9624067786	.9651064410	.9676253088	.9699742510
150+	.9721636047	.9742031883	.9761023163	.9778698143	.9795140354
155+	.9810428771	.9824637984	.9837838376	.9850096293	.9861474224
160+	.9872030973	.9881821832	.9890898745	.9899310481	.9907102785
* 190+	.9989629843	.9990500863	.9991300707	.9992035000	.9992708942
			P=1/ 9,	R= 1	
20+	.6588603543	.6926540430	.7235772494	.7517620311	.7773641173
25+	.8005528065	.8215033033	.8403910426	.8573876043	.8726578784
30+	.8863582012	.8986352300	.9096253727	.9194546259	.9282387053
35+	.9360833805	.9430849462	.9493307770	.9548999291	.9598637589
40+	.9642865407	.9682260668	.9717342225	.9748575285	.9776376490
45+	.9801118617	.9823134920	.9842723094	.9860148908	.9875649493
50+	.9889436326	.9901697929	.9912602296	.9922299085	.9930921578
* 70+	.9989496680	.9990663563	.9991700826	.9992622864	.9993442474
			P=1/ 9,	R= 2	
35+	.6828039194	.7086359840	.7327694253	.7552458829	.7761202823
40+	.7954574432	.8133291826	.8298118797	.8449844591	.8589267479
45+	.8717181604	.8834366670	.8941580029	.9039550820	.9128975799
50+	.9210516558	.9284797892	.9352407060	.9413893779	.9469770767
55+	.9520514714	.9566567565	.9608338027	.9646203218	.9680510410
60+	.9711578814	.9739701368	.9765146500	.9788159851	.9808965932
65+	.9827769715	.9844758140	.9860101545	.9873955003	.9886459579
70+	.9897743504	.9907923258	.9917104585	.9925383424	.9932846769
* 90+	.9987798646	.9989043504	.9990162376	.9991167887	.9992071413
			P=1/ 9,	R= 3	
50+	.7288848295	.7489595227	.7678225494	.7855071272	.8020522952
55+	.8175015296	.8319015316	.8453011788	.8577506284	.8693005607
60+	.8800015487	.8899035416	.8990554476	.9075048053	.9152975302
65+	.9224777270	.9290875576	.9351671555	.9407545796	.9458857997
70+	.9505947087	.9549131558	.9588709958	.9624961515	.9658146853
75+	.9688508784	.9716273136	.9741649609	.9764832653	.9786002327
80+	.9805325166	.9822955017	.9839033850	.9853692553	.9867051677
85+	.9879222161	.9890306011	.9900396950	.9909581030	.9917937200
* 110+	.9989758762	.9990741763	.9991631595	.9992436956	.9993165744
			P=1/ 9,	R= 4	
60+	.6798046899	.7005422995	.7202744313	.7390060411	.7567490479
65+	.7735212767	.7893454761	.8042484175	.8182600774	.8314129042
70+	.8437411664	.8552803801	.8660668102	.8761370411	.8855276107
75+	.8942747034	.9024138939	.9099799395	.9170066128	.9235265714
80+	.9295712598	.9351708382	.9403541359	.9451486243	.9495804075
85+	.9536742263	.9574534750	.9609402272	.9641552688	.9671181381
90+	.9698471695	.9723595409	.9746713233	.9767975319	.9787521778
95+	.9805483197	.9821981152	.9837128714	.9851030938	.9863785345
100+	.9875482383	.9886205869	.9896033417	.9905036843	.9913282551
* 125+	.9987759635	.9988872532	.9989885929	.9990808542	.9991648344
			P=1/ 9,	R= 5	
75+	.7380221545	.7545332079	.7702218616	.7851034076	.7991963916
80+	.8125220540	.8251038172	.8369668206	.8481375020	.8586432249
85+	.8685119482	.8777719378	.8864515158	.8945788447	.9021817445
90+	.9092875396	.9159229311	.9221138942	.9278855962	.9332623335
95+	.9382674847	.9429234791	.9472517763	.9512728574	.9550062252
100+	.9584704116	.9616829919	.9646606044	.9674189734	.9699729369
105+	.9723364757	.9745227451	.9765441081	.9784121689	.9801378071
110+	.9817312124	.9832019185	.9845588370	.9858102902	.9869640439

DIRICHLET 1 PROBABILITY, B= 4

N	0	1	2	3	4
			P=1/ 9, R= 5		
115+	.9880273382	.9890069182	.9899090630	.9907396138	.9915040009
* 140+	.9986794997	.9987941872	.9988991110	.9989950823	.9990828470
			P=1/ 9, R= 6		
85+	.7092399651	.7260065166	.7420463732	.7573648205	.7719705692
90+	.7858753069	.7990932750	.8116408733	.8235362934	.8347991815
95+	.8454503311	.8555114046	.8650046837	.8739528469	.8823787721
100+	.8903053642	.8977554055	.9047514262	.9113155944	.9174696243
105+	.9232346994	.9286314103	.9336797056	.9383988539	.9428074160
110+	.9469232257	.9507633790	.9543442292	.9576813885	.9607897337
115+	.9636834169	.9663758791	.9688798668	.9712074507	.9733700469
120+	.9753784386	.9772427993	.9789727168	.9805772172	.9820647895
125+	.9834434101	.9847205670	.9859032834	.9869981416	.9880113057
130+	.9889485436	.9898152492	.9906164627	.9913568909	.9920409263
* 155+	.9986631950	.9987748276	.9988773450	.9989714715	.9990578755
			P=1/ 9, R= 7		
100+	.7631701818	.7767491890	.7897096848	.8020635065	.8138241928
105+	.8250067153	.8356272282	.8457028394	.8552514003	.8642913159
110+	.8728413731	.8809205873	.8885480663	.8957428900	.9025240056
115+	.9089101360	.9149197027	.9205707588	.9258809339	.9308673891
120+	.9355467798	.9399352272	.9440482970	.9479009841	.9515077028
125+	.9548822822	.9580379655	.9609874135	.9637427103	.9663153726
130+	.9687163611	.9709560930	.9730444568	.9749908281	.9768040861
135+	.9784926311	.9800644021	.9815268952	.9828871817	.9841519263
140+	.9853274053	.9864195247	.9874338375	.9883755611	.9892495944
145+	.9900605337	.9908126889	.9915100987	.9921565454	.9927555691
* 170+	.9987029717	.9988076502	.9989040836	.9989929024	.9990746906
			P=1/ 9, R= 8		
110+	.7438001482	.7575537410	.7707390869	.7833632474	.7954350647
115+	.8069649322	.8179645787	.8284468656	.8384255987	.8479153532
120+	.8569313142	.8654891293	.8736047764	.8812944435	.8885744214
125+	.8954610077	.9019704226	.9081187350	.9139217975	.9193951916
130+	.9245541803	.9294136684	.9339881702	.9382917828	.9423381654
135+	.9461405237	.9497115985	.9530636587	.9562084977	.9591574328
140+	.9619213073	.9645104953	.9669349080	.9692040021	.9713267894
145+	.9733118477	.9751673328	.9769009907	.9785201711	.9800318410
150+	.9814425985	.9827586868	.9839860088	.9851301405	.9861963458
155+	.9871895897	.9881145523	.9889756422	.9897770093	.9905225576
* 185+	.9987786123	.9988742775	.9989626405	.9990442415	.9991195823
			P=1/ 9, R= 9		
120+	.7272367291	.7410382135	.7543179720	.7670795248	.7793281546
125+	.7910707149	.8023154461	.8130718007	.8233502774	.8331622660
130+	.8425199023	.8514359332	.8599235927	.8679964879	.8756684945
135+	.8829536632	.8898661342	.8964200608	.9026295423	.9085085633
140+	.9140709414	.9193302818	.9242999374	.9289929761	.9334221518
145+	.9375998821	.9415382292	.9452488852	.9487431614	.9520319804
150+	.9551258716	.9580349683	.9607690088	.9633373374	.9657489092
155+	.9680122946	.9701356865	.9721269075	.9739934189	.9757423299
160+	.9773804074	.9789140865	.9803494815	.9816923965	.9829483369
165+	.9841225204	.9852198886	.9862451179	.9872026308	.9880966067
170+	.9889309929	.9897095149	.9904356870	.9911128220	.9917440412
* 200+	.9988746958	.9989605269	.9990399852	.9991135286	.9991815835
			P=1/ 9, R=10		
130+	.7130366256	.7267997839	.7400832787	.7528880256	.7652166331
135+	.7770732427	.7884633738	.7993937740	.8098722760	.8199076612
140+	.8295095303	.8386881829	.8474545034	.8558198558	.8637959866
145+	.8713949343	.8786289482	.8855104127	.8920517803	.8982655102
150+	.9041640143	.9097596081	.9150644686	.9200905965	.9248497836

DIRICHLET 1   PROBABILITY,   B= 4

N	0	1	2	3	4
			P=1/ 9, R=10		
155+	.9293535855	.9336132974	.9376399347	.9414442165	.9450365530
160+	.9484270355	.9516254288	.9546411664	.9574833477	.9601607370
165+	.9626817643	.9650545277	.9672867966	.9693860165	.9713593147
170+	.9732135064	.9749551023	.9765903157	.9781250711	.9795650124
175+	.9809155116	.9821816781	.9833683676	.9844801911	.9855215244
180+	.9864965169	.9874091009	.9882630007	.9890617410	.9898086559
185+	.9905068972	.9911594430	.9917691052	.9923385379	.9928702446
* 215+	.9989801456	.9990561118	.9991265745	.9991919190	.9992525050
			P=1/10, R= 1		
25+	.7349745796	.7593168183	.7816463359	.8020832541	.8207509746
30+	.8377730708	.8532709649	.8673622394	.8801594580	.8917693875
35+	.9022925334	.9118229165	.9204480320	.9282489430	.9353004720
40+	.9416714596	.9474250689	.9526191154	.9573064109	.9615351106
45+	.9653490547	.9687881011	.9718884439	.9746829159	.9772012732
50+	.9794704621	.9815148668	.9833565394	.9850154119	.9865094911
55+	.9878550379	.9890667304	.9901578142	.9911402381	.9920247784
* 75+	.9985203696	.9986683003	.9988014444	.9989212792	.9990291347
			P=1/10, R= 2		
40+	.7055926068	.7273678420	.7478223359	.7669914956	.7849177937
45+	.8016490322	.8172368391	.8317353858	.8452003036	.8576877819
50+	.8692538281	.8799536684	.8898412730	.8989689878	.9073872573
55+	.9151444241	.9222865936	.9288575525	.9348987301	.9404491968
60+	.9455456902	.9502226638	.9545123530	.9584448541	.9620482122
65+	.9653485159	.9683699952	.9711351220	.9736647099	.9759780136
70+	.9780928258	.9800255721	.9817914012	.9834042724	.9848770388
75+	.9862215263	.9874486077	.9885682738	.9895896990	.9905213035
* 100+	.9987134686	.9988314784	.9989387531	.9990362594	.9991248780
			P=1/10, R= 3		
55+	.7153200002	.7339791311	.7516633215	.7683912089	.7841860198
60+	.7990747025	.8130871459	.8262554834	.8386134768	.8501959787
65+	.8610384655	.8711766383	.8806460839	.8894819905	.8977189139
70+	.9053905865	.9125297659	.9191681173	.9253361251	.9310630307
75+	.9363767915	.9413040591	.9458701722	.9500991638	.9540137775
80+	.9576354938	.9609845627	.9640800413	.9669398367	.9695807504
85+	.9720185255	.9742678954	.9763426321	.9782555953	.9800187808
90+	.9816433675	.9831397646	.9845176559	.9857860436	.9869532901
95+	.9880271580	.9890148487	.9899230385	.9907579134	.9915252018
* 120+	.9986764710	.9987902930	.9988944578	.9989897716	.9990769742
			P=1/10, R= 4		
70+	.7359107305	.7519214874	.7671594098	.7816389793	.7953774792
75+	.8083945158	.8207115792	.8323516453	.8433388181	.8536980094
80+	.8634546566	.8726344733	.8812632340	.8893665869	.8969698953
85+	.9040981034	.9107756244	.9170262495	.9228730742	.9283384416
90+	.9334438989	.9382101668	.9426571197	.9468037747	.9506682896
95+	.9542679661	.9576192602	.9607377967	.9636383872	.9663350519
100+	.9688410433	.9711688725	.9733303361	.9753365443	.9771979500
105+	.9789243778	.9805250528	.9820086297	.9833832214	.9846564263
110+	.9858353562	.9869266622	.9879365605	.9888708572	.9897349715
115+	.9905339592	.9912725339	.9919550879	.9925857120	.9931682143
* 140+	.9988213543	.9989177436	.9990063822	.9990878807	.9991628027
			P=1/10, R= 5		
85+	.7593737484	.7731800604	.7863497373	.7988958101	.8108330184
90+	.8221775228	.8329466407	.8431586051	.8528323451	.8619872893
95+	.8706431891	.8788199625	.8865375561	.8938158245	.9006744260
100+	.9071327329	.9132097559	.9189240802	.9242938134	.9293365431
105+	.9340693039	.9385085519	.9426701471	.9465693413	.9502207719
110+	.9536384602	.9568358144	.9598256351	.9626201247	.9652308987

DIRICHLET 1 PROBABILITY, B= 4

N	0	1	2	3	4
			P=1/10,	R= 5	
115+	.9676689995	.9699449112	.9720685765	.9740494142	.9758963372
120+	.9776177717	.9792216762	.9807155607	.9821065063	.9834011840
125+	.9846058739	.9857264838	.9867685673	.9877373417	.9886377053
130+	.9894742537	.9902512966	.9909728731	.9916427662	.9922645181
* 160+	.9990259261	.9991024250	.9991730296	.9992381837	.9992982986
			P=1/10,	R= 6	
95+	.7146402797	.7294080015	.7436020792	.7572262379	.7702863574
100+	.7827902236	.7947472909	.8061684590	.8170658618	.8274526714
105+	.8373429167	.8467513159	.8556931230	.8641839894	.8722398375
110+	.8798767480	.8871108589	.8939582766	.9004349969	.9065568369
115+	.9123393753	.9177979026	.9229473775	.9278023917	.9323771406
120+	.9366854004	.9407405096	.9445553563	.9481423685	.9515135095
125+	.9546802752	.9576536952	.9604443359	.9630623058	.9655172630
130+	.9678184235	.9699745715	.9719940702	.9738848739	.9756545404
135+	.9773102440	.9788587890	.9803066232	.9816598517	.9829242506
140+	.9841052809	.9852081015	.9862375834	.9871983222	.9880946511
145+	.9889306536	.9897101753	.9904368364	.9911140423	.9917449953
* 175+	.9988588850	.9989448156	.9990244162	.9990981410	.9991664122
			P=1/10,	R= 7	
110+	.7458406238	.7586544342	.7709689605	.7827902760	.7941258127
115+	.8049841953	.8153750835	.8253090243	.8347973129	.8438518648
120+	.8524850959	.8607098139	.8685391174	.8759863055	.8830647953
125+	.8897880477	.8961695017	.9022225154	.9079603147	.9133959476
130+	.9185422458	.9234117905	.9280168842	.9323695275	.9364813987
135+	.9403638395	.9440278419	.9474840401	.9507427041	.9538137367
140+	.9567066716	.9594306748	.9619945464	.9644067251	.9666752933
145+	.9688079831	.9708121840	.9726949509	.9744630123	.9761227801
150+	.9776803589	.9791415557	.9805118904	.9817966056	.9830006774
155+	.9841288249	.9851855213	.9861750032	.9871012813	.9879681496
160+	.9887791955	.9895378089	.9902471913	.9909103650	.9915301814
* 190+	.9987337612	.9988256200	.9989109763	.9989902768	.9990639388
			P=1/10,	R= 8	
120+	.7115020825	.7249090535	.7378608381	.7503584549	.7624044362
125+	.7740026862	.7851583441	.7958776522	.8061678287	.8160369481
130+	.8254938265	.8345479152	.8432091997	.8514881069	.8593954178
135+	.8669421886	.8741396768	.8809992748	.8875324490	.8937506851
140+	.8996654391	.9052880933	.9106299174	.9157020342	.9205153898
145+	.9250807279	.9294085676	.9335091848	.9373925966	.9410685490
150+	.9445465068	.9478356458	.9509448479	.9538826972	.9566574784
155+	.9592771765	.9617494784	.9640817743	.9662811620	.9683544509
160+	.9703081667	.9721485578	.9738816009	.9755130083	.9770482346
165+	.9784924846	.9798507202	.9811276690	.9823278316	.9834554898
170+	.9845147148	.9855093744	.9864431418	.9873195027	.9881417634
175+	.9889130578	.9896363555	.9903144685	.9909500583	.9915456429
* 210+	.9990621404	.9991285790	.9991904265	.9992479908	.9993015599
			P=1/10,	R= 9	
135+	.7446131243	.7563496462	.7676769295	.7785985906	.7891192442
140+	.7992444018	.8089803722	.8183341684	.8273134180	.8359262791
145+	.8441813610	.8520876499	.8596544401	.8668912701	.8738078630
150+	.8804140725	.8867198335	.8927351165	.8984698868	.9039340679
155+	.9091375083	.9140899524	.9188010145	.9232801559	.9275366655
160+	.9315796428	.9354179832	.9390603666	.9425152467	.9457908438
165+	.9488951386	.9518358674	.9546205196	.9572563357	.9597503071
170+	.9621091768	.9643394406	.9664473498	.9684389143	.9703199064
175+	.9720958649	.9737721001	.9753536988	.9768455299	.9782522500
180+	.9795783095	.9808279587	.9820052540	.9831140640	.9841580762
185+	.9851408032	.9860655888	.9869356146	.9877539061	.9885233388

DIRICHLET 1 PROBABILITY, B= 4

N	0	1	2	3	4
		P=1/10,	R= 9		
190+	.9892466441	.9899264155	.9905651140	.9911650743	.9917285097
* 225+	.9990169966	.9990845855	.9991476488	.9992064800	.9992613543
		P=1/10,	R=10		
145+	.7168939514	.7290590424	.7408454236	.7522536451	.7632853268
150+	.7739430707	.7842303740	.7941515452	.8037116236	.8129163006
155+	.8217718455	.8302850345	.8384630835	.8463135848	.8538444476
160+	.8610638421	.8679801480	.8746019065	.8809377758	.8869964903
165+	.8927868238	.8983175548	.9035974365	.9086351684	.9134393721
170+	.9180185685	.9223811586	.9265354063	.9304894230	.9342511553
175+	.9378283736	.9412286628	.9444594153	.9475278244	.9504408798
180+	.9532053642	.9558278509	.9583147028	.9606720718	.9629058991
185+	.9650219172	.9670256507	.9689224199	.9707173426	.9724153387
190+	.9740211329	.9755392596	.9769740667	.9783297207	.9796102112
195+	.9808193557	.9819608048	.9830380476	.9840544160	.9850130909
200+	.9859171065	.9867693561	.9875725965	.9883294539	.9890424281
205+	.9897138979	.9903461256	.9909412619	.9915013506	.9920283326
* 240+	.9989938455	.9990611736	.9991241206	.9991829621	.9992379571

DIRICHLET 1 PROBABILITY, B= 5

N	0	1	2	3	4
			P=1/ 6, R= 1		
15+	.6980043977	.7446324520	.7847071165	.8189230770	.8479875409
20+	.8725774848	.8933165344	.9107645583	.9254151646	.9376978598
25+	.9479827432	.9565863818	.9637780321	.9697857167	.9748018864
30+	.9789885405	.9824817617	.9853956751	.9878258670	.9898523142
35+	.9915418784	.9929504209	.9941245877	.9951033144	.9959190895
* 45+	.9986329761	.9988607936	.9990506481	.9992088646	.9993407146
			P=1/ 6, R= 2		
25+	.7054996871	.7434990347	.7772600485	.8070783480	.8332827937
30+	.8562135242	.8762066593	.8935841590	.9086475539	.9216745047
35+	.9329173694	.9426031514	.9509343585	.9580904270	.9642294651
40+	.9694901410	.9739935980	.9778453184	.9811368874	.9839476299
45+	.9863461060	.9883914617	.9901346385	.9916194481	.9928835217
* 60+	.9988465128	.9990239615	.9991743034	.9993016445	.9994094753
			P=1/ 6, R= 3		
35+	.7444984292	.7745227642	.8014980416	.8256257374	.8471218161
40+	.8662070461	.8830998763	.8980113927	.9111419254	.9226789330
45+	.9327958586	.9416517034	.9493911201	.9561448660	.9620304971
50+	.9671532101	.9716067640	.9754744318	.9788299468	.9817384190
55+	.9842572054	.9864367247	.9883212108	.9899494039	.9913551784
* 70+	.9984007658	.9986314602	.9988292837	.9989988507	.9991441399
			P=1/ 6, R= 4		
40+	.6282609018	.6651999367	.6993777532	.7308098285	.7595622538
45+	.7857386341	.8094689759	.8309005078	.8501902787	.8674993328
50+	.8829882441	.8968137909	.9091265714	.9200693786	.9297761771
55+	.9383715498	.9459705016	.9526785310	.9585918938	.9637980026
60+	.9683759143	.9723968712	.9759248681	.9790172253	.9817251537
65+	.9840942992	.9861652625	.9879740868	.9895527136	.9909294031
* 85+	.9991154732	.9992379619	.9993437266	.9994350126	.9995137698
			P=1/ 6, R= 5		
50+	.6982184026	.7275514846	.7546006194	.7794365547	.8021520136
55+	.8228551730	.8416642258	.8587029387	.8740970996	.8879717404
60+	.9004490213	.9116466715	.9216768881	.9306456072	.9386520734
65+	.9457886439	.9521407717	.9577871270	.9627998168	.9672446758
70+	.9711816042	.9746649334	.9777438063	.9804625599	.9828611032
75+	.9849752838	.9868372382	.9884757248	.9899164356	.9911822881
* 95+	.9990386359	.9991661617	.9992770551	.9993734419	.9994571831
			P=1/ 6, R= 6		
60+	.7555940643	.7789285408	.8004020497	.8200996545	.8381151590
65+	.8545477623	.8694993142	.8830721051	.8953671249	.9064827242
70+	.9165136167	.9255501666	.9336779105	.9409772691	.9475234125
75+	.9533862443	.9586304780	.9633157835	.9674969838	.9712242876
80+	.9745435437	.9774965096	.9801211252	.9824517855	.9845196080
85+	.9863526925	.9879763687	.9894134324	.9906843671	.9918075524
* 105+	.9990225427	.9991474839	.9992567572	.9993522830	.9994357529
			P=1/ 6, R= 7		
65+	.6843193626	.7114021317	.7366968556	.7602328122	.7820568391
70+	.8022293885	.8208210673	.8379096533	.8535775654	.8679097523
75+	.8809919614	.8929093429	.9037453503	.9135808934	.9224937079
80+	.9305759070	.9378436830	.9444171325	.9503401812	.9556705883
85+	.9604620126	.9647641254	.9686227591	.9720800806	.9751747806
90+	.9779422746	.9804149065	.9826221542	.9845908319	.9863452871
95+	.9879075901	.9892977160	.9905337167	.9916318830	.9926068975
* 115+	.9990497259	.9991673260	.9992706726	.9993614509	.9994411527
			P=1/ 6, R= 8		
75+	.7452215661	.7671208147	.7874879832	.8063761063	.8238461771
80+	.8399648911	.8548027149	.8684322567	.8809269122	.8923597574
85+	.9028026610	.9123255862	.9209960584	.9288787723	.9360353173

DIRICHLET 1 PROBABILITY, B= 5

N	0	1	2	3	4
			P=1/ 6, R= 8		
90+	.9425240016	.9483997571	.9537141102	.9585152050	.9628478680
95+	.9667537048	.9702712203	.9734359554	.9762806366	.9788353309
100+	.9811276049	.9831826845	.9850236117	.9866713984	.9881451752
105+	.9894623337	.9906386620	.9916884738	.9926247295	.9934591498
* 125+	.9991049497	.9992126026	.9993075914	.9993913668	.9994652190
			P=1/ 6, R= 9		
80+	.6852543309	.7099674796	.7332123839	.7550056361	.7753766997
85+	.7943654260	.8120198204	.8283940636	.8435467827	.8575395622
90+	.8704356778	.8822990344	.8931932902	.9031811447	.9123237718
95+	.9206803801	.9283078805	.9352606480	.9415903603	.9473459031
100+	.9525733275	.9573158526	.9616139022	.9655051700	.9690247047
105+	.9722050124	.9750761692	.9776659420	.9799999139	.9821016105
110+	.9839926277	.9856927564	.9872201052	.9885912186	.9898211911
115+	.9909237757	.9919114875	.9927957010	.9935867422	.9942939748
* 130+	.9984728716	.9986491009	.9988055131	.9989442680	.9990672989
			P=1/ 6, R=10		
90+	.7452287750	.7654736501	.7844254594	.8021222115	.8186081988
95+	.8339324679	.8481474790	.8613079466	.8734698484	.8846895899
100+	.8950233092	.9045263084	.9132525962	.9212545282	.9285825329
105+	.9352849097	.9414076898	.9469945490	.9520867645	.9567232072
110+	.9609403635	.9647723810	.9682511314	.9714062888	.9742654174
115+	.9768540670	.9791958731	.9813126600	.9832245445	.9849500390
120+	.9865061532	.9879084933	.9891713581	.9903078311	.9913298686
* 140+	.9986393773	.9987928577	.9989294805	.9990510386	.9991591418
			P=1/ 7, R= 1		
20+	.7827095736	.8120829165	.8377310832	.8600540611	.8794314380
25+	.8962152565	.9107265770	.9232545298	.9340569776	.9433621612
30+	.9513708970	.9582590294	.9641799456	.9692670274	.9736359675
35+	.9773869079	.9806063831	.9833690653	.9857393164	.9877725588
40+	.9895164800	.9910120856	.9922946160	.9933943442	.9943372657
* 55+	.9989604193	.9991089177	.9992362058	.9993453126	.9994388346
			P=1/ 7, R= 2		
30+	.7252105627	.7555323318	.7829671918	.8076862834	.8298772123
35+	.8497349398	.8674549572	.8832283362	.8972382825	.9096578718
40+	.9206486955	.9303601921	.9389294840	.9464815773	.9531298107
45+	.9589764698	.9641135009	.9686232744	.9725793659	.9760473264
50+	.9790854265	.9817453632	.9840729210	.9861085855	.9878881070
55+	.9894430152	.9908010863	.9919867652	.9930215452	.9939243084
* 70+	.9986958925	.9988674679	.9990166382	.9991463027	.9992589900
			P=1/ 7, R= 3		
40+	.7130834438	.7410082131	.7666832218	.7901987419	.8116620293
45+	.8311914252	.8489115583	.8649495242	.8794319137	.8924825601
50+	.9042208872	.9147607494	.9242096687	.9326683874	.9402306660
55+	.9469832699	.9530060970	.9583724059	.9631491176	.9673971615
60+	.9711718506	.9745232686	.9774966589	.9801328074	.9824684118
65+	.9845364355	.9863664408	.9879849029	.9894155007	.9906793867
* 85+	.9988341821	.9989786226	.9991053796	.9992165872	.9993141265
			P=1/ 7, R= 4		
50+	.7183976804	.7435577212	.7668872414	.7884469119	.8083101767
55+	.8265595380	.8432834234	.8585735868	.8725229932	.8852241295
60+	.8967676846	.9072415458	.9167300612	.9253135239	.9330678367
65+	.9400643244	.9463696630	.9520458992	.9571505416	.9617367030
70+	.9658532816	.9695451675	.9728534655	.9758157279	.9784661886
75+	.9808359968	.9829534449	.9848441889	.9865314593	.9880362613
80+	.9893775638	.9905724755	.9916364108	.9925832427	.9934254443
* 100+	.9990872996	.9991952934	.9992906948	.9993749466	.9994493297

DIRICHLET 1 PROBABILITY, B= 5

N	0	1	2	3	4
			P=1/ 7, R= 5		
60+	.7312723473	.7538312994	.7748524035	.7943840239	.8124834970
65+	.8292147207	.8446460813	.8588486997	.8718949688	.8838573539
70+	.8948074262	.9048151002	.9139480471	.9222712582	.9298467352
75+	.9367332856	.9429864053	.9486582314	.9537975514	.9584498574
80+	.9626574336	.9664594703	.9698921958	.9729890208	.9757806917
85+	.9782954460	.9805591703	.9825955556	.9844262497	.9860710047
90+	.9875478181	.9888730682	.9900616417	.9911270545	.9920815646
* 110+	.9987808928	.9989184449	.9990407568	.9991494806	.9992460938
			P=1/ 7, R= 6		
70+	.7474282376	.7676434641	.7865391761	.8041576156	.8205471791
75+	.8357608000	.8498545446	.8628864110	.8749153139	.8860002395
80+	.8961995535	.9055704430	.9141684776	.9220472727	.9292582408
85+	.9358504169	.9418703474	.9473620304	.9523668988	.9569238383
90+	.9610692322	.9648370284	.9682588223	.9713639518	.9741796007
95+	.9767309074	.9790410768	.9811314929	.9830218307	.9847301662
100+	.9862730838	.9876657795	.9889221602	.9900549387	.9910757235
* 120+	.9985023549	.9986643451	.9988091824	.9989386370	.9990543020
			P=1/ 7, R= 7		
80+	.7647783652	.7829045656	.7998828770	.8157517186	.8305536667
85+	.8443343394	.8571414254	.8690238469	.8800310477	.8902123960
90+	.8996166894	.9082917526	.9162841160	.9236387655	.9303989540
95+	.9366060658	.9422995258	.9475167472	.9522931105	.9566619678
100+	.9606546700	.9643006089	.9676272743	.9706603197	.9734236367
105+	.9759394337	.9782283184	.9803093818	.9822002827	.9839173321
110+	.9854755751	.9868888721	.9881699764	.9893306094	.9903815323
* 135+	.9990012627	.9991066635	.9992011782	.9992859023	.9993618249
			P=1/ 7, R= 8		
85+	.6848495013	.7065236517	.7270888688	.7465513779	.7649256636
90+	.7822331645	.7985010641	.8137611866	.8280490012	.8414027333
95+	.8538625799	.8654700251	.8762672494	.8862966242	.8956002872
100+	.9042197884	.9121958000	.9195678844	.9263743117	.9326519205
105+	.9384360180	.9437603111	.9486568662	.9531560919	.9572867411
110+	.9610759290	.9645491643	.9677303908	.9706420362	.9733050683
115+	.9757390548	.9779622258	.9799915395	.9818427470	.9835304585
120+	.9850682081	.9864685175	.9877429584	.9889022126	.9899561298
125+	.9909137838	.9917835250	.9925730312	.9932893558	.9939389724
* 145+	.9988679163	.9989834632	.9990874926	.9991811205	.9992653582
			P=1/ 7, R= 9		
95+	.7118433366	.7312906575	.7497349628	.7671891323	.7836717868
100+	.7992063380	.8138201153	.8275435721	.8404095726	.8524527559
105+	.8637089762	.8742148137	.8840071524	.8931228192	.9015982794
110+	.9094693829	.9167711563	.9235376353	.9298017336	.9355951430
115+	.9409482609	.9458901416	.9504484677	.9546495391	.9585182753
120+	.9620782314	.9653516221	.9683593548	.9711210685	.9736551778
125+	.9759789196	.9781084038	.9800586641	.9818437105	.9834765819
130+	.9849693981	.9863334115	.9875790573	.9887160024	.9897531929
135+	.9906988998	.9915607627	.9923458316	.9930606070	.9937110774
* 155+	.9987595671	.9988823761	.9989933329	.9990935466	.9991840265
			P=1/ 7, R=10		
105+	.7369115886	.7543838265	.7709484300	.7866222938	.8014262953
110+	.8153845947	.8285239963	.8408733705	.8524631357	.8633247986
115+	.8734905497	.8829929110	.8918644313	.9001374272	.9078437635
120+	.9150146716	.9216805999	.9278710947	.9336147063	.9389389184
125+	.9438700981	.9484334620	.9526530584	.9565517617	.9601512772
130+	.9634721552	.9665338129	.9693545611	.9719516369	.9743412393
135+	.9765385676	.9785578625	.9804124479	.9821147730	.9836764560
140+	.9851083259	.9864204654	.9876222521	.9887223984	.9897289914

DIRICHLET 1 PROBABILITY, B= 5

N	0	1	2	3	4
		P=1/ 7,	R=10		
145+	.9906495302	.9914909628	.9922597204	.9929617516	.9936025535
* 165+	.9986761431	.9988036081	.9989191290	.9990237889	.9991185765
		P=1/ 8,	R= 1		
20+	.6848459103	.7204811651	.7525763617	.7813548906	.8070634742
25+	.8299577435	.8502920119	.8683122989	.8842518192	.8983283006
30+	.9107426319	.9216784544	.9313024015	.9397647700	.9472004586
35+	.9537300575	.9594610067	.9644887632	.9688979406	.9727633938
40+	.9761512360	.9791197779	.9817203883	.9839982754	.9859931919
45+	.9877400680	.9892695781	.9906086445	.9917808866	.9928070177
* 60+	.9983428203	.9985499279	.9987311570	.9988897399	.9990285056
		P=1/ 8,	R= 2		
35+	.7388757826	.7640503077	.7871327966	.8082314105	.8274632779
40+	.8449502039	.8608152954	.8751803675	.8881640055	.8998801647
45+	.9104372094	.9199373006	.9284760584	.9361424374	.9430187624
50+	.9491808833	.9546984151	.9596350354	.9640488203	.9679925998
55+	.9715143222	.9746574166	.9774611478	.9799609588	.9821887956
60+	.9841734153	.9859406736	.9875137924	.9889136084	.9901588008
* 80+	.9985744815	.9987383101	.9988834576	.9990120339	.9991259135
		P=1/ 8,	R= 3		
45+	.6888043046	.7145586789	.7386178182	.7610193288	.7818149528
50+	.8010671301	.8188460282	.8352270169	.8502885576	.8641104655
55+	.8767725050	.8883532751	.8989293462	.9085746089	.9173598035
60+	.9253521967	.9326153813	.9392091737	.9451895906	.9506088873
65+	.9555156425	.9599548795	.9639682127	.9675940109	.9708675729
70+	.9738213078	.9764849177	.9788855787	.9810481181	.9829951861
75+	.9847474200	.9863236012	.9877408034	.9890145320	.9901588552
* 100+	.9990767000	.9991769358	.9992664159	.9993462783	.9994175428
		P=1/ 8,	R= 4		
55+	.6606330435	.6856907549	.7093644535	.7316597139	.7525955057
60+	.7722017933	.7905173547	.8075878305	.8234640050	.8382003109
65+	.8518535471	.8644817930	.8761435026	.8868967594	.8967986745
70+	.9059049096	.9142693086	.9219436224	.9289773123	.9354174207
75+	.9413084957	.9466925621	.9516091284	.9560952235	.9601854561
80+	.9639120918	.9673051431	.9703924687	.9731998788	.9757512438
85+	.9780686046	.9801722827	.9820809880	.9838119251	.9853808949
90+	.9868023937	.9880897062	.9892549955	.9903093872	.9912630491
* 115+	.9990669553	.9991633322	.9992499025	.9993276463	.9993974477
		P=1/ 8,	R= 5		
70+	.7534321160	.7716400742	.7887461373	.8047837560	.8197906218
75+	.8338076179	.8468778986	.8590460911	.8703576102	.8808580785
80+	.8905928407	.8996065639	.9079429143	.9156442996	.9227516707
85+	.9293043740	.9353400461	.9408945461	.9460019189	.9506943833
90+	.9550023424	.9589544107	.9625774545	.9658966437	.9689355114
95+	.9717160195	.9742586285	.9765823696	.9787049187	.9806426701
100+	.9824108097	.9840233874	.9854933874	.9868327958	.9880526667
105+	.9891631847	.9901737250	.9910929102	.9919286646	.9926882647
* 125+	.9985516676	.9986938245	.9988222940	.9989383634	.9990432027
		P=1/ 8,	R= 6		
80+	.7415796962	.7593730152	.7761859515	.7920412927	.8069659599
85+	.8209901744	.8341467102	.8464702279	.8579966893	.8687628464
90+	.8788058018	.8881626323	.8968700733	.9049642553	.9124804890
95+	.9194530925	.9259152563	.9318989406	.9374348017	.9425521418
100+	.9472788806	.9516415443	.9556652685	.9593738147	.9627895952
105+	.9659337075	.9688259734	.9714849842	.9739281486	.9761717433
110+	.9782309652	.9801199843	.9818519964	.9834392760	.9848932280
115+	.9862244381	.9874427219	.9885571728	.9895762071	.9905076082
* 140+	.9987559357	.9988738392	.9989808002	.9990778087	.9991657683

DIRICHLET 1 PROBABILITY, B= 5

N	0	1	2	3	4
		P=1/ 8,	R= 7		
90+	.7345815458	.7518160659	.7681707340	.7836617650	.7983091435
95+	.8121359639	.8251678285	.8374323027	.8489584266	.8597762810
100+	.8699166062	.8794104684	.8882889728	.8965830175	.9043230863
105+	.9115390760	.9182601545	.9245146472	.9303299472	.9357324482
110+	.9407474951	.9453993519	.9497111828	.9537050466	.9574018995
115+	.9608216077	.9639829669	.9669037263	.9696006188	.9720893935
120+	.9743848512	.9765008825	.9784505056	.9802459071	.9818984808
125+	.9834188677	.9848169956	.9861021174	.9872828487	.9883672048
130+	.9893626358	.9902760609	.9911139011	.9918821106	.9925862068
* 155+	.9989652691	.9990604593	.9991470839	.9992258935	.9992975749
		P=1/ 8,	R= 8		
100+	.7310641577	.7476777506	.7634945520	.7785267011	.7927896737
105+	.8063017571	.8190835638	.8311575865	.8425477942	.8532792683
110+	.8633778772	.8728699892	.8817822192	.8901412090	.8979734374
115+	.9053050584	.9121617654	.9185686772	.9245502460	.9301301831
120+	.9353314016	.9401759729	.9446850970	.9488790826	.9527773378
125+	.9563983686	.9597597840	.9628783077	.9657697940	.9684492480
130+	.9709308490	.9732279764	.9753532372	.9773184956	.9791349029
135+	.9808129282	.9823623895	.9837924845	.9851118216	.9863284495
140+	.9874498875	.9884831536	.9894347932	.9903109057	.9911171706
* 165+	.9986579546	.9987769063	.9988855737	.9989848197	.9990754373
		P=1/ 8,	R= 9		
110+	.7300868409	.7460599422	.7613054204	.7758328740	.7896548144
115+	.8027862431	.8152442558	.8270476775	.8382167280	.8487727173
120+	.8587377710	.8681345840	.8769862015	.8853158264	.8931466496
125+	.9005017042	.9074037397	.9138751148	.9199377086	.9256128467
130+	.9309212411	.9358829435	.9405173094	.9448429716	.9488778234
135+	.9526390084	.9561429173	.9594051904	.9624407245	.9652636842
140+	.9678875162	.9703249669	.9725881016	.9746883260	.9766364086
145+	.9784425041	.9801161779	.9816664307	.9831017232	.9844300004
150+	.9856587170	.9867948607	.9878449762	.9888151886	.9897112255
155+	.9905384387	.9913018257	.9920060497	.9926554587	.9932541049
* 180+	.9989366676	.9990288439	.9991132292	.9991904624	.9992611316
		P=1/ 8,	R=10		
120+	.7309880045	.7463234740	.7609895803	.7749943348	.7883482766
125+	.8010641262	.8131564618	.8246414170	.8355364030	.8458598531
130+	.8556309905	.8648696188	.8735959329	.8818303522	.8895933713
135+	.8969054301	.9037868007	.9102574895	.9163371542	.9220450331
140+	.9273998879	.9324199556	.9371229108	.9415258373	.9456452062
145+	.9494968616	.9530960118	.9564572264	.9595944369	.9625209420
150+	.9652494158	.9677919189	.9701599116	.9723642695	.9744152996
155+	.9763227592	.9780958738	.9797433573	.9812734315	.9826938467
160+	.9840119012	.9852344624	.9863679858	.9874185356	.9883918030
165+	.9892931258	.9901275064	.9908996296	.9916138797	.9922743572
* 190+	.9987037617	.9988126338	.9989126097	.9990043926	.9990886322
		P=1/ 9,	R= 1		
25+	.7551646752	.7803803037	.8032264146	.8238734302	.8424924158
30+	.8592510061	.8743105589	.8878242651	.8999359916	.9107796773
35+	.9204791373	.9291481613	.9368908196	.9438019070	.9499674748
40+	.9554654104	.9603660384	.9647327209	.9686224440	.9720863798
45+	.9751704182	.9779156646	.9803589020	.9825330173	.9844673915
50+	.9861882569	.9877190212	.9890805610	.9902914880	.9913683883
* 70+	.9986871422	.9988329899	.9989626379	.9990778849	.9991803302
		P=1/ 9,	R= 2		
40+	.7488975388	.7703935144	.7903041983	.8087023686	.8256660676
45+	.8412763774	.8556156069	.8687658425	.8808078111	.8918200094
50+	.9018780579	.9110542406	.9194171985	.9270317460	.9339587874

DIRICHLET 1 PROBABILITY,   B= 5

N	0	1	2	3	4
			P=1/ 9, R= 2		
55+	.9402553118	.9459744474	.9511655628	.9558744013	.9601432393
60+	.9640110606	.9675137395	.9706842289	.9735527486	.9761469708
65+	.9784922012	.9806115528	.9825261126	.9842550997	.9858160148
70+	.9872247796	.9884958689	.9896424316	.9906764050	.9916086190
* 90+	.9984749023	.9986304948	.9987703420	.9988960217	.9990089550
			P=1/ 9, R= 3		
50+	.6696016217	.6933319951	.7157552338	.7368837797	.7567408591
55+	.7753584658	.7927755460	.8090363870	.8241892076	.8382849379
60+	.8513761780	.8635163202	.8747588182	.8851565893	.8947615317
65+	.9036241447	.9117932358	.9193157036	.9262363842	.9325979515
70+	.9384408618	.9438033347	.9487213645	.9532287542	.9573571690
75+	.9611362037	.9645934603	.9677546337	.9706436021	.9732825203
80+	.9756919143	.9778907753	.9798966527	.9817257454	.9833929893
85+	.9849121426	.9862958665	.9875558032	.9887026488	.9897462231
90+	.9906955348	.9915588430	.9923437148	.9930570793	.9937052774
* 110+	.9987198915	.9988427575	.9989539793	.9990546435	.9991457374
			P=1/ 9, R= 4		
65+	.7224889130	.7414261852	.7593328166	.7762301815	.7921446304
70+	.8071065091	.8211492776	.8343087288	.8466223006	.8581284790
75+	.8688662840	.8788748321	.8881929686	.8968589632	.9049102612
80+	.9123832840	.9193132741	.9257341769	.9316785556	.9371775343
85+	.9422607644	.9469564111	.9512911565	.9552902163	.9589773679
90+	.9623749862	.9655040877	.9683843787	.9710343077	.9734711201
95+	.9757109151	.9777687020	.9796584583	.9813931854	.9829849647
100+	.9844450115	.9857837278	.9870107528	.9881350114	.9891647610
105+	.9901076360	.9909706894	.9917604331	.9924828759	.9931435586
* 130+	.9990516009	.9991385240	.9992176014	.9992895285	.9993549403
			P=1/ 9, R= 5		
75+	.6802462875	.6998195908	.7185028425	.7362989899	.7532166233
80+	.7692691856	.7844742331	.7988527501	.8124285205	.8252275573
85+	.8372775902	.8486076073	.8592474524	.8692274702	.8785782006
90+	.8873301149	.8955133926	.9031577336	.9102922037	.9169451083
95+	.9231438924	.9289150634	.9342841330	.9392755777	.9439128129
100+	.9482181806	.9522129474	.9559173124	.9593504212	.9625303878
105+	.9654743198	.9681983493	.9707176651	.9730465481	.9751984084
110+	.9771858228	.9790205733	.9807136853	.9822754655	.9837155403
115+	.9850428917	.9862658938	.9873923474	.9884295138	.9893841469
120+	.9902625245	.9910704779	.9918134199	.9924963723	.9931239910
* 145+	.9989538909	.9990455766	.9991293725	.9992059428	.9992758971
			P=1/ 9, R= 6		
90+	.7371909552	.7530628566	.7681792963	.7825533885	.7962010410
95+	.8091404994	.8213919269	.8329770218	.8439186707	.8542406372
100+	.8639672851	.8731233331	.8817336404	.8898230204	.8974160801
105+	.9045370839	.9112098389	.9174575995	.9233029905	.9287679453
110+	.9338736582	.9386405490	.9430882389	.9472355349	.9511004232
115+	.9547000690	.9580508224	.9611682291	.9640670456	.9667612574
120+	.9692641001	.9715880826	.9737450117	.9757460187	.9776015857
125+	.9793215730	.9809152470	.9823913072	.9837579140	.9850227149
130+	.9861928716	.9872750848	.9882756202	.9892003315	.9900546848
* 160+	.9989214965	.9990124452	.9990958813	.9991724102	.9992425899
			P=1/ 9, R= 7		
100+	.7100248016	.7262358713	.7417636563	.7566133673	.7707931983
105+	.7843139425	.7971886299	.8094321879	.8210611274	.8320932527
110+	.8425473977	.8524431851	.8618008111	.8706408511	.8789840886
115+	.8868513637	.8942634408	.9012408941	.9078040092	.9139726997
120+	.9197664378	.9252041963	.9303044024	.9350849013	.9395629280
125+	.9437550878	.9476773428	.9513450052	.9547727349	.9579745422

DIRICHLET 1 PROBABILITY,   B= 5

N	0	1	2	3	4
			P=1/ 9, R= 7		
130+	.9609637942	.9637532244	.9663549448	.9687804603	.9710406853
135+	.9731459611	.9751060750	.9769302797	.9786273134	.9802054209
140+	.9816723735	.9830354903	.9843016585	.9854773537	.9865686598
145+	.9875812885	.9885205985	.9893916137	.9901990411	.9909472880
* 175+	.9989375134	.9990241456	.9991038714	.9991772267	.9992447071
			P=1/ 9, R= 8		
115+	.7624991676	.7757780071	.7884652915	.8005723309	.8121119575
120+	.8230982861	.8335464917	.8434726051	.8528933268	.8618258572
125+	.8702877441	.8782967454	.8858707078	.8930274591	.8997847136
130+	.9061599903	.9121705424	.9178332970	.9231648047	.9281811974
135+	.9328981548	.9373308771	.9414940642	.9454019014	.9490680487
140+	.9525056357	.9557272595	.9587449866	.9615703571	.9642143918
145+	.9666876009	.9689999953	.9711610983	.9731799589	.9750651666
150+	.9768248658	.9784667714	.9799981851	.9814260108	.9827567710
155+	.9839966231	.9851513748	.9862265005	.9872271564	.9881581957
160+	.9890241836	.9898294114	.9905779108	.9912734671	.9919196322
* 190+	.9989864348	.9990666618	.9991406889	.9992089818	.9992719721
			P=1/ 9, R= 9		
125+	.7433577350	.7568805049	.7698518098	.7822788253	.7941703522
130+	.8055366047	.8163890118	.8267400308	.8366029743	.8459918509
135+	.8549212183	.8634060499	.8714616135	.8791033619	.8863468350
140+	.8932075729	.8997010383	.9058425495	.9116472213	.9171299145
145+	.9223051926	.9271872856	.9317900599	.9361269939	.9402111586
150+	.9440552030	.9476713436	.9510713572	.9542665775	.9572678941
155+	.9600857538	.9627301647	.9652107018	.9675365144	.9697163339
160+	.9717584841	.9736708918	.9754610974	.9771362675	.9787032068
165+	.9801683708	.9815378783	.9828175249	.9840127952	.9851288760
170+	.9861706689	.9871428029	.9880496466	.9888953203	.9896837075
175+	.9904184670	.9911030430	.9917406767	.9923344164	.9928871271
* 205+	.9990557036	.9991285016	.9991958264	.9992580774	.9993156258
			P=1/ 9, R=10		
135+	.7265619584	.7402116478	.7533484486	.7659763003	.7781007899
140+	.7897289677	.8008691726	.8115308658	.8217244742	.8314612440
145+	.8407531046	.8496125421	.8580524833	.8660861893	.8737271589
150+	.8809890405	.8878855536	.8944304182	.9006372917	.9065197138
155+	.9120910577	.9173644883	.9223529256	.9270690140	.9315250965
160+	.9357331932	.9397049845	.9434517971	.9469845946	.9503139696
165+	.9534501402	.9564029472	.9591818548	.9617959522	.9642539572
170+	.9665642213	.9687347356	.9707731383	.9726867220	.9744824433
175+	.9761669313	.9777464976	.9792271462	.9806145839	.9819142307
180+	.9831312302	.9842704603	.9853365436	.9863338576	.9872665454
185+	.9881385255	.9889535019	.9897149737	.9904262445	.9910904318
* 215+	.9987252200	.9988201719	.9989082452	.9989899216	.9990656506
			P=1/10, R= 1		
25+	.6775030547	.7062388172	.7327778178	.7572120768	.7796476655
30+	.8001993201	.8189861954	.8361285802	.8517454098	.8659524339
35+	.8788609107	.8905767221	.9011998193	.9108239257	.9195364344
40+	.9274184532	.9345449564	.9409850127	.9468020640	.9520542375
45+	.9567946740	.9610718644	.9649299827	.9684092133	.9715460646
50+	.9743736689	.9769220655	.9792184662	.9812875028	.9831514573
55+	.9848304742	.9863427560	.9877047432	.9889312787	.9900357584
* 80+	.9989078042	.9990170061	.9991152913	.9992037509	.9992833668
			P=1/10, R= 2		
45+	.7565573010	.7752990044	.7927981504	.8091060758	.8242773988
50+	.8383687743	.8514378546	.8635424322	.8747397444	.8850859181
55+	.8946355374	.9034413141	.9115538471	.9190214555	.9258900726
60+	.9322031915	.9380018505	.9433246517	.9482078056	.9526851954

DIRICHLET 1 PROBABILITY, B= 5

N	0	1	2	3	4
			P=1/10,	R= 2	
65+	.9567884562	.9605470656	.9639884408	.9671380413	.9700194738
70+	.9726545967	.9750636255	.9772652346	.9792766571	.9811137809
75+	.9827912411	.9843225073	.9857199678	.9869950083	.9881580860
80+	.9892188002	.9901859571	.9910676318	.9918712249	.9926035160
* 105+	.9990067969	.9990982637	.9991813702	.9992568735	.9993254631
			P=1/10,	R= 3	
60+	.7533968961	.7702379732	.7861197508	.8010716216	.8151258905
65+	.8283170385	.8406810776	.8522549901	.8630762480	.8731824039
70+	.8826107472	.8913980203	.8995801857	.9071922405	.9142680709
75+	.9208403417	.9269404168	.9325983051	.9378426285	.9427006091
80+	.9471980703	.9513594528	.9552078383	.9587649839	.9620513608
85+	.9650861995	.9678875375	.9704722699	.9728562013	.9750540990
90+	.9770797453	.9789459903	.9806648032	.9822473229	.9837039068
95+	.9850441780	.9862770707	.9874108740	.9884532730	.9894113889
100+	.9902918161	.9911006579	.9918435603	.9925257434	.9931520313
* 125+	.9989459668	.9990371641	.9991205676	.9991968337	.9992665642
			P=1/10,	R= 4	
75+	.7645469772	.7793874136	.7934543836	.8067691067	.8193547909
80+	.8312361921	.8424392217	.8529905966	.8629175333	.8722474799
85+	.8810078854	.8892260014	.8969287160	.9041424135	.9108928596
90+	.9172051094	.9231034336	.9286112629	.9337511473	.9385447283
95+	.9430127228	.9471749158	.9510501627	.9546563971	.9580106453
100+	.9611290449	.9640268675	.9667185444	.9692176945	.9715371540
105+	.9736890069	.9756846172	.9775346602	.9792491550	.9808374960
110+	.9823084844	.9836703590	.9849308265	.9860970899	.9871758777
115+	.9881734702	.9890957265	.9899481090	.9907357074	.9914632618
* 145+	.9990396137	.9991187214	.9991914136	.9992582008	.9993195541
			P=1/10,	R= 5	
85+	.7059143375	.7223391636	.7380654990	.7530993149	.7674496356
90+	.7811281283	.7941487197	.8065272376	.8182810824	.8294289247
95+	.8399904313	.8499860179	.8594366279	.8683635346	.8767881685
100+	.8847319642	.8922162286	.8992620276	.9058900892	.9121207221
105+	.9179737491	.9234684517	.9286235268	.9334570534	.9379864678
110+	.9422285466	.9461993964	.9499144497	.9533884657	.9566355351
115+	.9596690888	.9625019102	.9651461486	.9676133365	.9699144067
120+	.9720597121	.9740590457	.9759216614	.9776562952	.9792711870
125+	.9807741015	.9821723504	.9834728128	.9846819562	.9858058572
130+	.9868502208	.9878204002	.9887214149	.9895579691	.9903344691
* 160+	.9987824526	.9988780691	.9989663187	.9990477562	.9991228955
			P=1/10,	R= 6	
100+	.7337785487	.7480944763	.7618111828	.7749366100	.7874806275
105+	.7994547681	.8108719810	.8217464020	.8320931412	.8419280892
110+	.8512677398	.8601290298	.8685291949	.8764856411	.8840158300
115+	.8911371789	.8978669718	.9042222841	.9102199167	.9158763402
120+	.9212076491	.9262295229	.9309571954	.9354054303	.9395885028
125+	.9435201868	.9472137459	.9506819294	.9539369712	.9569905924
130+	.9598540056	.9625379228	.9650525638	.9674076675	.9696125032
135+	.9716758841	.9736061810	.9754113368	.9770988813	.9786759466
140+	.9801492824	.9815252715	.9828099451	.9840089982	.9851278045
145+	.9861714313	.9871446537	.9880519693	.9888976111	.9896855615
150+	.9904195648	.9911031396	.9917395910	.9923320218	.9928833435
* 180+	.9990370554	.9991101964	.9991778948	.9992405456	.9992985161
			P=1/10,	R= 7	
115+	.7603660055	.7728785608	.7848715929	.7963540762	.8073362038
120+	.8178292106	.8278452092	.8373970368	.8464981153	.8551623222
125+	.8634038735	.8712372176	.8786769396	.8857376757	.8924340376
130+	.8987805450	.9047915667	.9104812700	.9158635761	.9209521233

DIRICHLET 1 PROBABILITY, B= 5

N	0	1	2	3	4
			$P=1/10$,	$R= 7$	
135+	.9257602350	.9303008942	.9345867219	.9386299613	.9424424644
140+	.9460356837	.9494206660	.9526080497	.9556080642	.9584305316
145+	.9610848700	.9635800995	.9659248479	.9681273587	.9701955004
150+	.9721367753	.9739583298	.9756669657	.9772691503	.9787710286
155+	.9801784343	.9814969017	.9827316770	.9838877301	.9849697659
160+	.9859822358	.9869293487	.9878150819	.9886431916	.9894172235
165+	.9901405226	.9908162433	.9914473584	.9920366687	.9925868112
* 195+	.9989154731	.9989948772	.9990685993	.9991370347	.9992005525
			$P=1/10$,	$R= 8$	
125+	.7232383070	.7365652435	.7494087731	.7617723146	.7736607747
130+	.7850803874	.7960385605	.8065437297	.8166052209	.8262331210
135+	.8354381572	.8442315852	.8526250854	.8606306686	.8682605882
140+	.8755272619	.8824432000	.8890209414	.8952729957	.9012117926
145+	.9068496370	.9121986689	.9172708302	.9220778346	.9266311427
150+	.9309419415	.9350211269	.9388792899	.9425267061	.9459733271
155+	.9492287756	.9523023418	.9552029816	.9579393176	.9605196405
160+	.9629519125	.9652437714	.9674025362	.9694352130	.9713485020
165+	.9731488049	.9748422329	.9764346146	.9779315054	.9793381954
170+	.9806597190	.9819008634	.9830661782	.9841599839	.9851863812
175+	.9861492598	.9870523070	.9878990163	.9886926961	.9894364774
180+	.9901333223	.9907860314	.9913972517	.9919694834	.9925050875
* 210+	.9988277144	.9989107569	.9989880614	.9990600127	.9991269705
			$P=1/10$,	$R= 9$	
140+	.7534346216	.7651428772	.7764190126	.7872684719	.7976976734
145+	.8077138949	.8173251644	.8265401573	.8353681006	.8438186822
150+	.8519019679	.8596283238	.8670083454	.8740527926	.8807725298
155+	.8871784730	.8932815405	.8990926097	.9046224784	.9098818302
160+	.9148812047	.9196309705	.9241413030	.9284221647	.9324832887
165+	.9363341655	.9399840314	.9434418601	.9467163561	.9498159496
170+	.9527487941	.9555227643	.9581454565	.9606241892	.9629660055
175+	.9651776764	.9672657043	.9692363276	.9710955263	.9728490271
180+	.9745023099	.9760606142	.9775289455	.9789120825	.9802145841
185+	.9814407966	.9825948606	.9836807187	.9847021224	.9856626395
190+	.9865656608	.9874144077	.9882119385	.9889611555	.9896648117
195+	.9903255168	.9909457441	.9915278358	.9920740100	.9925863654
* 225+	.9987712882	.9988557682	.9989345922	.9990081267	.9990767158
			$P=1/10$,	$R=10$	
150+	.7231320428	.7354224320	.7473077995	.7587900930	.7698723678
155+	.7805586833	.7908540026	.8007640962	.8102954501	.8194551783
160+	.8282509399	.8366908619	.8447834655	.8525375988	.8599623732
165+	.8670671052	.8738612626	.8803544154	.8865561904	.8924762311
170+	.8981241602	.9035095471	.9086418778	.9135305291	.9181847450
175+	.9226136173	.9268260669	.9308308301	.9346364449	.9382512409
180+	.9416833306	.9449406026	.9480307162	.9509610982	.9537389396
185+	.9563711953	.9588645833	.9612255859	.9634604511	.9655751950
190+	.9675756051	.9694672436	.9712554516	.9729453537	.9745418628
195+	.9760496851	.9774733257	.9788170939	.9800851094	.9812813075
200+	.9824094452	.9834731074	.9844757124	.9854205180	.9863106273
205+	.9871489946	.9879384309	.9886816097	.9893810723	.9900392336
* 240+	.9987423508	.9988265048	.9989051834	.9989787307	.9990474705

DIRICHLET 1 PROBABILITY, B= 6

N	0	1	2	3	4
			P=1/ 7, R= 1		
20+	.7427272334	.7769978107	.8070707585	.8333510369	.8562393434
25+	.8761182816	.8933442601	.9082433889	.9211100811	.9322074209
30+	.9417686268	.9499991441	.9570790458	.9631655314	.9683953851
35+	.9728873129	.9767441086	.9800546296	.9828955745	.9853330675
40+	.9874240600	.9892175629	.9907557252	.9920747745	.9932058342
* 55+	.9987525307	.9989307209	.9990834610	.9992143851	.9993266086
			P=1/ 7, R= 2		
30+	.6759468339	.7109745596	.7428528673	.7717171800	.7977371806
35+	.8211032408	.8420159329	.8606781489	.8772893622	.8920416063
40+	.9051168043	.9166851368	.9269041942	.9359187055	.9438606825
45+	.9508498508	.9569942712	.9623910771	.9671272734	.9712805566
50+	.9749201292	.9781074868	.9808971668	.9833374505	.9854710133
55+	.9873355234	.9889641875	.9903862469	.9916274245	.9927103279
* 70+	.9984351103	.9986409903	.9988199870	.9989755788	.9991107993
			P=1/ 7, R= 3		
40+	.6618494548	.6940421213	.7238061732	.7511973322	.7763009509
45+	.7992236075	.8200861100	.8390177981	.8561520008	.8716224956
50+	.8855608180	.8980942790	.9093445624	.9194267889	.9284489506
55+	.9365116337	.9437079623	.9501237059	.9558375079	.9609211972
60+	.9654401546	.9694537131	.9730155736	.9761742243	.9789733532
65+	.9814522484	.9836461797	.9855867602	.9873022845	.9888180445
70+	.9901566202	.9913381485	.9923805691	.9932998486	.9941101856
* 85+	.9986010467	.9987743682	.9989264713	.9990599164	.9991769606
			P=1/ 7, R= 4		
50+	.6678645065	.6969090846	.7239744774	.7490949509	.7723262057
55+	.7937401575	.8134204455	.8314586390	.8479510894	.8629963640
60+	.8766931943	.8891388696	.9004280104	.9106516622	.9198966537
65+	.9282451726	.9357745168	.9425569838	.9486598707	.9541455558
70+	.9590716446	.9634911588	.9674527582	.9710009816	.9741764988
75+	.9770163667	.9795542844	.9818208422	.9838437633	.9856481344
80+	.9872566247	.9886896929	.9899657799	.9911014897	.9921117566
* 100+	.9989047749	.9990343638	.9991488426	.9992499425	.9993392007
			P=1/ 7, R= 5		
60+	.6826428866	.7087569569	.7331961487	.7559901451	.7771835966
65+	.7968327394	.8150024396	.8317636494	.8471912527	.8613622666
70+	.8743543656	.8862446899	.8971089034	.9070204651	.9160500839
75+	.9242653268	.9317303545	.9385057627	.9446485076	.9502119002
80+	.9552456535	.9597959714	.9639056675	.9676143077	.9709583666
85+	.9739713942	.9766841882	.9791249661	.9813195379	.9832914730
90+	.9850622638	.9866514813	.9880769243	.9893547604	.9904996589
* 110+	.9985370992	.9987021552	.9988489246	.9989793892	.9990953221
			P=1/ 7, R= 6		
70+	.7012876156	.7247651000	.7467919345	.7673977550	.7866225647
75+	.8045144571	.8211276114	.8365205521	.8507546568	.8638928958
80+	.8759987811	.8871355042	.8973652399	.9067485958	.9153441886
85+	.9232083271	.9303947872	.9369546632	.9429362825	.9483851728
90+	.9533440701	.9578529614	.9619491529	.9656673574	.9690397977
95+	.9720963190	.9748645088	.9773698199	.9796356954	.9816836930
100+	.9835336065	.9852035857	.9867102505	.9880688017	.9892931260
105+	.9903958958	.9913886635	.9922819500	.9930853282	.9938075004
* 125+	.9989891448	.9990998318	.9991986284	.9992867830	.9993654166
			P=1/ 7, R= 7		
80+	.7213999547	.7425213144	.7623683758	.7809714667	.7983680673
85+	.8146012328	.8297182000	.8437691708	.8568062659	.8688826317
90+	.8800516916	.8903665242	.8998793563	.9086411571	.9167013196
95+	.9241074196	.9309050382	.9371376415	.9428465059	.9480706823
100+	.9528469924	.9572100512	.9611923092	.9648241117	.9681337702

DIRICHLET 1 PROBABILITY, B= 6

N	0	1	2	3	4
			P=1/ 7, R= 7		
105+	.9711476427	.9738902208	.9763842206	.9786506760	.9807090333
110+	.9825772441	.9842718578	.9858081114	.9872000168	.9884604440
115+	.9896012019	.9906331140	.9915660916	.9924092015	.9931707310
* 135+	.9988015318	.9989280092	.9990414239	.9991430907	.9992341960
			P=1/ 7, R= 8		
90+	.7417144444	.7607306488	.7786177207	.7954063105	.8111319729
95+	.8258340494	.8395546805	.8523379434	.8642291066	.8752739929
100+	.8855184418	.8950078617	.9037868618	.9118989565	.9193863302
105+	.9262896591	.9326479777	.9384985875	.9438769984	.9488169000
110+	.9533501560	.9575068191	.9613151606	.9648017141	.9679913281
115+	.9709072266	.9735710751	.9760030503	.9782219123	.9802450777
120+	.9820886931	.9837677077	.9852959449	.9866861722	.9879501679
125+	.9890987870	.9901420230	.9910890674	.9919483663	.9927276740
* 145+	.9986415209	.9987801725	.9989050043	.9990173549	.9991184379
			P=1/ 7, R= 9		
100+	.7615392010	.7786719802	.7947988048	.8099500568	.8241594407
105+	.8374631796	.8498993056	.8615070398	.8723262567	.8823970256
110+	.8917592230	.9004522094	.9085145630	.9159838642	.9228965265
115+	.9292876649	.9351910000	.9406387909	.9456617934	.9502892393
120+	.9545488338	.9584667678	.9620677421	.9653750011	.9684103749
125+	.9711943270	.9737460063	.9760833025	.9782229032	.9801803528
130+	.9819701110	.9836056117	.9850993203	.9864627909	.9877067210
135+	.9888410047	.9898747843	.9908164992	.9916739332	.9924542590
* 155+	.9985115060	.9986588715	.9987920155	.9989122686	.9990208418
			P=1/ 7, R=10		
105+	.6889875817	.7092458258	.7285155821	.7468040194	.7641245570
110+	.7804959075	.7959411910	.8104871250	.8241632903	.8370014724
115+	.8490350763	.8602986112	.8708272414	.8806563976	.8898214454
120+	.8983574050	.9062987169	.9136790500	.9205311463	.9268866987
125+	.9327762575	.9382291619	.9432734932	.9479360468	.9522423197
130+	.9562165114	.9598815363	.9632590446	.9663694509	.9692319692
135+	.9718646517	.9742844314	.9765071676	.9785476923	.9804198584
140+	.9821365876	.9837099183	.9851510535	.9864704071	.9876776495
145+	.9887817518	.9897910285	.9907131786	.9915553248	.9923240511
* 170+	.9990452823	.9991384855	.9992228131	.9992990858	.9993680507
			P=1/ 8, R= 1		
20+	.6306755522	.6713033207	.7081867040	.7414789473	.7713859048
25+	.7981440818	.8220043281	.8432200910	.8620392469	.8786986761
30+	.8934208945	.9064121976	.9178618902	.9279422737	.9368091459
35+	.9446026304	.9514482007	.9574578052	.9627310237	.9673562114
40+	.9714115982	.9749663265	.9780814150	.9808106450	.9832013678
45+	.9852952358	.9871288600	.9887343985	.9901400830	.9913706857
* 65+	.9989799385	.9991074320	.9992189923	.9993166103	.9994020280
			P=1/ 8, R= 2		
35+	.6921909824	.7212530804	.7480348401	.7726220836	.7951191468
40+	.8156425650	.8343159657	.8512660240	.8666193284	.8805000146
45+	.8930280335	.9043179356	.9144780698	.9236101084	.9318088256
50+	.9391620691	.9457508757	.9516496895	.9569266541	.9616439507
55+	.9658581647	.9696206654	.9729779875	.9759722066	.9786413024
60+	.9810019061	.9831376280	.9850233666	.9867015947	.9881946258
65+	.9895224591	.9907030049	.9917522905	.9926846481	.9935128859
* 85+	.9990721299	.9991792791	.9992741400	.9993581101	.9994324299
			P=1/ 8, R= 3		
50+	.7641906374	.7849687524	.8041724921	.8218777395	.8381650259
55+	.8531174011	.8668187022	.8793521710	.8907993744	.9012393815
60+	.9107481576	.9193981387	.9272579536	.9343922654	.9408617097
65+	.9467229076	.9520285368	.9568274465	.9611648045	.9650822665

DIRICHLET 1   PROBABILITY,   B= 6

N	0	1	2	3	4
			P=1/ 8, R= 3		
70+	.9686181599	.9718076755	.9746830636	.9772738282	.9796069190
75+	.9817069171	.9835962138	.9852951810	.9868223326	.9881944770
80+	.9894268590	.9905332936	.9915262892	.9924171627	.9932161455
* 100+	.9988920623	.9990123404	.9991197127	.9992155446	.9993010596
			P=1/ 8, R= 4		
60+	.7305458093	.7518422906	.7717589779	.7903387849	.8076315825
65+	.8236923549	.8385796122	.8523540400	.8650773691	.8768114403
70+	.8876174454	.8975553205	.9066832733	.9150574243	.9227315451
75+	.9297568785	.9361820266	.9420528945	.9474126797	.9523018970
80+	.9567584333	.9608176230	.9645123404	.9678731041	.9709281892
85+	.9737037442	.9762239103	.9785109413	.9805853218	.9824658828
90+	.9841699143	.9857132730	.9871104858	.9883748478	.9895185159
95+	.9905525958	.9914872251	.9923316498	.9930942973	.9937828427
* 115+	.9988803681	.9989960158	.9990998965	.9991931861	.9992769455
			P=1/ 8, R= 5		
70+	.7087525675	.7298498826	.7497410687	.7684496057	.7860061751
75+	.8024471963	.8178135215	.8321492852	.8455009039	.8579162134
80+	.8694437369	.8801320690	.8900293668	.8991829337	.9076388877
85+	.9154419007	.9226350019	.9292594339	.9353545549	.9409577797
90+	.9461045025	.9508283457	.9551606810	.9591311669	.9627675503
95+	.9660957778	.9691400654	.9719229732	.9744654838	.9767870821
100+	.9789058371	.9808384817	.9826004929	.9842061692	.9856687056
105+	.9870002668	.9882120566	.9893143843	.9903167279	.9912277942
* 130+	.9989654670	.9990680266	.9991605945	.9992441241	.9993194799
			P=1/ 8, R= 6		
80+	.6950201402	.7155900343	.7350963001	.7535513127	.7709741259
85+	.7873893322	.8028260237	.8173168524	.8308971936	.8436044051
90+	.8554771808	.8665549910	.8768776029	.8864846756	.8954154214
95+	.9037083267	.9114009268	.9185296268	.9251295639	.9312345053
100+	.9368767762	.9420872149	.9468951485	.9513283887	.9554132414
105+	.9591745303	.9626356289	.9658185021	.9687437529	.9714306738
110+	.9738973016	.9761604747	.9782358904	.9801381642	.9818808873
115+	.9834766843	.9849372696	.9862735015	.9874954357	.9886123760
120+	.9896329229	.9905650207	.9914160012	.9921926265	.9929011285
* 145+	.9990946159	.9991813258	.9992598888	.9993310529	.9993954996
			P=1/ 8, R= 7		
90+	.6869155629	.7068124354	.7257603792	.7437655627	.7608400829
95+	.7770010759	.7922698911	.8066713331	.8202329741	.8329845345
100+	.8449573316	.8561837930	.8666970301	.8765304698	.8857175376
105+	.8942913897	.9022846878	.9097294139	.9166567190	.9230968043
110+	.9290788278	.9346308364	.9397797172	.9445511676	.9489696798
115+	.9530585385	.9568398291	.9603344549	.9635621610	.9665415653
120+	.9692901922	.9718245115	.9741599787	.9763110776	.9782913640
125+	.9801135093	.9817893451	.9833299066	.9847454759	.9860456239
130+	.9872392513	.9883346283	.9893394332	.9902607882	.9911052956
* 155+	.9987583474	.9988725709	.9989765167	.9990710851	.9991571003
			P=1/ 8, R= 8		
100+	.6828334194	.7020007433	.7203116959	.7377690982	.7543809372
105+	.7701596627	.7851215270	.7992859721	.8126750658	.8253129879
110+	.8372255665	.8484398622	.8589837987	.8688858378	.8781746955
115+	.8868790963	.8950275629	.9026482381	.9097687356	.9164160174
120+	.9226162944	.9283949480	.9337764700	.9387844187	.9434413887
125+	.9477689932	.9517878564	.9555176152	.9589769274	.9621834872
130+	.9651540448	.9679044310	.9704495844	.9728035810	.9749796668
135+	.9769902901	.9788471361	.9805611608	.9821426257	.9836011320
140+	.9849456545	.9861845745	.9873257125	.9883763595	.9893433071
145+	.9902328775	.9910509509	.9918029931	.9924940810	.9931289272

DIRICHLET 1 PROBABILITY, B= 6

N	0	1	2	3	4
		P=1/ 8,	R= 8		
* 170+	.9989898014	.9990803836	.9991630274	.9992384098	.9993071516
		P=1/ 8,	R= 9		
110+	.6816818884	.7001076551	.7177525670	.7346177694	.7507088745
115+	.7660353974	.7806102232	.7944491071	.8075702108	.8199936762
120+	.8317412356	.8428358594	.8533014397	.8631625077	.8724439841
125+	.8811709598	.8893685056	.8970615073	.9042745263	.9110316806
130+	.9173565473	.9232720816	.9288005526	.9339634932	.9387816628
135+	.9432750199	.9474627059	.9513630363	.9549934990	.9583707588
140+	.9615106676	.9644282781	.9671378616	.9696529282	.9719862496
145+	.9741498840	.9761552011	.9780129097	.9797330845	.9813251935
150+	.9827981262	.9841602208	.9854192911	.9865826533	.9876571522
155+	.9886491862	.9895647323	.9904093691	.9911883003	.9919063766
* 180+	.9987240259	.9988346330	.9989358917	.9990285684	.9991133691
		P=1/ 8,	R=10		
125+	.7640098726	.7781534022	.7916148622	.8044096152	.8165548318
130+	.8280691632	.8389724428	.8492854154	.8590294940	.8682265418
135+	.8768986795	.8850681157	.8927569982	.8999872860	.9067806392
140+	.9131583263	.9191411455	.9247493614	.9300026526	.9349200714
145+	.9395200128	.9438201919	.9478376300	.9515886461	.9550888546
150+	.9583531686	.9613958061	.9642303008	.9668695158	.9693256589
155+	.9716103011	.9737343955	.9757082982	.9775417896	.9792440967
160+	.9808239157	.9822894340	.9836483536	.9849079128	.9860749090
165+	.9871557199	.9881563251	.9890823266	.9899389688	.9907311584
* 195+	.9989991289	.9990842162	.9991622519	.9992338030	.9992993923
		P=1/ 9,	R= 1		
25+	.7114886689	.7406060064	.7671222046	.7911916779	.8129795128
30+	.8326547814	.8503855979	.8663355719	.8806613674	.8935111168
35+	.9050234905	.9153272550	.9245411925	.9327742772	.9401260304
40+	.9466869934	.9525392714	.9577571161	.9624075193	.9665508024
45+	.9702411873	.9735273413	.9764528902	.9790568981	.9813743109
50+	.9834363646	.9852709595	.9869030006	.9883547076	.9896458945
55+	.9907942237	.9918154332	.9927235426	.9935310369	.9942490320
* 70+	.9984246394	.9985996413	.9987552070	.9988934942	.9990164214
		P=1/ 9,	R= 2		
40+	.7040925818	.7288968591	.7519739098	.7733817117	.7931890588
45+	.8114723177	.8283127191	.8437941305	.8580012582	.8710182198
50+	.8829274357	.8938087904	.9037390195	.9127912819	.9210348841
55+	.9285351270	.9353532487	.9415464434	.9471679383	.9522671137
60+	.9568896540	.9610777212	.9648701405	.9683025953	.9714078238
65+	.9742158144	.9767539978	.9790474329	.9811189851	.9829894959
70+	.9846779437	.9862015954	.9875761472	.9888158577	.9899336698
75+	.9909413253	.9918494696	.9926677491	.9934049003	.9940688322
* 95+	.9989325343	.9990419153	.9991401692	.9992284181	.9993076721
		P=1/ 9,	R= 3		
55+	.7345617868	.7547748819	.7737104133	.7914096605	.8079194307
60+	.8232905506	.8375765715	.8508326691	.8631147221	.8744785505
65+	.8849792973	.8946709328	.9036058689	.9118346651	.9194058139
70+	.9263655927	.9327579715	.9386245659	.9440046278	.9489350661
75+	.9534504910	.9575832767	.9613636380	.9648197174	.9679776785
80+	.9708618042	.9734945973	.9758968812	.9780879004	.9800854188
85+	.9819058152	.9835641759	.9850743836	.9864492016	.9877003548
90+	.9888386052	.9898738246	.9908150614	.9916706040	.9924480404
* 115+	.9990738077	.9991632767	.9992442017	.9993173876	.9993835645
		P=1/ 9,	R= 4		
65+	.6735922223	.6953291027	.7159687815	.7355188570	.7539948840
70+	.7714190556	.7878189951	.8032266656	.8176773946	.8312090120
75+	.8438610966	.8556743237	.8666899072	.8769491270	.8864929346

DIRICHLET 1 PROBABILITY, B= 6

N	0	1	2	3	4
			P=1/ 9,	R= 4	
80+	.8953616280	.9035945882	.9112300702	.9183050413	.9248550603
85+	.9309141917	.9365149501	.9416882697	.9464634945	.9508683863
90+	.9549291469	.9586704509	.9621154886	.9652860147	.9682024027
95+	.9708837028	.9733477020	.9756109855	.9776889989	.9795961095
100+	.9813456673	.9829500644	.9844207926	.9857684998	.9870030431
105+	.9881335407	.9891684208	.9901154683	.9909818693	.9917742526
* 130+	.9988619489	.9989662513	.9990611401	.9991474491	.9992259404
			P=1/ 9,	R= 5	
80+	.7274157932	.7450460631	.7617683780	.7776006919	.7925648084
85+	.8066856673	.8199906991	.8325092460	.8442720486	.8553107940
90+	.8656577218	.8753452839	.8844058543	.8928714835	.9007736935
95+	.9081433102	.9150103271	.9214037984	.9273517573	.9328811556
100+	.9380178228	.9427864412	.9472105348	.9513124698	.9551134648
105+	.9586336090	.9618918871	.9649062089	.9676934434	.9702694556
110+	.9726491459	.9748464907	.9768745836	.9787456778	.9804712278
115+	.9820619304	.9835277660	.9848780379	.9861214114	.9872659505
120+	.9883191547	.9892879929	.9901789369	.9909979927	.9917507309
* 145+	.9987447022	.9988547191	.9989552693	.9990471496	.9991310915
			P=1/ 9,	R= 6	
90+	.6903694159	.7086513185	.7261222593	.7427871163	.7586551556
95+	.7737394229	.7880561739	.8016243468	.8144650790	.8266012668
100+	.8380571682	.8488580480	.8590298620	.8685989799	.8775919424
105+	.8860352521	.8939551935	.9013776810	.9083281317	.9148313605
110+	.9209114951	.9265919092	.9318951706	.9368430041	.9414562656
115+	.9457549268	.9497580690	.9534838845	.9569496841	.9601719102
120+	.9631661543	.9659471780	.9685289370	.9709246072	.9731466124
125+	.9752066532	.9771157372	.9788842085	.9805217783	.9820375552
130+	.9834400747	.9847373286	.9859367934	.9870454587	.9880698535
135+	.9890160727	.9898898021	.9906963422	.9914406318	.9921272696
* 160+	.9987058302	.9988149626	.9989150811	.9990069117	.9990911238
			P=1/ 9,	R= 7	
105+	.7448087421	.7597844836	.7740613810	.7876525787	.8005734390
110+	.8128411749	.8244745130	.8354933861	.8459186551	.8557718589
115+	.8650749922	.8738503079	.8821201443	.8899067737	.8972322728
120+	.9041184111	.9105865572	.9166576001	.9223518854	.9276891634
125+	.9326885489	.9373684910	.9417467512	.9458403900	.9496657592
130+	.9532385008	.9565735505	.9596851451	.9625868342	.9652914939
135+	.9678113435	.9701579643	.9723423188	.9743747724	.9762651152
140+	.9780225842	.9796558866	.9811732219	.9825823054	.9838903905
145+	.9851042908	.9862304023	.9872747242	.9882428800	.9891401373
150+	.9899714272	.9907413631	.9914542586	.9921141446	.9927247859
* 175+	.9987250486	.9988290017	.9989246682	.9990126907	.9990936641
			P=1/ 9,	R= 8	
115+	.7195013311	.7348808971	.7496148642	.7637100263	.7771756112
120+	.7900229571	.8022652077	.8139170299	.8249943516	.8355141227
125+	.8454940954	.8549526272	.8639085018	.8723807708	.8803886119
130+	.8879512048	.8950876222	.9018167362	.9081571366	.9141270630
135+	.9197443471	.9250263651	.9299899997	.9346516097	.9390270069
140+	.9431314399	.9469795832	.9505855310	.9539627961	.9571243114
145+	.9600824361	.9628489630	.9654351294	.9678516290	.9701086263
150+	.9722157709	.9741822141	.9760166257	.9777272110	.9793217286
155+	.9808075084	.9821914691	.9834801364	.9846796602	.9857958324
160+	.9868341036	.9877996000	.9886971391	.9895312460	.9903061686
* 190+	.9987837505	.9988800181	.9989688468	.9990507949	.9991263805
			P=1/ 9,	R= 9	
125+	.6973868856	.7129869283	.7279937105	.7424091174	.7562375278
130+	.7694855354	.7821616826	.7942762078	.8058408093	.8168684229

DIRICHLET 1 PROBABILITY, B= 6

N	0	1	2	3	4
		P=1/ 9,	R= 9		
135+	.8273730175	.8373694057	.8468730701	.8559000065	.8644665803
140+	.8725893986	.8802851951	.8875707288	.8944626937	.9009776403
145+	.9071319079	.9129415656	.9184223626	.9235896867	.9284585298
150+	.9330434601	.9373585999	.9414176094	.9452336747	.9488194997
155+	.9521873028	.9553488153	.9583152842	.9610974758	.9637056828
160+	.9661497325	.9684389962	.9705824004	.9725884387	.9744651844
165+	.9762203041	.9778610711	.9793943798	.9808267599	.9821643905
170+	.9834131146	.9845784536	.9856656206	.9866795350	.9876248358
175+	.9885058945	.9893268287	.9900915139	.9908035962	.9914665036
* 205+	.9988668685	.9989542222	.9990350087	.9991097072	.9991787629
		P=1/ 9,	R=10		
140+	.7510553685	.7640295169	.7764726997	.7883930797	.7998001815
145+	.8107047034	.8211183430	.8310536336	.8405237941	.8495425901
150+	.8581242082	.8662831404	.8740340810	.8813918325	.8883712233
155+	.8949870333	.9012539289	.9071864064	.9127987427	.9181049529
160+	.9231187546	.9278535379	.9323223408	.9365378296	.9405122829
165+	.9442575806	.9477851956	.9511061889	.9542312072	.9571704834
170+	.9599338387	.9625306864	.9649700381	.9672605105	.9694103332
175+	.9714273582	.9733190695	.9750925935	.9767547101	.9783118635
180+	.9797701745	.9811354513	.9824132018	.9836086454	.9847267241
185+	.9857721146	.9867492395	.9876622786	.9885151796	.9893116692
190+	.9900552632	.9907492768	.9913968342	.9920008784	.9925641802
* 220+	.9989630412	.9990411140	.9991134595	.9991804849	.9992425695
		P=1/10,	R= 1		
30+	.7637746747	.7856294619	.8056435020	.8239345489	.8406210654
35+	.8558198865	.8696445010	.8822038337	.8936014250	.9039349254
40+	.9132958308	.9217694042	.9294347355	.9363649018	.9426271992
45+	.9482834222	.9533901733	.9579991879	.9621576648	.9659085939
50+	.9692910763	.9723406319	.9750894921	.9775668765	.9797992523
55+	.9818105753	.9836225147	.9852546595	.9867247097	.9880486511
60+	.9892409155	.9903145272	.9912812369	.9921516430	.9929353025
* 80+	.9986894180	.9988204497	.9989383835	.9990445282	.9991400618
		P=1/10,	R= 2		
45+	.7131827546	.7348019047	.7550675623	.7740207927	.7917093807
50+	.8081860317	.8235068312	.8377299461	.8509145416	.8631198915
55+	.8744046590	.8848263253	.8944407454	.9033018128	.9114612161
60+	.9189682718	.9258698210	.9322101773	.9380311178	.9433719060
65+	.9482693424	.9527578341	.9568694796	.9606341650	.9640796661
70+	.9672317564	.9701143162	.9727494417	.9751575542	.9773575057
75+	.9793666816	.9812011005	.9828755082	.9844034685	.9857974490
80+	.9870689016	.9882283387	.9892854050	.9902489439	.9911270604
* 105+	.9988081958	.9989179487	.9990176705	.9991082696	.9991905732
		P=1/10,	R= 3		
60+	.7094599891	.7288785476	.7472567089	.7646153147	.7809803565
65+	.7963819513	.8108534259	.8244305083	.8371506202	.8490522643
70+	.8601744994	.8705564956	.8802371619	.8892548388	.8976470478
75+	.9054502919	.9126999005	.9194299120	.9256729893	.9314603642
80+	.9368218040	.9417855993	.9463785674	.9506260694	.9545520379
85+	.9581790139	.9615281896	.9646194567	.9674714582	.9701016427
90+	.9725263197	.9747607165	.9768190339	.9787145022	.9804594357
95+	.9820652860	.9835426935	.9849015374	.9861509834	.9872995293
100+	.9883550488	.9893248326	.9902156282	.9910336765	.9917847471
* 125+	.9987352027	.9988446320	.9989447101	.9990362243	.9991198967
		P=1/10,	R= 4		
75+	.7222558099	.7394113072	.7557230163	.7712063186	.7858800334
80+	.7997658097	.8128875742	.8252710336	.8369432309	.8479321515
85+	.8582663792	.8679747952	.8770863197	.8856296912	.8936332803

DIRICHLET 1 PROBABILITY, B= 6

N	0	1	2	3	4
			P=1/10, R= 4		
90+	.9011249347	.9081318516	.9146804758	.9207964188	.9265043975
95+	.9318281898	.9367906045	.9414134634	.9457175949	.9497228355
100+	.9534480393	.9569110942	.9601289423	.9631176046	.9658922097
105+	.9684670235	.9708554821	.9730702247	.9751231283	.9770253420
110+	.9787873215	.9804188633	.9819291389	.9833267272	.9846196470
115+	.9858153881	.9869209416	.9879428288	.9888871290	.9897595062
120+	.9905652347	.9913092230	.9919960371	.9926299223	.9932148238
* 145+	.9988475700	.9989424936	.9990297195	.9991098603	.9991834810
			P=1/10, R= 5		
90+	.7413930644	.7564986366	.7708970011	.7846019058	.7976293268
95+	.8099970883	.8217245139	.8328321111	.8433412864	.8532740908
100+	.8626529938	.8715006843	.8798398969	.8876932604	.8950831695
105+	.9020316740	.9085603870	.9146904089	.9204422653	.9258358578
110+	.9308904268	.9356245228	.9400559883	.9442019456	.9480787925
115+	.9517022027	.9550871314	.9582478253	.9611978348	.9639500305
120+	.9665166208	.9689091718	.9711386280	.9732153351	.9751490620
125+	.9769490246	.9786239089	.9801818947	.9816306785	.9829774964
130+	.9842291473	.9853920138	.9864720849	.9874749756	.9884059480
135+	.9892699299	.9900715339	.9908150753	.9915045889	.9921438456
* 165+	.9990306779	.9991073985	.9991781534	.9992433967	.9993035490
			P=1/10, R= 6		
105+	.7626431315	.7759384778	.7886297440	.8007299533	.8122535527
110+	.8232161643	.8336343586	.8435254474	.8529072968	.8617981593
115+	.8702165239	.8781809834	.8857101171	.8928223894	.8995360612
120+	.9058691141	.9118391866	.9174635205	.9227589164	.9277416982
125+	.9324276856	.9368321720	.9409699102	.9448551021	.9485013936
130+	.9519218734	.9551290754	.9581349837	.9609510409	.9635881578
135+	.9660567252	.9683666270	.9705272548	.9725475230	.9744358845
140+	.9762003476	.9778484924	.9793874878	.9808241082	.9821647509
145+	.9834154522	.9845819040	.9856694702	.9866832020	.9876278537
150+	.9885078971	.9893275366	.9900907223	.9908011642	.9914623445
* 180+	.9988444983	.9989322626	.9990134964	.9990886738	.9991582354
			P=1/10, R= 7		
115+	.7173372091	.7317929084	.7456850519	.7590186006	.7718004502
120+	.7840391949	.7957449046	.8069289163	.8176036390	.8277823732
125+	.8374791449	.8467085532	.8554856320	.8638257251	.8717443729
130+	.8792572123	.8863798869	.8931279687	.8995168882	.9055618756
135+	.9112779082	.9166796675	.9217815018	.9265973961	.9311409475
140+	.9354253455	.9394633576	.9432673183	.9468491225	.9502202217
145+	.9533916231	.9563738913	.9591771520	.9618110979	.9642849956
150+	.9666076946	.9687876362	.9708328648	.9727510386	.9745494419
155+	.9762349966	.9778142756	.9792935148	.9806786262	.9819752105
160+	.9831885702	.9843237217	.9853854081	.9863781111	.9873060632
165+	.9881732592	.9889834672	.9897402404	.9904469268	.9911066802
* 195+	.9986986080	.9987938869	.9988823482	.9989644663	.9990406840
			P=1/10, R= 8		
130+	.7459098163	.7586358506	.7708625718	.7825969421	.7938471823
135+	.8046226084	.8149334766	.8247908393	.8342064118	.8431924487
140+	.8517616301	.8599269588	.8677016657	.8750991249	.8821327774
145+	.8888160627	.8951623585	.9011849272	.9068968689	.9123110814
150+	.9174402251	.9222966934	.9268925881	.9312396988	.9353494867
155+	.9392330716	.9429012224	.9463643501	.9496325038	.9527153687
160+	.9556222663	.9583621561	.9609436388	.9633749616	.9656640233
165+	.9678183814	.9698452598	.9717515568	.9735438538	.9752284249
170+	.9768112459	.9782980047	.9796941103	.9810047036	.9822346670
175+	.9833886346	.9844710016	.9854859349	.9864373819	.9873290805
180+	.9881645680	.9889471904	.9896801110	.9903663190	.9910086377

DIRICHLET 1 PROBABILITY, B= 6

N	0	1	2	3	4
		P=1/10'	R= 8		
* 215+	.9990271469	.9990966943	.9991613846	.9992215476	.9992774915
		P=1/10,	R= 9		
140+	.7093061853	.7228139207	.7358558837	.7484340055	.7605517137
145+	.7722137818	.7834261847	.7941959602	.8045310785	.8144403183
150+	.8239331511	.8330196326	.8417103027	.8500160922	.8579482372
155+	.8655182016	.8727376051	.8796181591	.8861716082	.8924096783
160+	.8983440303	.9039862184	.9093476543	.9144395759	.9192730195
165+	.9238587969	.9282074756	.9323293624	.9362344904	.9399326079
170+	.9434331708	.9467453367	.9498779606	.9528395937	.9556384822
175+	.9582825686	.9607794940	.9631366011	.9653609386	.9674592662
180+	.9694380605	.9713035207	.9730615760	.9747178923	.9762778798
185+	.9777467008	.9791292773	.9804302990	.9816542315	.9828053241
190+	.9838876178	.9849049536	.9858609800	.9867591608	.9876027827
195+	.9883949632	.9891386572	.9898366644	.9904916366	.9911060839
* 230+	.9989688415	.9990404118	.9991071372	.9991693358	.9992273059
		P=1/10,	R=10		
155+	.7406384287	.7525848128	.7641081742	.7752128758	.7859042847
160+	.7961886617	.8060730549	.8155652003	.8246734273	.8334065701
165+	.8417738860	.8497849785	.8574497268	.8647782207	.8717807009
170+	.8784675049	.8848490177	.8909356272	.8967376841	.9022654667
175+	.9075291486	.9125387714	.9173042199	.9218352012	.9261412263
180+	.9302315950	.9341153831	.9378014315	.9412983384	.9446144524
185+	.9477578681	.9507364226	.9535576944	.9562290022	.9587574062
190+	.9611497094	.9634124601	.9655519554	.9675742450	.9694851355
195+	.9712901959	.9729947624	.9746039445	.9761226308	.9775554953
200+	.9789070037	.9801814196	.9813828117	.9825150598	.9835818614
205+	.9845867387	.9855330446	.9864239693	.9872625469	.9880516612
210+	.9887940521	.9894923216	.9901489393	.9907662486	.9913464717
* 245+	.9989340731	.9990060917	.9990733710	.9991362133	.9991949020

DIRICHLET 1 PROBABILITY,   B= 7

N	0	1	2	3	4
			P=1/ 8, R= 1		
25+	.7670438184	.7942565489	.8185358557	.8401345615	.8593011613
30+	.8762739868	.8912775310	.9045203642	.9161941994	.9264737644
35+	.9355172232	.9434669545	.9504505450	.9565818977	.9619623830
40+	.9666819828	.9708203963	.9744480853	.9776272488	.9804127204
45+	.9828527869	.9849899297	.9868614937	.9885002854	.9899351083
50+	.9911912385	.9922908477	.9932533771	.9940958690	.9948332583
* 65+	.9988099547	.9989586905	.9990888393	.9992027232	.9993023744
			P=1/ 8, R= 2		
35+	.6472742770	.6798669326	.7100595100	.7379034064	.7634798644
40+	.7868914890	.8082552267	.8276966775	.8453455807	.8613323090
45+	.8757852141	.8888286740	.9005817143	.9111570875	.9206607145
50+	.9291914077	.9368408076	.9436934793	.9498271252	.9553128772
55+	.9602156438	.9645944878	.9685030207	.9719898000	.9750987209
60+	.9778693965	.9803375206	.9825352116	.9844913352	.9862318052
65+	.9877798619	.9891563290	.9903798494	.9914671009	.9924329934
* 85+	.9989175040	.9990425070	.9991531746	.9992511371	.9993378415
			P=1/ 8, R= 3		
50+	.7282587841	.7518582182	.7737389222	.7939687715	.8126244020
55+	.8297882612	.8455461669	.8599853190	.8731927108	.8852538830
60+	.8962519729	.9062670083	.9153754078	.9236496485	.9311580698
65+	.9379647867	.9441296874	.9497084983	.9547528975	.9593106654
70+	.9634258612	.9671390152	.9704873317	.9735048961	.9762228809
75+	.9786697499	.9808714558	.9828516300	.9846317641	.9862313819
80+	.9876682012	.9889582853	.9901161846	.9911550684	.9920868463
* 100+	.9987074321	.9988477508	.9989730139	.9990848143	.9991845791
			P=1/ 8, R= 4		
60+	.6901506401	.7142118671	.7367936286	.7579257799	.7776490868
65+	.7960127739	.8130723803	.8288879096	.8435222573	.8570398899
70+	.8695057532	.8809843831	.8915391940	.9012319218	.9101221992
75+	.9182672428	.9257216356	.9325371863	.9387628549	.9444447286
80+	.9496260412	.9543472244	.9586459846	.9625573986	.9661140238
85+	.9693460166	.9722812570	.9749454767	.9773623869	.9795538055
90+	.9815397812	.9833387143	.9849674726	.9864415026	.9877749348
95+	.9889806843	.9900705443	.9910552748	.9919446859	.9927477143
* 115+	.9986937880	.9988287050	.9989498950	.9990587295	.9991564462
			P=1/ 8, R= 5		
70+	.6655648091	.6893145086	.7117890451	.7329972498	.7529586091
75+	.7717013664	.7892608045	.8056777094	.8209970165	.8352666293
80+	.8485364028	.8608572776	.8722805515	.8828572760	.8926377627
85+	.9016711873	.9100052783	.9176860791	.9247577731	.9312625623
90+	.9372405905	.9427299038	.9477664426	.9523840575	.9566145461
95+	.9604877052	.9640313950	.9672716123	.9702325705	.9729367830
100+	.9754051502	.9776570463	.9797104065	.9815818129	.9832865789
105+	.9848388301	.9862515834	.9875368221	.9887055678	.9897679493
110+	.9907332665	.9916100523	.9924061297	.9931286658	.9937842225
* 130+	.9987930653	.9989127139	.9990207065	.9991181550	.9992060681
			P=1/ 8, R= 6		
85+	.7545075579	.7721110890	.7886728966	.8042254059	.8188045072
90+	.8324486753	.8451981997	.8570945168	.8681796387	.8784956680
95+	.8880843933	.8969869540	.9052435685	.9128933185	.9199739804
100+	.9265218996	.9325719006	.9381572275	.9433095111	.9480587567
105+	.9524333513	.9564600838	.9601641788	.9635693380	.9666977905
110+	.9695703469	.9722064589	.9746242799	.9768407280	.9788715497
115+	.9807313825	.9824338176	.9839914611	.9854159931	.9867182256
120+	.9879081579	.9889950293	.9899873703	.9908930506	.9917193248
* 145+	.9989437332	.9990448919	.9991365464	.9992195693	.9992947556

DIRICHLET 1 PROBABILITY, B= 7

N	0	1	2	3	4
			P=1/ 8, R= 7		
95+	.7426617019	.7600484998	.7764847767	.7919944441	.8066049163
100+	.8203463697	.8332510817	.8453528474	.8566864676	.8672873057
105+	.8771909066	.8864326735	.8950475960	.9030700257	.9105334923
110+	.9174705575	.9239127011	.9298902346	.9354322404	.9405665310
115+	.9453196265	.9497167475	.9537818200	.9575374915	.9610051550
120+	.9642049805	.9671559511	.9698759040	.9723815741	.9746886402
125+	.9768117716	.9787646766	.9805601501	.9822101215	.9837257018
130+	.9851172291	.9863943139	.9875658823	.9886402175	.9896250001
135+	.9905273465	.9913538455	.9921105932	.9928032256	.9934369505
* 155+	.9985514338	.9986846892	.9988059548	.9989162811	.9990166293
			P=1/ 8, R= 8		
105+	.7348646027	.7518854939	.7680356288	.7833331679	.7977995938
110+	.8114590937	.8243380007	.8364642928	.8478671470	.8585765471
115+	.8686229395	.8780369365	.8868490607	.8950895286	.9027880688
120+	.9099737723	.9166749699	.9229191353	.9287328097	.9341415460
125+	.9391698694	.9438412523	.9481781022	.9522017592	.9559325033
130+	.9593895682	.9625911615	.9655544902	.9682957894	.9708303542
135+	.9731725744	.9753359701	.9773332286	.9791762419	.9808761445
140+	.9824433507	.9838875917	.9852179518	.9864429039	.9875703442
145+	.9886076253	.9895615882	.9904385936	.9912445510	.9919849473
* 170+	.9988214527	.9989271286	.9990235437	.9991114876	.9991916846
			P=1/ 8, R= 9		
115+	.7301620891	.7467332607	.7625026768	.7774846191	.7916964186
120+	.8051579406	.8178911144	.8299195064	.8412679368	.8519621363
125+	.8620284434	.8714935378	.8803842079	.8887271509	.8965488012
130+	.9038751855	.9107318021	.9171435213	.9231345056	.9287281462
135+	.9339470156	.9388128320	.9433464367	.9475677806	.9514959191
140+	.9551490150	.9585443471	.9616983237	.9646265010	.9673436043
145+	.9698635524	.9721994835	.9743637839	.9763681161	.9782234495
150+	.9799400897	.9815277096	.9829953786	.9843515932	.9856043058
155+	.9867609535	.9878284858	.9888133915	.9897217251	.9905591316
* 185+	.9990561115	.9991388806	.9992145596	.9992837391	.9993469623
			P=1/ 8, R=10		
125+	.7278465169	.7439232053	.7592577935	.7738619465	.7877500776
130+	.8009389178	.8134471190	.8252948907	.8365036713	.8470958315
135+	.8570944089	.8665228726	.8754049157	.8837642739	.8916245682
140+	.8990091697	.9059410852	.9124428604	.9185365005	.9242434054
145+	.9295843180	.9345792846	.9392476259	.9436079168	.9476779743
150+	.9514748520	.9550148415	.9583134776	.9613855485	.9642451098
155+	.9669055003	.9693793616	.9716786587	.9738147023	.9757981718
160+	.9776391401	.9793470976	.9809309773	.9823991794	.9837595963
165+	.9850196366	.9861862493	.9872659470	.9882648289	.9891886028
170+	.9900426065	.9908318286	.9915609279	.9922342529	.9928558602
* 195+	.9988323337	.9989315992	.9990226385	.9991061127	.9991826319
			P=1/ 9, R= 1		
25+	.6694868366	.7021549284	.7320613224	.7593308974	.7841115045
30+	.8065643998	.8268569188	.8451570025	.8616292232	.8764320020
35+	.8897157580	.9016217722	.9122815922	.9218168369	.9303392905
40+	.9379511993	.9447457058	.9508073680	.9562127276	.9610308989
45+	.9653241572	.9691485140	.9725542685	.9755865298	.9782857051
50+	.9806879540	.9828256067	.9847275478	.9864195667	.9879246762
55+	.9892634016	.9904540410	.9915129016	.9924545113	.9932918090
* 75+	.9989799991	.9990933173	.9991940480	.9992835890	.9993631831
			P=1/ 9, R= 2		
40+	.6610055206	.6888126942	.7148020708	.7390089858	.7614861146
45+	.7822991587	.8015231656	.8192394415	.8355330018	.8504905013
50+	.8641985835	.8767425893	.8882055715	.8986675657	.9082050719

DIRICHLET 1 PROBABILITY,   B= 7

N	0	1	2	3	4
			P=1/ 9, R= 2		
55+	.9168907102	.9247930155	.9319763430	.9385008618	.9444226142
60+	.9497936270	.9546620576	.9590723680	.9630655137	.9666791439
65+	.9699478052	.9729031450	.9755741128	.9779871545	.9801663999
70+	.9821338417	.9839095046	.9855116049	.9869567003	.9882598292
75+	.9894346404	.9904935140	.9914476722	.9923072822	.9930815506
* 95+	.9987546549	.9988822595	.9989968839	.9990998368	.9991922966
			P=1/ 9, R= 3		
55+	.6950886578	.7178826552	.7393107991	.7594028128	.7781972957
60+	.7957397188	.8120806718	.8272743504	.8413772689	.8544471762
65+	.8665421579	.8777199021	.8880371096	.8975490294	.9063091022
70+	.9143686958	.9217769176	.9285804929	.9348236950	.9405483208
75+	.9457937001	.9505967338	.9549919531	.9590115963	.9626856964
80+	.9660421785	.9691069621	.9719040665	.9744557176	.9767824542
85+	.9789032325	.9808355284	.9825954357	.9841977610	.9856561146
90+	.9869829963	.9881898769	.9892872755	.9902848315	.9911913723
* 115+	.9989194641	.9990238403	.9991182494	.9992036301	.9992808343
			P=1/ 9, R= 4		
70+	.7366665054	.7552802528	.7728136332	.7892970711	.8047652688
75+	.8192561889	.8328101608	.8454691066	.8572758748	.8682736766
80+	.8785056114	.8880142752	.8968414412	.9050278039	.9126127801
85+	.9196343565	.9261289803	.9321314845	.9376750427	.9427911490
90+	.9475096189	.9518586066	.9558646354	.9595526401	.9629460161
95+	.9660666760	.9689351100	.9715704493	.9739905316	.9762119671
100+	.9782502043	.9801195955	.9818334605	.9834041490	.9848431008
105+	.9861609038	.9873673495	.9884714860	.9894816685	.9904056072
* 130+	.9986723061	.9987939862	.9989046848	.9990053747	.9990969446
			P=1/ 9, R= 5		
80+	.6869385619	.7068013055	.7256998330	.7436433482	.7606467886
85+	.7767299044	.7919164128	.8062332307	.8197097848	.8323773971
90+	.8442687430	.8554173778	.8658573263	.8756227317	.8847475572
95+	.8932653365	.9012089676	.9086105446	.9155012248	.9219111245
100+	.9278692416	.9334034008	.9385402175	.9433050787	.9477221377
105+	.9518143206	.9556033417	.9591097272	.9623528456	.9653509419
110+	.9681211766	.9706796667	.9730415291	.9752209249	.9772311047
115+	.9790844532	.9807925340	.9823661341	.9838153062	.9851494115
120+	.9863771594	.9875066473	.9885453979	.9895003949	.9903781179
* 150+	.9990757418	.9991574339	.9992320264	.9993001241	.9993622814
			P=1/ 9, R= 6		
95+	.7392360483	.7554955328	.7709399595	.7855868699	.7994566279
100+	.8125718920	.8249571377	.8366382292	.8476420382	.8579961074
105+	.8677283558	.8768668233	.8854394514	.8934738957	.9009973696
110+	.9080365137	.9146172895	.9207648950	.9265036977	.9318571860
115+	.9368479333	.9414975759	.9458268004	.9498553415	.9536019866
120+	.9570845870	.9603200749	.9633244846	.9661129765	.9686998650
125+	.9710986474	.9733220351	.9753819851	.9772897331	.9790558254
130+	.9806901523	.9822019798	.9835999821	.9848922721	.9860864323
135+	.9871895439	.9882082155	.9891486100	.9900164717	.9908171507
* 165+	.9990297390	.9991123104	.9991879880	.9992573344	.9993208675
			P=1/ 9, R= 7		
105+	.7064739530	.7234012106	.7395796932	.7550172970	.7697252169
110+	.7837174721	.7970104649	.8096225743	.8215737850	.8328853500
115+	.8435794874	.8536791089	.8632075795	.8721885051	.8806455479
120+	.8886022654	.8960819737	.9031076301	.9097017357	.9158862542
125+	.9216825464	.9271113183	.9321925811	.9369456216	.9413889823
130+	.9455404492	.9494170468	.9530350391	.9564099355	.9595565010
135+	.9624887701	.9652200637	.9677630078	.9701295550	.9723310064
140+	.9743780354	.9762807118	.9780485263	.9796904153	.9812147858

DIRICHLET 1 PROBABILITY, B= 7

N	0	1	2	3	4
			P=1/ 9, R= 7		
145+	.9826295401	.9839420999	.9851594300	.9862880620	.9873341166
150+	.9883033259	.9892010541	.9900323185	.9908018090	.9915139068
* 180+	.9990295094	.9991094036	.9991828569	.9992503757	.9993124278
			P=1/ 9, R= 8		
120+	.7576973978	.7716389737	.7849330468	.7975935519	.8096360788
125+	.8210775811	.8319361103	.8422305741	.8519805183	.8612059322
130+	.8699270744	.8781643204	.8859380276	.8932684186	.9001754811
135+	.9066788812	.9127978918	.9185513323	.9239575192	.9290342270
140+	.9337986577	.9382674181	.9424565039	.9463812901	.9500565261
145+	.9534963360	.9567142225	.9597230737	.9625351735	.9651622139
150+	.9676153090	.9699050112	.9720413277	.9740337390	.9758912170
155+	.9776222445	.9792348344	.9807365492	.9821345203	.9834354674
160+	.9846457174	.9857712233	.9868175822	.9877900532	.9886935747
165+	.9895327810	.9903120187	.9910353623	.9917066286	.9923293919
* 195+	.9990621115	.9991370868	.9992062011	.9992699003	.9993285978
			P=1/ 9, R= 9		
130+	.7343555450	.7487449813	.7625249277	.7757042245	.7882935937
135+	.8003053709	.8117532562	.8226520832	.8330176076	.8428663137
140+	.8522152391	.8610818171	.8694837348	.8774388074	.8849648665
145+	.8920796630	.8988007813	.9051455666	.9111310617	.9167739544
150+	.9220905332	.9270966513	.9318076980	.9362385760	.9404036849
155+	.9443169096	.9479916135	.9514406349	.9546762879	.9577103649
160+	.9605541433	.9632183927	.9657133850	.9680489057	.9702342663
165+	.9722783178	.9741894649	.9759756812	.9776445238	.9792031493
170+	.9806583294	.9820164663	.9832836084	.9844654659	.9855674257
175+	.9865945664	.9875516732	.9884432515	.9892735413	.9900465304
* 205+	.9986780415	.9987799495	.9988741967	.9989613417	.9990419041
			P=1/ 9, R=10		
140+	.7134739272	.7281571945	.7422702593	.7558177831	.7688064214
145+	.7812445827	.7931422010	.8045105231	.8153619089	.8257096478
150+	.8355677886	.8449509841	.8538743495	.8623533342	.8704036064
155+	.8780409497	.8852811714	.8921400211	.8986331198	.9047758979
160+	.9105835418	.9160709482	.9212526862	.9261429647	.9307556073
165+	.9351040308	.9392012298	.9430597649	.9466917549	.9501088725
170+	.9533223421	.9563429416	.9591810052	.9618464286	.9643486757
175+	.9666967874	.9688993902	.9709647068	.9729005674	.9747144211
180+	.9764133482	.9780040726	.9794929747	.9808861038	.9821891915
185+	.9834076641	.9845466555	.9856110197	.9866053431	.9875339567
190+	.9884009479	.9892101718	.9899652628	.9906696450	.9913265429
* 220+	.9987902376	.9988813190	.9989657190	.9990439128	.9991163426
			P=1/10, R= 1		
30+	.7284774740	.7531855145	.7758962777	.7967193347	.8157699847
35+	.8331657412	.8490236429	.8634582436	.8765801566	.8884950445
40+	.8993029611	.9090979699	.9179679739	.9259947070	.9332538434
45+	.9398151924	.9457429532	.9510960073	.9559282337	.9602888355
50+	.9642226664	.9677705534	.9709696083	.9738535269	.9764528716
55+	.9787953381	.9809060041	.9828075615	.9845205301	.9860634558
60+	.9874530925	.9887045686	.9898315396	.9908463277	.9917600484
* 85+	.9990970725	.9991873531	.9992686081	.9993417395	.9994075593
			P=1/10, R= 2		
45+	.6714856278	.6957158660	.7185215187	.7399280185	.7599715259
50+	.7786965549	.7961539020	.8123988640	.8274897239	.8414864812
55+	.8544498025	.8664401651	.8775171720	.8877390121	.8971620474
60+	.9058405058	.9138262636	.9211687018	.9279146238	.9341082222
65+	.9397910855	.9450022357	.9497781912	.9541530462	.9581585654
70+	.9618242867	.9651776307	.9682440128	.9710469567	.9736082079
75+	.9759478443	.9780843846	.9800348928	.9818150783	.9834393915

DIRICHLET 1 PROBABILITY, B= 7

N	0	1	2	3	4
			P=1/10, R= 2		
80+	.9849211150	.9862724491	.9875045929	.9886278203	.9896515512
85+	.9905844182	.9914343282	.9922085204	.9929136200	.9935556876
* 105+	.9986096080	.9987376445	.9988539796	.9989596729	.9990556890
			P=1/10, R= 3		
65+	.7652226956	.7816856583	.7971665190	.8117004471	.8253249514
70+	.8380791514	.8500031500	.8611374990	.8715227492	.8811990763
75+	.8902059738	.8985820060	.9063646129	.9135899601	.9202928277
80+	.9265065330	.9322628809	.9375921376	.9425230250	.9470827305
85+	.9512969304	.9551898230	.9587841712	.9621013500	.9651613997
90+	.9679830817	.9705839366	.9729803431	.9751875779	.9772198742
95+	.9790904801	.9808117157	.9823950279	.9838510439	.9851896225
100+	.9864199036	.9875503543	.9885888142	.9895425374	.9904182325
* 130+	.9990624735	.9991440445	.9992185984	.9992867307	.9993489871
			P=1/10, R= 4		
75+	.6814889951	.7007580951	.7191378005	.7366349907	.7532616456
80+	.7690340690	.7839721715	.7980988163	.8114392261	.8240204528
85+	.8358709045	.8470199301	.8574974550	.8673336672	.8765587468
90+	.8852026371	.8932948530	.9008643216	.9079392536	.9145470394
95+	.9207141696	.9264661748	.9318275830	.9368218920	.9414715546
100+	.9457979749	.9498215134	.9535614997	.9570362517	.9602630998
105+	.9632584143	.9660376373	.9686153155	.9710051356	.9732199601
110+	.9752718642	.9771721726	.9789314960	.9805597676	.9820662783
115+	.9834597119	.9847481784	.9859392468	.9870399765	.9880569475
120+	.9889962892	.9898637082	.9906645147	.9914036471	.9920856961
* 145+	.9986555374	.9987662751	.9988680332	.9989615262	.9990474132
			P=1/10, R= 5		
90+	.7029486023	.7199774780	.7362528441	.7517828556	.7665790070
95+	.7806556429	.7940295034	.8067193075	.8187453727	.8301292715
100+	.8408935226	.8510613158	.8606562689	.8697022144	.8782230134
105+	.8862423962	.8937838246	.9008703771	.9075246521	.9137686887
110+	.9196239036	.9251110403	.9302501315	.9350604718	.9395605993
115+	.9437682855	.9477005320	.9513735731	.9548028832	.9580031887
120+	.9609884825	.9637720425	.9663664515	.9687836188	.9710348035
125+	.9731306388	.9750811562	.9768958108	.9785835067	.9801526216
130+	.9816110318	.9829661367	.9842248823	.9853937851	.9864789543
135+	.9874861136	.9884206229	.9892874978	.9900914301	.9908368057
* 165+	.9988691504	.9989586535	.9990411973	.9991173116	.9991874867
			P=1/10, R= 6		
105+	.7268912440	.7419373041	.7563322891	.7700857034	.7832092319
110+	.7957164176	.8076223624	.8189434524	.8296971060	.8399015467
115+	.8495755972	.8587384954	.8674097312	.8756089013	.8833555833
120+	.8906692250	.8975690497	.9040739752	.9102025455	.9159728739
125+	.9214025967	.9265088358	.9313081701	.9358166135	.9400496001
130+	.9440219748	.9477479883	.9512412976	.9545149689	.9575814845
135+	.9604527518	.9631401150	.9656543680	.9680057695	.9702040587
140+	.9722584721	.9741777609	.9759702087	.9776436502	.9792054888
145+	.9806627151	.9820219253	.9832893384	.9844708145	.9855718715
150+	.9865977018	.9875531893	.9884429241	.9892712189	.9900421227
* 180+	.9986519519	.9987543378	.9988491057	.9989368085	.9990179602
			P=1/10, R= 7		
120+	.7511035308	.7643995468	.7771271703	.7892967327	.8009199783
125+	.8120098462	.8225802678	.8326459809	.8422223599	.8513252623
130+	.8599708891	.8681756610	.8759561064	.8833287640	.8903100957
135+	.8969164116	.9031638046	.9090680947	.9146447817	.9199090053
140+	.9248755134	.9295586352	.9339722613	.9381298280	.9420443067
145+	.9457281971	.9491935236	.9524518350	.9555142073	.9583912481
150+	.9610931028	.9636294638	.9660095791	.9682422639	.9703359119

DIRICHLET 1 PROBABILITY, B= 7

N	0	1	2	3	4
			P=1/10, R= 7		
155+	.9722985076	.9741376397	.9758605141	.9774739676	.9789844819
160+	.9803981973	.9817209267	.9829581690	.9841151229	.9851967003
165+	.9862075389	.9871520156	.9880342586	.9888581594	.9896273845
170+	.9903453868	.9910154167	.9916405321	.9922236089	.9927673505
* 200+	.9989633368	.9990398983	.9991109253	.9991768078	.9992379090
			P=1/10, R= 8		
130+	.7079612018	.7223225805	.7361515901	.7494516711	.7622280966
135+	.7744877629	.7862389920	.7974913446	.8082554450	.8185428182
140+	.8283657382	.8377370896	.8466702394	.8551789207	.8632771279
145+	.8709790210	.8782988408	.8852508330	.8918491812	.8981079474
150+	.9040410212	.9096620747	.9149845247	.9200215006	.9247858169
155+	.9292899521	.9335460303	.9375658079	.9413606631	.9449415896
160+	.9483191915	.9515036824	.9545048854	.9573322355	.9599947836
165+	.9625012023	.9648597918	.9670784880	.9691648709	.9711261735
170+	.9729692920	.9747007953	.9763269357	.9778536599	.9792866192
175+	.9806311809	.9818924386	.9830752236	.9841841154	.9852234518
180+	.9861973400	.9871096663	.9879641062	.9887641339	.9895130318
185+	.9902138997	.9908696636	.9914830839	.9920567642	.9925931587
* 215+	.9988650294	.9989461644	.9990216333	.9990918208	.9991570864
			P=1/10, R= 9		
145+	.7376674030	.7503794515	.7626125516	.7743725362	.7856664551
150+	.7965024252	.8068894888	.8168374819	.8263569100	.8354588345
155+	.8441547666	.8524565707	.8603763759	.8679264955	.8751193542
160+	.8819674229	.8884831601	.8946789601	.9005671066	.9061597329
165+	.9114687870	.9165060009	.9212828654	.9258106084	.9301001768
170+	.9341622221	.9380070891	.9416448071	.9450850837	.9483373008
175+	.9514105129	.9543134462	.9570545007	.9596417519	.9620829556
180+	.9643855517	.9665566707	.9686031394	.9705314885	.9723479594
185+	.9740585129	.9756688365	.9771843536	.9786102318	.9799513913
190+	.9812125141	.9823980525	.9835122380	.9845590895	.9855424224
195+	.9864658565	.9873328244	.9881465797	.9889102044	.9896266171
200+	.9902985802	.9909287070	.9915194686	.9920732012	.9925921119
* 230+	.9987970092	.9988805040	.9989583469	.9990309091	.9990985383
			P=1/10, R=10		
155+	.7019890504	.7154590661	.7284799712	.7410530772	.7531811293
160+	.7648681664	.7761193880	.7869410254	.7973402205	.8073249104
165+	.8169037192	.8260858565	.8348810235	.8432993252	.8513511896
170+	.8590472935	.8663984951	.8734157718	.8801101645	.8864927274
175+	.8925744823	.8983663794	.9038792604	.9091238285	.9141106201
180+	.9188499816	.9233520493	.9276267322	.9316836979	.9355323607
185+	.9391818731	.9426411180	.9459187042	.9490229628	.9519619455
190+	.9547434241	.9573748916	.9598635640	.9622163831	.9644400201
195+	.9665408807	.9685251091	.9703985945	.9721669765	.9738356514
200+	.9754097795	.9768942908	.9782938930	.9796130780	.9808561294
205+	.9820271293	.9831299659	.9841683403	.9851457739	.9860656150
210+	.9869310459	.9877450898	.9885106170	.9892303518	.9899068783
215+	.9905426471	.9911399808	.9917010801	.9922280289	.9927228004
* 245+	.9987564478	.9988404653	.9989189544	.9989922673	.9990607349

DIRICHLET 1   PROBABILITY,   B= 8

N	0	1	2	3	4
		P=1/ 9,	R= 1		
25♦	.6291216362	.6650018143	.6980268046	.7282797093	.7558807654
30♦	.7809747557	.8037211060	.8242862730	.8428380326	.8595413131
35♦	.8745552588	.8880312581	.9001117150	.9109293836	.9206071199
40♦	.9292579382	.9369852814	.9438834364	.9500380422	.9555266514
45♦	.9604193158	.9647791748	.9686630318	.9721219089	.9752015719
50♦	.9779430237	.9803829618	.9825542017	.9844860649	.9862047331
55♦	.9877335715	.9890934204	.9903028609	.9913784527	.9923349492
* 75♦	.9988343107	.9989638115	.9990789277	.9991812568	.9992722188
		P=1/ 9,	R= 2		
45♦	.7537501064	.7752421263	.7950983610	.8134006265	.8302351479
50♦	.8456902980	.8598547894	.8728162579	.8846601779	.8954690561
55♦	.9053218548	.9142936030	.9224551604	.9298731008	.9366096915
60♦	.9427229447	.9482667246	.9532908945	.9578414919	.9619609230
65♦	.9656881679	.9690589904	.9721061494	.9748596059	.9773467256
70♦	.9795924726	.9816195959	.9834488042	.9850989325	.9865870968
75♦	.9879288392	.9891382628	.9902281566	.9912101112	.9920946250
* 95♦	.9985767844	.9987226108	.9988536042	.9989712600	.9990769246
		P=1/ 9,	R= 3		
60♦	.7687187520	.7870250077	.8040812189	.8199422992	.8346660007
65♦	.8483117012	.8609393981	.8726088842	.8833790844	.8933075320
70♦	.9024499650	.9108600242	.9185890379	.9256858783	.9321968773
75♦	.9381657909	.9436338028	.9486395584	.9532192239	.9574065635
80♦	.9612330298	.9647278644	.9679182047	.9708291940	.9734840934
85♦	.9759043932	.9781099229	.9801189586	.9819483263	.9836135022
90♦	.9851287076	.9865069996	.9877603571	.9888997616	.9899352732
95♦	.9908761021	.9917306750	.9925066977	.9932112126	.9938506527
* 115♦	.9987651267	.9988844089	.9989923011	.9990898759	.9991781067
		P=1/ 9,	R= 4		
70♦	.7028366195	.7235238114	.7430626704	.7614760952	.7787933199
75♦	.7950486162	.8102801405	.8245289212	.8378379794	.8502515732
80♦	.8618145545	.8725718305	.8825679155	.8918465663	.9004504887
85♦	.9084211073	.9157983898	.9226207185	.9289248001	.9347456111
90♦	.9401163695	.9450685305	.9496318015	.9538341709	.9577019508
95♦	.9612598277	.9645309210	.9675368460	.9702977809	.9728325355
100♦	.9751586211	.9772923201	.9792487556	.9810419584	.9826849339
105♦	.9841897252	.9855674754	.9868284860	.9879822734	.9890376224
110♦	.9900026368	.9908847877	.9916909585	.9924274872	.9931002069
* 135♦	.9990631057	.9991496103	.9992282393	.9992996980	.9993646294
		P=1/ 9,	R= 5		
85♦	.7474040476	.7643799626	.7804157505	.7955378904	.8097758865
90♦	.8231615642	.8357284486	.8475112205	.8585452451	.8688661677
95♦	.8785095701	.8875106823	.8959041438	.9037238089	.9110025909
100♦	.9177723406	.9240637541	.9299063064	.9353282074	.9403563764
105♦	.9450164321	.9493326961	.9533282074	.9570247454	.9604428604
110♦	.9636019097	.9665200975	.9692145189	.9717012048	.9739951697
115♦	.9761104590	.9780601971	.9798566352	.9815111979	.9830345294
120♦	.9844365384	.9857264410	.9869128027	.9880035787	.9890061518
125♦	.9899273695	.9907735784	.9915506578	.9922640505	.9929187928
* 150♦	.9989437213	.9990370807	.9991223269	.9992001508	.9992711861
		P=1/ 9,	R= 6		
95♦	.7056193710	.7237014248	.7409171837	.7572788568	.7728026981
100♦	.7875083419	.8014181945	.8145568835	.8269507631	.8386274726
105♦	.8496155476	.8599440782	.8696424123	.8787399011	.8872656811
110♦	.8952484912	.9027165204	.9096972827	.9162175173	.9223031096
115♦	.9279790320	.9332693000	.9381969442	.9427839932	.9470514680
120♦	.9510193842	.9547067628	.9581316466	.9613111207	.9642613391
125♦	.9669975521	.9695341380	.9718846354	.9740617770	.9760775245

DIRICHLET 1 PROBABILITY, B= 8

N	0	1	2	3	4
		P=1/ 9,	R= 6		
130+	.9779431035	.9796690380	.9812651857	.9827407714	.9841044209
135+	.9853641934	.9865276131	.9876016996	.9885929974	.9895076039
140+	.9903511967	.9911290592	.9918461052	.9925069022	.9931156936
* 165+	.9988911479	.9989855121	.9990719983	.9991512492	.9992238567
		P=1/ 9,	R= 7		
110+	.7551710724	.7700469453	.7841851658	.7976037290	.8103225993
115+	.8223633179	.8337486506	.8445022713	.8546484821	.8642119651
120+	.8732175668	.8816901100	.8896542329	.8971342518	.9041540457
125+	.9107369611	.9169057332	.9226824244	.9280883757	.9331441700
130+	.9378696076	.9422836892	.9464046083	.9502497502	.9538356959
135+	.9571782326	.9602923669	.9631923420	.9658916575	.9684030917
140+	.9707387250	.9729099645	.9749275706	.9768016824	.9785418451
145+	.9801570359	.9816556908	.9830457307	.9843345868	.9855292260
150+	.9866361750	.9876615440	.9886110497	.9894900371	.9903035010
* 180+	.9988908850	.9989821898	.9990661340	.9991432964	.9992142114
		P=1/ 9,	R= 8		
120+	.7261137074	.7416615059	.7565156664	.7706869717	.7841885938
125+	.7970357253	.8092452405	.8208353841	.8318254889	.8422357209
130+	.8520868507	.8614000501	.8701967117	.8784982915	.8863261708
135+	.8937015379	.9006452864	.9071779290	.9133195267	.9190896289
140+	.9245072268	.9295907159	.9343578678	.9388258102	.9430110134
145+	.9469292835	.9505957605	.9540249210	.9572305845	.9602259234
150+	.9630234757	.9656351592	.9680722885	.9703455926	.9724652344
155+	.9744408300	.9762814697	.9779957382	.9795917361	.9810771001
160+	.9824590246	.9837442813	.9849392401	.9860498884	.9870818503
165+	.9880404055	.9889305071	.9897567992	.9905236337	.9912350864
* 195+	.9989281430	.9990138265	.9990928121	.9991656094	.9992326907
		P=1/ 9,	R= 9		
130+	.7001355308	.7161299775	.7314790185	.7461874692	.7602627586
135+	.7737145970	.7865546625	.7987963097	.8104542997	.8215445519
140+	.8320839168	.8420899704	.8515808279	.8605749774	.8690911309
145+	.8771480929	.8847646452	.8919594456	.8987509408	.9051572923
150+	.9111963124	.9168854117	.9222415556	.9272812289	.9320204084
155+	.9364745421	.9406585341	.9445867347	.9482729355	.9517303684
160+	.9549717074	.9580090742	.9608540458	.9635176644	.9660104493
165+	.9683424094	.9705230579	.9725614274	.9744660856	.9762451515
170+	.9779063125	.9794568406	.9809036098	.9822531126	.9835114770
175+	.9846844824	.9857775762	.9867958893	.9877442518	.9886272073
180+	.9894490283	.9902137294	.9909250813	.9915866235	.9922016772
* 210+	.9989901466	.9990687860	.9991414437	.9992085621	.9992705521
		P=1/ 9,	R=10		
145+	.7523676089	.7656799770	.7784198187	.7905978537	.8022261797
150+	.8133180556	.8238877020	.8339501179	.8435209132	.8526161580
155+	.8612522451	.8694457684	.8772134138	.8845718627	.8915377073
160+	.8981273772	.9043570749	.9102427219	.9157999122	.9210438736
165+	.9259894361	.9306510067	.9350425490	.9391775688	.9430691034
170+	.9467297153	.9501714886	.9534060295	.9564444684	.9592974649
175+	.9619752142	.9644874558	.9668434831	.9690521536	.9711219017
180+	.9730607501	.9748763234	.9765758613	.9781662322	.9796539472
185+	.9810451737	.9823457496	.9835611967	.9846967347	.9857572941
190+	.9867475296	.9876718330	.9885343454	.9893389695	.9900893814
* 225+	.9990667779	.9991377594	.9992034720	.9992642954	.9993205834
		P=1/10,	R= 1		
30+	.6942862749	.7216392342	.7468762584	.7700922738	.7913939225
35+	.8108947722	.8287115474	.8449612238	.8597588397	.8732158932
40+	.8854392143	.8965302118	.9065844183	.9156912628	.9239340182
45+	.9313898783	.9381301296	.9442203885	.9497208835	.9546867637

DIRICHLET 1   PROBABILITY,   B= 8

N	0	1	2	3	4
			P=1/10, R= 1		
50+	.9591684211	.9632118173	.9668588061	.9701474479	.9731123111
55+	.9757847595	.9781932224	.9803634478	.9823187388	.9840801718
60+	.9856667990	.9870958338	.9883828216	.9895417960	.9905854214
* 85+	.9989681060	.9990712792	.9991641383	.9992477141	.9993229344
			P=1/10, R= 2		
50+	.7498913538	.7693724065	.7875442582	.8044616504	.8201830005
55+	.8347689873	.8482813753	.8607820544	.8723322652	.8829919879
60+	.8928194691	.9018708673	.9101999973	.9178581570	.9248940220
65+	.9313535961	.9372802054	.9427145278	.9476946503	.9522561463
70+	.9564321692	.9602535559	.9637489381	.9669448574	.9698658824
75+	.9725347252	.9749723569	.9771981198	.9792298361	.9810839125
80+	.9827754397	.9843182877	.9857251950	.9870078537	.9881769885
85+	.9892424316	.9902131920	.9910975205	.9919029702	.9926364523
* 110+	.9990204832	.9991110144	.9991932355	.9992679030	.9993357053
			P=1/10, R= 3		
65+	.7348288281	.7531697935	.7704569318	.7867210881	.8019969343
70+	.8163220204	.8297359476	.8422796538	.8539948036	.8649232732
75+	.8751067202	.8845862299	.8934020284	.9015932549	.9091977845
80+	.9162520959	.9227911761	.9288484576	.9344557823	.9396433875
85+	.9444399106	.9488724091	.9529663919	.9567458607	.9602333582
90+	.9634500225	.9664156439	.9691487259	.9716665462	.9739852193
95+	.9761197583	.9780841358	.9798913441	.9815534535	.9830816679
100+	.9844863800	.9857772225	.9869631180	.9880523258	.9890524873
105+	.9899706678	.9908133965	.9915867046	.9922961602	.9929469017
* 130+	.9989285627	.9990217829	.9991069842	.9991848471	.9992559952
			P=1/10, R= 4		
80+	.7390315827	.7556856911	.7714682410	.7864010832	.8005089385
85+	.8138187890	.8263593361	.8381605207	.8492531042	.8596683039
90+	.8694374804	.8785918707	.8871623642	.8951793166	.9026723968
95+	.9096704649	.9162014758	.9222924058	.9279691998	.9332567351
100+	.9381788001	.9427580856	.9470161866	.9509736130	.9546498068
105+	.9580631660	.9612310728	.9641699250	.9668951710	.9694213459
110+	.9717621093	.9739302843	.9759378958	.9777962095	.9795157706
115+	.9811064409	.9825774364	.9839373626	.9851942496	.9863555852
120+	.9874283475	.9884190353	.9893336981	.9901779634	.9909570639
* 150+	.9990015056	.9990843029	.9991603329	.9992301390	.9992942224
			P=1/10, R= 5		
95+	.7519586148	.7668950654	.7810935368	.7945720993	.8073507987
100+	.8194512581	.8308963204	.8417097281	.8519158404	.8615393837
105+	.8706052347	.8791382321	.8871630156	.8947038893	.9017847076
110+	.9084287802	.9146587956	.9204967604	.9259639518	.9310808836
115+	.9358672816	.9403420687	.9445233588	.9484284562	.9520738613
120+	.9554752817	.9586476461	.9616051227	.9643611394	.9669284063
125+	.9693189401	.9715440889	.9736145585	.9755404391	.9773312313
130+	.9789958735	.9805427679	.9819798066	.9833143972	.9845534877
135+	.9857035906	.9867708065	.9877608466	.9886790548	.9895304281
140+	.9903196371	.9910510450	.9917287258	.9923564819	.9929378605
* 165+	.9987076304	.9988099148	.9989042465	.9989912308	.9990714281
			P=1/10, R= 6		
105+	.6921824593	.7088550424	.7248432389	.7401517134	.7547881629
110+	.7687629189	.7820885751	.7947796403	.8068522194	.8183237206
115+	.8292125897	.8395380692	.8493199832	.8585785452	.8673341875
120+	.8756074125	.8834186618	.8907882036	.8977360367	.9042818083
125+	.9104447456	.9162436001	.9216966011	.9268214205	.9316351452
130+	.9361542572	.9403946206	.9443714735	.9480994265	.9515924638
135+	.9548639488	.9579266330	.9607926669	.9634736136	.9659804637
140+	.9683236516	.9705130735	.9725581054	.9744676221	.9762500164

DIRICHLET 1 PROBABILITY, B= 8

N	0	1	2	3	4
			P=1/10,	R= 6	
145+	.9779132188	.9794647165	.9809115732	.9822604480	.9835176144
150+	.9846889785	.9857800974	.9867961963	.9877421857	.9886226779
155+	.9894420027	.9902042225	.9909131475	.9915723491	.9921851735
* 185+	.9989634908	.9990428581	.9991162636	.9991841448	.9992469083
			P=1/10,	R= 7	
120+	.7190107666	.7337998308	.7479842489	.7615712784	.7745701863
125+	.7869919758	.7988491313	.8101553816	.8209254827	.8311750185
130+	.8409202200	.8501778008	.8589648109	.8672985048	.8751962248
135+	.8826752985	.8897529489	.8964462163	.9027718911	.9087464573
140+	.9143860437	.9197063848	.9247227881	.9294501087	.9339027296
145+	.9380945470	.9420389607	.9457488687	.9492366649	.9525142401
150+	.9555929862	.9584838017	.9611971000	.9637428194	.9661304333
155+	.9683689631	.9704669907	.9724326722	.9742737522	.9759975785
160+	.9776111161	.9791209630	.9805333641	.9818542265	.9830891341
165+	.9842433614	.9853218882	.9863294129	.9872703662	.9881489239
170+	.9889690197	.9897343574	.9904484223	.9911144932	.9917356526
* 200+	.9988152657	.9989027610	.9989839317	.9990592233	.9991290510
			P=1/10,	R= 8	
135+	.7450258773	.7581417992	.7707214294	.7827735491	.7943082590
140+	.8053367885	.8158713182	.8259248156	.8355108853	.8446436306
145+	.8533375295	.8616073214	.8694679062	.8769342537	.8840213233
150+	.8907439936	.8971170003	.9031548825	.9088719368	.9142821777
155+	.9193993048	.9242366755	.9288072828	.9331237379	.9371982570
160+	.9410426513	.9446683212	.9480862525	.9513070159	.9543407682
165+	.9571972558	.9598858193	.9624154008	.9647945507	.9670314368
170+	.9691338535	.9711092323	.9729646523	.9747068511	.9763422365
175+	.9778768976	.9793166170	.9806668818	.9819328958	.9831195908
180+	.9842316379	.9852734589	.9862492370	.9871629278	.9880182697
185+	.9888187939	.9895678343	.9902685375	.9909238713	.9915366342
* 215+	.9987029189	.9987956405	.9988818872	.9989620984	.9990366851
			P=1/10,	R= 9	
145+	.7040612215	.7181737675	.7317806811	.7448848411	.7574908026
150+	.7696046128	.7812336358	.7923863874	.8030723788	.8133019722
155+	.8230862449	.8324368648	.8413659759	.8498860924	.8580100035
160+	.8657506869	.8731212299	.8801347602	.8868043830	.8931431259
165+	.8991638901	.9048794084	.9103022082	.9154445801	.9203185513
170+	.9249358631	.9293079532	.9334459407	.9373606146	.9410624256
175+	.9445614805	.9478675380	.9509900079	.9539379516	.9567200833
180+	.9593447741	.9618200566	.9641536298	.9663528662	.9684248191
185+	.9703762295	.9722135356	.9739428806	.9755701217	.9771008399
190+	.9785403487	.9798937036	.9811657119	.9823609414	.9834837304
195+	.9845381965	.9855282457	.9864575814	.9873297132	.9881479651
200+	.9889154842	.9896352485	.9903100746	.9909426258	.9915354187
* 235+	.9990417977	.9991089038	.9991714196	.9992296502	.9992838815
			P=1/10,	R=10	
160+	.7342806558	.7468225542	.7589048796	.7705329566	.7817132378
165+	.7924531684	.8027610587	.8126459643	.8221175746	.8311861087
170+	.8398622196	.8481569058	.8560814305	.8636472477	.8708659350
175+	.8777491339	.8843084945	.8905556279	.8965020626	.9021592067
180+	.9075383146	.9126504582	.9175065020	.9221170819	.9264925877
185+	.9306431480	.9345786189	.9383085744	.9418422997	.9451887859
190+	.9483567276	.9513545207	.9541902633	.9568717566	.9594065075
195+	.9618017320	.9640643598	.9662010391	.9682181424	.9701217727
200+	.9719177701	.9736117189	.9752089545	.9767145714	.9781334301
205+	.9794701655	.9807291941	.9819147221	.9830307529	.9840810946
210+	.9850693681	.9859990140	.9868733004	.9876953294	.9884680450
215+	.9891942392	.9898765590	.9905175128	.9911194769	.9916847010

DIRICHLET 1 PROBABILITY, B= 8

N	0	1	2	3	4
		P=1/10,	R=10		
* 250*	.9989996359	.9990678500	.9991315270	.9991909599	.9992464233

DIRICHLET 1 PROBABILITY, B= 9

N	0	1	2	3	4
			P=1/10, R= 1		
30+	.6611798504	.6909756324	.7185728740	.7440459200	.7674876384
35+	.7890032941	.8087056244	.8267109551	.8431361970	.8580965754
40+	.8717039615	.8840656890	.8952837595	.9054543524	.9146675717
45+	.9230073735	.9305516279	.9373722794	.9435355776	.9491023529
50+	.9541283226	.9586644113	.9627570767	.9664486332	.9697775664
55+	.9727788367	.9754841673	.9779223169	.9801193347	.9820987983
60+	.9838820345	.9854883226	.9869350827	.9882380477	.9894114212
65+	.9904680214	.9914194129	.9922760263	.9930472674	.9937416156
* 85+	.9988391453	.9989552100	.9990596723	.9991536917	.9992383119
			P=1/10, R= 2		
50+	.7217614932	.7431557198	.7631612242	.7818266950	.7992067720
55+	.8153602390	.8303485015	.8442343225	.8570807856	.8689504600
60+	.8799047379	.8900033214	.8993038362	.9078615506	.9157291835
65+	.9229567854	.9295916781	.9356784422	.9412589426	.9463723836
70+	.9510553853	.9553420781	.9592642079	.9628512493	.9661305241
75+	.9691273207	.9718650147	.9743651872	.9766477406	.9787310109
80+	.9806318749	.9823658539	.9839472110	.9853890438	.9867033719
85+	.9879012186	.9889926883	.9899870376	.9908927428	.9917175619
* 110+	.9988980649	.9989999085	.9990924039	.9991764023	.9992526777
			P=1/10, R= 3		
65+	.7051899699	.7252978937	.7442956859	.7622079228	.7790646956
70+	.7949004326	.8097528587	.8236620874	.8366698365	.8488187568
75+	.8601518649	.8707120693	.8805417795	.8896825884	.8981750199
80+	.9060583310	.9133703636	.9201474366	.9264242733	.9322339577
85+	.9376079164	.9425759193	.9471660974	.9514049742	.9553175063
90+	.9589271332	.9622558319	.9653241773	.9681514035	.9707554685
95+	.9731531184	.9753599522	.9773904851	.9792582114	.9809756649
100+	.9825544777	.9840054370	.9853385393	.9865630416	.9876875114
105+	.9887198730	.9896674516	.9905370154	.9913348144	.9920666178
* 130+	.9987946573	.9988995257	.9989953737	.9990829666	.9991630058
			P=1/10, R= 4		
80+	.7097490164	.7280208470	.7453736300	.7618243840	.7773941757
85+	.7921073691	.8059909492	.8190739194	.8313867674	.8429609986
90+	.8538287296	.8640223386	.8735741679	.8825162729	.8908802126
95+	.8986968785	.9059963561	.9128078161	.9191594300	.9250783094
100+	.9305904632	.9357207714	.9404929731	.9449296655	.9490523134
105+	.9528812655	.9564357785	.9597340456	.9627932287	.9656294949
110+	.9682580534	.9706931959	.9729483364	.9750360517	.9769681228
115+	.9787505744	.9804087150	.9819371758	.9833499478	.9846554191
120+	.9858614094	.9869752042	.9880035872	.9889528711	.9898289271
125+	.9906372133	.9913828007	.9920703988	.9927043795	.9932887993
* 150+	.9988767138	.9989698574	.9990553883	.9991339179	.9992060097
			P=1/10, R= 5		
95+	.7238985602	.7403153050	.7559501376	.7708177852	.7849357693
100+	.7983239141	.8110038999	.8229988618	.8343330301	.8450314117
105+	.8551195095	.8646320776	.8735679089	.8819796530	.8898836619
110+	.8973048598	.9042676358	.9107957560	.9169122934	.9226395740
115+	.9279991354	.9330116987	.9376971492	.9420745279	.9461620287
120+	.9499770032	.9535359707	.9568546325	.9599478898	.9628298649
125+	.9655139243	.9680127039	.9703381354	.9725014736	.9745133240
130+	.9763836711	.9781219067	.9797368571	.9812368114	.9826295478
135+	.9839223604	.9851220845	.9862351215	.9872674627	.9882247128
140+	.9891121113	.9899345542	.9906966141	.9914025597	.9920563737
* 170+	.9990385406	.9991151967	.9991858412	.9992509363	.9993109096
			P=1/10, R= 6		
110+	.7423501110	.7570285471	.7710304535	.7843697960	.7970624154
115+	.8091256968	.8205782689	.8314397325	.8417304156	.8514711560

DIRICHLET 1   PROBABILITY,   B= 9

N	0	1	2	3	4
			P=1/10,	R= 6	
120+	.8606831081	.8693875730	.8776058502	.8853591093	.8926682799
125+	.8995539579	.9060363273	.9121350950	.9178694381	.9232579622
130+	.9283186686	.9330689312	.9375254794	.9417043886	.9456210760
135+	.9492903007	.9527261690	.9559421423	.9589510481	.9617650936
140+	.9643958815	.9668544268	.9691511751	.9712960218	.9732983323
145+	.9751669620	.9769102767	.9785361737	.9800521019	.9814650820
150+	.9827817268	.9840082607	.9851505386	.9862140645	.9872040100
155+	.9881252310	.9889822852	.9897794478	.9905207272	.9912098799
* 185+	.9988339476	.9989232326	.9990058111	.9990821752	.9991527821
			P=1/10,	R= 7	
125+	.7624215460	.7755119234	.7880112370	.7999330013	.8112919584
130+	.8221038518	.8323852203	.8421532113	.8514254130	.8602197045
135+	.8685541226	.8764467439	.8839155820	.8909784980	.8976531230
140+	.9039567929	.9099064922	.9155188081	.9208098929	.9257954334
145+	.9304906275	.9349101670	.9390682249	.9429784482	.9466539543
150+	.9501073308	.9533506389	.9563954189	.9592526982	.9619330008
155+	.9644463589	.9668023252	.9690099865	.9710779781	.9730144986
160+	.9748273252	.9765238294	.9781109924	.9795954213	.9809833643
165+	.9822807264	.9834930846	.9846257029	.9856835470	.9866712983
170+	.9875933682	.9884539110	.9892568373	.9900058267	.9907043390
* 200+	.9986672005	.9987656287	.9988569423	.9989416424	.9990201961
			P=1/10,	R= 8	
135+	.7162292448	.7306355929	.7444759419	.7577565198	.7704853444
140+	.7826719913	.7943273761	.8054635522	.8160935238	.8262310741
145+	.8358906076	.8450870074	.8538355060	.8621515693	.8700507921
150+	.8775488064	.8846611999	.8914034441	.8977908330	.9038384289
155+	.9095610173	.9149730679	.9200887029	.9249216707	.9294853250
160+	.9337926090	.9378560431	.9416877169	.9452992845	.9487019623
165+	.9519065300	.9549233332	.9577622879	.9604328875	.9629442099
170+	.9653049266	.9675233129	.9696072577	.9715642754	.9734015173
175+	.9751257837	.9767435359	.9782609086	.9796837227	.9810174971
180+	.9822674615	.9834385679	.9845355033	.9855627007	.9865243513
185+	.9874244148	.9882666310	.9890545299	.9897914421	.9904805085
* 220+	.9989942991	.9990668185	.9991342221	.9991968616	.9992550651
			P=1/10,	R= 9	
150+	.7432345561	.7560478317	.7683535346	.7801596394	.7914753069
155+	.8023107179	.8126769185	.8225856772	.8320493528	.8410807736
160+	.8496931276	.8578998625	.8657145954	.8731510322	.8802228956
165+	.8869438609	.8933274994	.8993872300	.9051362755	.9105876263
170+	.9157540091	.9206478608	.9252813065	.9296661422	.9338138210
175+	.9377354428	.9414417462	.9449431044	.9482495226	.9513706374
180+	.9543157182	.9570936707	.9597130409	.9621820207	.9645084548
185+	.9666998478	.9687633723	.9707058778	.9725338997	.9742536687
190+	.9758711206	.9773919061	.9788214010	.9801647158	.9814267062
195+	.9826119827	.9837249206	.9847696694	.9857501628	.9866701277
200+	.9875330932	.9883423999	.9891012081	.9898125063	.9904791196
* 235+	.9989220386	.9989975307	.9990678589	.9991333667	.9991943754
			P=1/10,	R=10	
160+	.7044173741	.7181752790	.7314507261	.7442466318	.7565674096
165+	.7684188079	.7798077552	.7907422140	.8012310439	.8112838731
170+	.8209109791	.8301231782	.8389317235	.8473482119	.8553844985
175+	.8630526192	.8703647210	.8773329989	.8839696394	.8902867707
180+	.8962964184	.9020104667	.9074406244	.9125983959	.9174950564
185+	.9221416304	.9265488747	.9307272641	.9346869797	.9384379008
190+	.9419895980	.9453513297	.9485320391	.9515403542	.9543845883
195+	.9570727425	.9596125085	.9620112735	.9642761248	.9664138560
200+	.9684309732	.9703337020	.9721279948	.9738195386	.9754137626

DIRICHLET 1 PROBABILITY, B= 9

N	0	1	2	3	4
		P=1/10,	R=10		
205+	.9769158465	.9783307287	.9796631142	.9809174834	.9820980997
210+	.9832090184	.9842540941	.9852369891	.9861611814	.9870299717
215+	.9878464916	.9886137107	.9893344436	.9900113571	.9906469767
# 250+	.9988746079	.9989513463	.9990229809	.9990898410	.9991522357

DIRICHLET 1 PROBABILITY,   B=10

N	0	1	2	3	4
			P=1/11, R= 1		
35+	.6825376155	.7080612285	.7318439260	.7539459302	.7744378672
40+	.7933974494	.8109067552	.8270500345	.8419119656	.8555762979
45+	.8681248162	.8796365743	.8901873471	.8998492622	.9086905740
50+	.9167755511	.9241644520	.9309135693	.9370753241	.9426983978
55+	.9478278909	.9525054980	.9567696942	.9606559252	.9641967997
60+	.9674222776	.9703598544	.9730347387	.9754700217	.9776868380
65+	.9797045185	.9815407322	.9832116209	.9847319230	.9861150901
70+	.9873733940	.9885180261	.9895591896	.9905061839	.9913674821
* 95+	.9988315584	.9989377589	.9990343087	.9991220844	.9992018831
			P=1/11, R= 2		
55+	.6909583931	.7120948153	.7320545379	.7508632701	.7685530559
60+	.7851608771	.8007274308	.8152960708	.8289119018	.8416210120
65+	.8534698309	.8645045985	.8747709327	.8843134826	.8931756563
70+	.9013994128	.9090251082	.9160913880	.9226351182	.9286913487
75+	.9342933026	.9394723871	.9442582216	.9486786792	.9527599389
80+	.9565265452	.9600014734	.9632061990	.9661607687	.9688838734
85+	.9713929203	.9737041048	.9758324808	.9777920294	.9795957248
90+	.9812555983	.9827827992	.9841876530	.9854797168	.9866678318
95+	.9877601727	.9887642942	.9896871757	.9905352617	.9913145008
* 120+	.9985980836	.9987156997	.9988235191	.9989223506	.9990129371
			P=1/11, R= 3		
75+	.7381566483	.7547514867	.7704704200	.7853373061	.7993786207
80+	.8126228549	.8250999842	.8368410034	.8478775237	.8582414261
85+	.8679645666	.8770785287	.8856144164	.8936026854	.9010730068
90+	.9080541599	.9145739497	.9206591462	.9263354432	.9316274321
95+	.9365585899	.9411512782	.9454267518	.9494051752	.9531056453
100+	.9565462190	.9597439448	.9627148974	.9654742146	.9680361350
105+	.9704140378	.9726204816	.9746672439	.9765653606	.9783251640
110+	.9799563204	.9814678668	.9828682464	.9841653424	.9853665111
115+	.9864786134	.9875080446	.9884607632	.9893423183	.9901578752
* 145+	.9988065582	.9989010098	.9989880703	.9990683102	.9991422569
			P=1/11, R= 4		
90+	.7129652675	.7294299245	.7451407401	.7601102727	.7743538565
95+	.7878891370	.8007356472	.8129144249	.8244476692	.8353584340
100+	.8456703576	.8554074249	.8645937599	.8732534474	.8814103798
105+	.8890881271	.8963098287	.9030981037	.9094749778	.9154618261
110+	.9210793280	.9263474344	.9312853448	.9359114929	.9402435403
115+	.9442983760	.9480921215	.9516401405	.9549570528	.9580567503
120+	.9609524167	.9636565479	.9661809751	.9685368878	.9707348588
125+	.9727848688	.9746963312	.9764781173	.9781385813	.9796855846
130+	.9811265199	.9824683345	.9837175535	.9848803019	.9859623256
135+	.9869690126	.9879054122	.9887762548	.9895859693	.9903387012
* 165+	.9986862844	.9987851413	.9988766683	.9989613993	.9990398302
			P=1/11, R= 5		
105+	.7011505151	.7170303911	.7322546587	.7468303943	.7607671480
110+	.7740765978	.7867722275	.7988690305	.8103832389	.8213320765
115+	.8317335360	.8416061788	.8509689566	.8598410529	.8682417438
120+	.8761902758	.8837057610	.8908070853	.8975128322	.9038412169
125+	.9098100330	.9154366085	.9207377703	.9257298175	.9304285008
130+	.9348490088	.9390059589	.9429133937	.9465847809	.9500330167
135+	.9532704321	.9563088024	.9591593576	.9618327954	.9643392952
140+	.9666885332	.9688896987	.9709515103	.9728822329	.9746896956
145+	.9763813082	.9779640792	.9794446327	.9808292255	.9821237639
150+	.9833338199	.9844646472	.9855211971	.9865081328	.9874298450
155+	.9882904649	.9890938785	.9898437395	.9905434814	.9911963298
* 185+	.9986910546	.9987854813	.9988732132	.9989547150	.9990304203

DIRICHLET 1 PROBABILITY, B=10

N	0	1	2	3	4
		P=1/11,	R= 6		
120+	.6977934468	.7129142169	.7274558312	.7414228806	.7548220333
125+	.7676617752	.7799521666	.7917046150	.8029316645	.8136468024
130+	.8238642812	.8335989575	.8428661455	.8516814852	.8600608246
135+	.8680201148	.8755753169	.8827423212	.8895368754	.8959745236
140+	.9020705537	.9078399529	.9132973701	.9184570857	.9233329865
145+	.9279385464	.9322868120	.9363903915	.9402614488	.9439116994
150+	.9473524099	.9505944002	.9536480471	.9565232903	.9592296399
155+	.9617761849	.9641716030	.9664241713	.9685417775	.9705319318
160+	.9724017794	.9741581124	.9758073833	.9773557171	.9788089243
165+	.9801725139	.9814517051	.9826514408	.9837763987	.9848310037
170+	.9858194396	.9867456598	.9876133987	.9884261823	.9891873380
175+	.9899000050	.9905671434	.9911915434	.9917758345	.9923224931
* 205+	.9987736081	.9988589578	.9989384807	.9990125655	.9990815758
		P=1/11,	R= 7		
135+	.6999201077	.7142261097	.7280131378	.7412847880	.7540463592
140+	.7663046556	.7780678006	.7893450609	.8001466837	.8104837439
145+	.8203680048	.8298117887	.8388278593	.8474293145	.8556294895
150+	.8634418698	.8708800127	.8779574779	.8846877662	.8910842653
155+	.8971602026	.9029286046	.9084022617	.9135936988	.9185151506
160+	.9231785412	.9275954676	.9317771872	.9357346081	.9394782826
165+	.9430184028	.9463647989	.9495269393	.9525139324	.9553345297
170+	.9579971308	.9605097889	.9628802174	.9651157974	.9672235859
175+	.9692103241	.9710824467	.9728460911	.9745071068	.9760710654
180+	.9775432701	.9789287655	.9802323474	.9814585726	.9826117684
185+	.9836960418	.9847152891	.9856732046	.9865732898	.9874188618
190+	.9882130615	.9889588620	.9896590765	.9903163654	.9909332443
* 225+	.9988969912	.9989714117	.9990409133	.9991058134	.9991664093
		P=1/11,	R= 8		
150+	.7056544347	.7191426306	.7321604195	.7447111224	.7567994416
155+	.7684313080	.7796137361	.7903546872	.8006629415	.8105479781
160+	.8200198640	.8290891518	.8377667850	.8460640117	.8539923058
165+	.8615632953	.8687886984	.8756802646	.8822497241	.8885087407
170+	.8944688721	.9001415339	.9055379687	.9106692197	.9155461078
175+	.9201792125	.9245788559	.9287550903	.9327176874	.9364761311
180+	.9400396113	.9434170211	.9466169541	.9496477044	.9525172676
185+	.9552333430	.9578033367	.9602343657	.9625332630	.9647065828
190+	.9667606068	.9687013506	.9705345707	.9722657717	.9739002135
195+	.9754429188	.9768986813	.9782720727	.9795674511	.9807889681
200+	.9819405770	.9830260399	.9840489355	.9850126665	.9859204665
205+	.9867754070	.9875804049	.9883382283	.9890515039	.9897227225
210+	.9903542458	.9909483118	.9915070409	.9920324411	.9925264136
* 245+	.9990359636	.9990992720	.9991585118	.9992139380	.9992657898
		P=1/11,	R= 9		
165+	.7137668633	.7264581415	.7387193033	.7505537213	.7619658831
170+	.7729612709	.7835462479	.7937279497	.8035141832	.8129133318
175+	.8219342670	.8305862662	.8388789370	.8468221476	.8544259629
180+	.8617005862	.8686563067	.8753034515	.8816523425	.8877132585
185+	.8934964005	.8990118617	.9042696013	.9092794207	.9140509441
190+	.9185936014	.9229166136	.9270289808	.9309394728	.9346566214
195+	.9381887140	.9415437898	.9447296372	.9477537920	.9506235371
200+	.9533459038	.9559276731	.9583753782	.9606953081	.9628935111
205+	.9649757991	.9669477530	.9688147271	.9705818551	.9722540555
210+	.9738360379	.9753323087	.9767471773	.9780847627	.9793489993
215+	.9805436434	.9816722794	.9827383259	.9837450419	.9846955330
220+	.9855927571	.9864395300	.9872385318	.9879923117	.9887032939
225+	.9893737830	.9900059684	.9906019301	.9911636432	.9916929823
* 260+	.9988524413	.9989254568	.9989939350	.9990581499	.9991183597

DIRICHLET 1 PROBABILITY, B=10

N	0	1	2	3	4
		P=1/11,	R=10		
180+	.7234255495	.7353516047	.7468818443	.7580198191	.7687699771
185+	.7791375677	.7891285506	.7987495102	.8080075746	.8169103394
190+	.8254657973	.8336822718	.8415683567	.8491328595	.8563847496
195+	.8633331115	.8699871011	.8763559068	.8824487142	.8882746741
200+	.8938428745	.8991623146	.9042418833	.9090903389	.9137162922
205+	.9181281921	.9223343126	.9263427423	.9301613760	.9337979071
210+	.9372598224	.9405543977	.9436886948	.9466695600	.9495036229
215+	.9521972966	.9547567786	.9571880524	.9594968895	.9616888525
220+	.9637692979	.9657433802	.9676160554	.9693920856	.9710760433
225+	.9726723163	.9741851125	.9756184647	.9769762362	.9782621251
230+	.9794796703	.9806322561	.9817231178	.9827553464	.9837318939
235+	.9846555784	.9855290888	.9863549900	.9871357274	.9878736315
240+	.9885709227	.9892297157	.9898520237	.9904397628	.9909947557
* 280+	.9990398360	.9990996489	.9991558245	.9992085776	.9992581110

TABLE C

(Inverse for n)

This table gives the n-value for which the corresponding I - value reaches a specified P^*

for $P^* = .75, .90, .95, .975, .99, .999, .9999,$

for $p = 1/j,$ $j = b + 1(1)20,$

for $r = 1(1)10,$

for $b = 1(1)10.$

The entry given is obtained by linear interpolation and can be used with the help of randomization to obtain P^* exactly; otherwise we use the next largest integer.

DIRICHLET 1, N-VALUES FOR INVERSE PROBLEM (LI)   B= 1

P*-VALUES

R	.7500	.9000	.9500	.9750	.9900	.9990	.9999
			P=1/ 2				
1	2.00	3.40	4.40	5.40	6.72	9.98	13.36
2	4.50	6.20	7.46	8.65	10.15	13.89	17.58
3	6.80	8.81	10.21	11.57	13.26	17.34	21.21
4	9.05	11.33	12.86	14.33	16.15	20.53	24.64
5	11.30	13.77	15.44	16.97	18.93	23.57	27.85
6	13.52	16.15	17.92	19.61	21.68	26.50	30.94
7	15.70	18.53	20.41	22.14	24.33	29.36	33.96
8	17.87	20.85	22.83	24.67	26.91	32.13	36.93
9	20.03	23.17	25.24	27.14	29.52	34.88	39.84
10	22.20	25.48	27.63	29.61	32.01	37.62	42.71
			P=1/ 3				
1	3.47	5.72	7.44	9.12	11.41	17.04	22.76
2	7.20	10.14	12.26	14.30	16.91	23.33	29.56
3	10.76	14.21	16.62	18.89	21.81	28.77	35.44
4	14.22	18.07	20.73	23.22	26.38	33.81	40.84
5	17.62	21.85	24.71	27.39	30.75	38.62	45.95
6	20.97	25.55	28.59	31.44	34.96	43.23	50.89
7	24.32	29.16	32.40	35.40	39.10	47.72	55.68
8	27.63	32.75	36.14	39.28	43.16	52.09	60.33
9	30.91	36.30	39.84	43.11	47.15	56.41	64.86
10	34.18	39.82	43.51	46.90	51.07	60.63	69.34
			P=1/ 4				
1	4.84	8.00	10.45	12.84	16.01	24.01	32.02
2	9.89	14.05	17.01	19.90	23.62	32.64	41.41
3	14.69	19.56	22.93	26.16	30.28	40.07	49.48
4	19.35	24.80	28.53	32.04	36.50	46.96	56.88
5	23.91	29.88	33.90	37.70	42.43	53.52	63.89
6	28.43	34.85	39.15	43.17	48.17	59.80	70.62
7	32.89	39.75	44.30	48.53	53.77	65.90	77.11
8	37.33	44.58	49.36	53.78	59.26	71.86	83.46
9	41.74	49.35	54.35	58.95	64.66	77.70	89.65
10	46.12	54.08	59.29	64.07	69.96	83.44	95.72
			P=1/ 5				
1	6.23	10.34	13.45	16.56	20.66	30.96	41.30
2	12.60	17.95	21.79	25.51	30.28	41.90	53.21
3	18.62	24.89	29.26	33.43	38.72	51.37	63.47
4	24.46	31.50	36.31	40.84	46.58	60.08	72.86
5	30.20	37.89	43.09	47.96	54.07	68.37	81.75
6	35.86	44.15	49.70	54.87	61.33	76.32	90.27
7	41.47	50.31	56.17	61.63	68.40	84.02	98.50
8	47.03	56.38	62.54	68.25	75.32	91.57	106.51
9	52.56	62.38	68.82	74.77	82.11	98.92	114.34
10	58.05	68.32	75.02	81.21	88.82	106.17	122.00
			P=1/ 6				
1	7.63	12.65	16.45	20.25	25.28	37.90	50.54
2	15.30	21.86	26.55	31.08	36.93	51.16	64.99
3	22.54	30.23	35.57	40.66	47.14	62.63	77.43
4	29.58	38.19	44.07	49.64	56.64	73.19	88.83
5	36.48	45.90	52.27	58.23	65.71	83.20	99.59
6	43.29	53.45	60.23	66.57	74.47	92.82	109.88
7	50.03	60.85	68.03	74.71	82.99	102.14	119.84
8	56.72	68.16	75.71	82.70	91.35	111.23	129.53
9	63.37	75.39	83.28	90.57	99.56	120.13	138.98
10	69.98	82.55	90.76	98.33	107.64	128.88	148.27

DIRICHLET 1, N-VALUES FOR INVERSE PROBLEM (LI)   B= 1

P#-VALUES

R	.7500	.9000	.9500	.9750	.9900	.9990	.9999
			P=1/ 7				
1	8.99	14.94	19.45	23.94	29.88	44.82	59.76
2	17.98	25.76	31.30	36.67	43.59	60.42	76.79
3	26.47	35.56	41.87	47.89	55.56	73.87	91.39
4	34.69	44.88	51.84	58.42	66.70	86.28	104.77
5	42.76	53.90	61.43	68.49	77.33	98.01	117.40
6	50.72	62.73	70.76	78.25	87.60	109.30	129.49
7	58.60	71.40	79.89	87.79	97.59	120.23	141.17
8	66.42	79.94	88.87	97.14	107.37	130.89	152.54
9	74.18	88.40	97.73	106.35	116.98	141.33	163.63
10	81.90	96.76	106.49	115.44	126.45	151.58	174.51
			P=1/ 8				
1	10.40	17.26	22.45	27.64	34.50	51.74	68.98
2	20.69	29.65	36.04	42.24	50.23	69.67	88.56
3	30.39	40.88	48.17	55.13	63.97	85.11	105.33
4	39.80	51.57	59.60	67.19	76.76	99.36	120.71
5	49.03	61.90	70.60	78.74	88.94	112.83	135.21
6	58.14	72.01	81.28	89.93	100.72	125.78	149.08
7	67.16	81.93	91.75	100.86	112.17	138.32	162.49
8	76.10	91.73	102.03	111.58	123.39	150.55	175.53
9	84.98	101.40	112.18	122.12	134.41	162.51	188.26
10	93.82	110.98	122.20	132.54	145.25	174.26	200.74
			P=1/ 9				
1	11.78	19.56	25.45	31.33	39.10	58.66	78.21
2	23.38	33.54	40.79	47.82	56.87	78.90	100.33
3	34.31	46.21	54.48	62.36	72.39	96.36	119.28
4	44.91	58.25	67.36	75.96	86.81	112.43	136.64
5	55.31	69.90	79.75	88.98	100.56	127.64	153.00
6	65.57	81.29	91.80	101.61	113.84	142.25	168.67
7	75.72	92.48	103.60	113.93	126.76	156.40	183.80
8	85.79	103.50	115.19	126.01	139.40	170.19	198.51
9	95.79	114.40	126.63	137.90	151.82	183.69	212.87
10	105.74	125.19	137.92	149.63	164.04	196.93	226.95
			P=1/10				
1	13.16	21.86	28.45	35.01	43.72	65.58	87.43
2	26.07	37.44	45.54	53.40	63.52	88.15	112.09
3	38.23	51.54	60.78	69.59	80.80	107.60	133.21
4	50.02	64.93	75.12	84.74	96.86	125.50	152.57
5	61.59	77.90	88.91	99.24	112.17	142.45	170.80
6	72.99	90.57	102.32	113.29	126.95	158.71	188.25
7	84.28	103.01	115.45	126.99	141.34	174.47	205.10
8	95.48	115.28	128.35	140.45	155.41	189.83	221.49
9	106.60	127.40	141.06	153.68	169.23	204.85	237.49
10	117.66	139.41	153.63	166.73	182.84	219.61	253.16
			P=1/11				
1	14.56	24.17	31.44	38.71	48.33	72.49	96.65
2	28.77	41.33	50.29	58.97	70.16	97.39	123.86
3	42.15	56.86	67.07	76.82	89.21	138.58	147.15
4	55.13	71.62	82.87	93.51	106.91	138.58	168.49
5	67.86	85.90	98.07	109.49	123.79	157.25	188.60
6	80.42	99.85	112.84	124.96	140.07	175.18	207.83
7	92.84	113.55	127.30	140.06	155.91	192.55	226.41
8	105.17	127.05	141.50	154.87	171.42	209.47	244.47
9	117.40	140.40	155.51	169.45	186.65	226.02	262.10
10	129.57	153.62	169.35	183.82	201.64	242.28	279.38

DIRICHLET 1, N-VALUES FOR INVERSE PROBLEM (LI)   B= 1

P#-VALUES

R	.7500	.9000	.9500	.9750	.9900	.9990	.9999
				P=1/12			
1	15.94	26.47	34.44	42.41	52.93	79.40	105.86
2	31.46	45.22	55.03	64.55	76.80	106.63	135.63
3	46.07	62.18	73.37	84.04	97.63	130.06	161.09
4	60.25	78.30	90.63	102.28	116.96	151.64	184.42
5	74.14	93.90	107.23	119.73	135.40	172.05	206.39
6	87.84	109.13	123.36	136.64	153.18	191.64	227.41
7	101.41	124.08	139.14	153.13	170.50	210.62	247.71
8	114.85	138.83	154.66	169.30	187.43	229.10	267.44
9	128.21	153.40	169.94	185.21	204.05	247.19	286.71
10	141.49	167.83	185.05	200.91	220.43	264.95	305.58
				P=1/13			
1	17.33	28.77	37.44	46.09	57.54	86.31	115.07
2	34.15	49.10	59.78	70.12	83.45	115.86	147.39
3	49.99	67.51	79.67	91.28	106.03	141.30	175.02
4	65.36	84.98	98.39	111.05	127.01	164.71	200.34
5	80.42	101.89	116.39	129.97	147.00	186.85	224.18
6	95.27	118.41	133.87	148.31	166.30	208.10	246.98
7	109.96	134.62	150.98	166.19	185.07	228.68	269.01
8	124.54	150.60	167.81	183.73	203.43	248.74	290.42
9	139.01	166.40	184.39	200.98	221.46	268.36	311.31
10	153.40	182.03	200.76	218.00	239.21	287.62	331.78
				P=1/14			
1	18.71	31.07	40.43	49.78	62.15	93.22	124.29
2	36.84	52.99	64.53	75.69	90.08	125.10	159.15
3	53.92	72.83	85.97	98.50	114.44	152.54	188.95
4	70.47	91.67	106.14	119.82	137.06	177.78	216.26
5	86.69	109.89	125.54	140.22	158.62	201.65	241.96
6	102.69	127.68	144.39	159.98	179.41	224.57	266.56
7	118.53	145.15	162.83	179.26	199.65	246.75	290.31
8	134.22	162.38	180.96	198.15	219.44	268.37	313.39
9	149.82	179.40	198.82	216.75	238.87	289.52	335.91
10	165.32	196.24	216.47	235.09	258.00	310.28	357.98
				P=1/15			
1	20.10	33.38	43.43	53.48	66.75	100.13	133.51
2	39.54	56.89	69.27	81.27	96.73	134.34	170.91
3	57.84	78.16	92.27	105.73	122.85	163.77	202.89
4	75.58	98.35	113.90	128.59	147.11	190.84	232.17
5	92.97	117.88	134.70	150.47	170.22	216.45	259.76
6	110.12	136.96	154.90	171.65	192.53	241.02	286.13
7	127.08	155.69	174.68	192.32	214.22	264.82	311.61
8	143.91	174.15	194.11	212.58	235.44	288.00	336.36
9	160.62	192.39	213.26	232.52	256.28	310.68	360.52
10	177.23	210.45	232.18	252.18	276.79	332.94	384.19
				P=1/16			
1	21.49	35.68	46.43	57.16	71.36	107.03	142.72
2	42.23	60.78	74.01	86.84	103.37	143.58	182.67
3	61.76	83.48	98.57	112.95	131.26	174.99	216.82
4	80.69	105.03	121.66	137.36	157.16	203.91	248.09
5	99.24	125.88	143.85	160.71	181.83	231.25	277.55
6	117.54	146.23	165.42	183.32	205.64	257.48	305.71
7	135.64	166.22	186.52	205.38	228.79	282.88	332.90
8	153.60	185.92	207.26	227.00	251.45	307.63	359.33
9	171.42	205.39	227.70	248.28	273.69	331.84	385.12
10	189.15	224.66	247.89	269.27	295.58	355.61	410.39

C 4

DIRICHLET 1, N-VALUES FOR INVERSE PROBLEM (LI)   B= 1

P#-VALUES

R	.7500	.9000	.9500	.9750	.9900	.9990	.9999
				P=1/17			
1	22.87	37.98	49.42	60.85	75.96	113.94	151.93
2	44.92	64.67	78.76	92.41	110.00	152.81	194.43
3	65.68	88.80	104.86	120.18	139.67	186.23	230.76
4	85.80	111.72	129.41	146.13	167.21	216.97	264.01
5	105.52	133.87	153.00	170.95	193.44	246.04	295.33
6	124.96	155.51	175.93	194.99	218.75	273.94	325.28
7	144.20	176.76	198.37	218.44	243.37	300.95	354.20
8	163.28	197.69	220.41	241.43	267.45	327.26	382.29
9	182.23	218.39	242.13	264.05	291.09	353.00	409.72
10	201.06	238.87	263.60	286.35	314.37	378.27	436.59
				P=1/18			
1	24.26	40.29	52.42	64.54	80.58	120.86	161.14
2	47.62	68.56	83.50	97.98	116.65	162.04	206.19
3	69.60	94.13	111.16	127.41	148.07	197.46	244.69
4	90.91	118.40	137.17	154.90	177.26	230.04	279.93
5	111.79	141.87	162.16	181.20	205.04	260.84	313.12
6	132.39	164.79	186.45	206.66	231.86	290.40	344.85
7	152.76	187.29	210.21	231.51	257.94	319.01	375.49
8	172.96	209.47	233.56	255.86	283.45	346.89	405.26
9	193.03	231.38	256.57	279.81	308.50	374.16	434.32
10	212.98	253.08	279.31	303.44	333.15	400.93	462.79
				P=1/19			
1	25.65	42.59	55.41	68.23	85.18	127.77	170.36
2	50.31	72.45	88.25	103.56	123.29	171.28	217.94
3	73.52	99.45	117.46	134.64	156.48	208.69	258.62
4	96.01	125.08	144.92	163.67	187.30	243.10	295.85
5	118.07	149.86	171.32	191.44	216.65	275.64	330.90
6	139.81	174.06	196.96	218.33	244.97	306.86	364.43
7	161.32	197.82	222.05	244.57	272.52	337.08	396.78
8	182.65	221.24	246.71	270.28	299.46	366.52	428.23
9	203.83	244.38	271.01	295.58	325.90	395.32	458.92
10	224.89	267.28	295.01	320.53	351.94	423.59	488.98
				P=1/20			
1	27.03	44.89	58.41	71.92	89.79	134.68	179.57
2	53.00	76.34	92.99	109.13	129.92	180.52	229.71
3	77.44	104.77	123.75	141.86	164.89	219.92	272.55
4	101.13	131.76	152.68	172.44	197.35	256.17	311.76
5	124.34	157.86	180.47	201.68	228.26	290.44	348.69
6	147.23	183.34	207.48	230.00	258.08	323.32	383.99
7	169.88	208.36	233.90	257.63	287.09	355.15	418.08
8	192.34	233.01	259.86	284.71	315.46	386.15	451.19
9	214.64	257.37	285.45	311.35	343.31	416.48	483.52
10	236.81	281.49	310.72	337.62	370.72	446.25	515.18
			P=1/ 2,	B = 2			
1	3.00	4.40	5.40	6.40	7.72	10.98	14.36
2	5.80	7.46	8.65	9.80	11.32	14.97	18.65
3	8.36	10.21	11.57	12.84	14.53	18.54	22.38
4	10.80	12.86	14.33	15.73	17.53	21.77	25.82
5	13.19	15.44	16.97	18.50	20.39	24.86	29.08
6	15.56	17.92	19.61	21.17	23.16	27.85	32.28
7	17.87	20.41	22.14	23.81	25.88	30.78	35.36
8	20.18	22.83	24.67	26.40	28.58	33.64	38.35
9	22.48	25.24	27.14	28.92	31.18	36.44	41.28
10	24.74	27.63	29.61	31.47	33.79	39.17	44.15

C 5

DIRICHLET 1, N-VALUES FOR INVERSE PROBLEM (LI)   B= 2
P#-VALUES

R	.7500	.9000	.9500	.9750	.9900	.9990	.9999
			P=1/ 3				
1	5.11	7.43	9.11	10.84	13.08	18.78	24.47
2	9.42	12.25	14.30	16.30	18.87	25.21	31.41
3	13.36	16.61	18.88	21.09	23.93	30.80	37.40
4	17.14	20.72	23.22	25.63	28.68	35.94	42.90
5	20.83	24.70	27.39	29.93	33.18	40.86	48.11
6	24.45	28.59	31.44	34.13	37.56	45.59	53.12
7	28.00	32.39	35.40	38.23	41.80	50.16	57.96
8	31.54	36.13	39.28	42.24	45.94	54.64	62.71
9	35.01	39.83	43.10	46.19	50.03	58.99	67.33
10	38.49	43.51	46.89	50.08	54.06	63.32	71.85
			P=1/ 4				
1	7.15	10.42	12.83	15.25	18.45	26.46	34.46
2	12.95	16.98	19.89	22.72	26.37	35.30	43.99
3	18.29	22.90	26.15	29.29	33.29	42.93	52.24
4	23.40	28.50	32.02	35.43	39.73	49.99	59.79
5	28.36	33.87	37.69	41.30	45.86	56.70	66.92
6	33.22	39.12	43.16	46.96	51.79	63.11	73.77
7	38.01	44.26	48.52	52.53	57.56	69.36	80.39
8	42.75	49.32	53.77	57.95	63.20	75.44	86.81
9	47.44	54.31	58.94	63.30	68.74	81.39	93.09
10	52.09	59.25	64.06	68.58	74.19	87.23	99.26
			P=1/ 5				
1	9.17	13.40	16.54	19.66	23.76	34.07	44.41
2	16.50	21.74	25.49	29.11	33.82	45.34	56.58
3	23.20	29.20	33.40	37.44	42.61	55.04	67.01
4	29.63	36.24	40.82	45.19	50.75	63.98	76.63
5	35.86	43.02	47.94	52.61	58.51	72.47	85.67
6	41.97	49.64	54.85	59.78	65.98	80.61	94.33
7	47.99	56.10	61.60	66.77	73.27	88.49	102.70
8	53.93	62.47	68.22	73.63	80.40	96.16	110.84
9	59.82	68.75	74.75	80.37	87.38	103.69	118.79
10	65.66	74.95	81.18	86.99	94.26	111.06	126.59
			P=1/ 6				
1	11.19	16.38	20.22	24.02	29.06	41.71	54.34
2	20.01	26.47	31.04	35.51	41.27	55.37	69.11
3	28.11	35.48	40.63	45.57	51.89	67.13	81.80
4	35.84	43.97	49.60	54.94	61.75	77.96	93.42
5	43.35	52.16	58.19	63.90	71.12	88.22	104.38
6	50.71	60.12	66.53	72.57	80.17	98.05	114.85
7	57.95	67.92	74.67	81.00	88.96	107.59	124.98
8	65.10	75.60	82.66	89.28	97.56	116.87	134.83
9	72.19	83.16	90.52	97.41	105.99	125.95	144.45
10	79.21	90.64	98.28	105.42	114.30	134.87	153.86
			P=1/ 7				
1	13.20	19.35	23.90	28.43	34.38	49.32	64.26
2	23.55	31.18	36.62	41.89	48.72	65.39	81.66
3	33.01	41.75	47.85	53.70	61.18	79.21	96.56
4	42.05	51.71	58.36	64.70	72.74	91.92	110.20
5	50.84	61.29	68.44	75.19	83.74	103.95	123.06
6	59.44	70.62	78.19	85.35	94.34	115.50	135.37
7	67.90	79.74	87.73	95.23	104.65	126.68	147.25
8	76.27	88.72	97.08	104.91	114.72	137.56	158.80
9	84.54	97.57	106.29	114.43	124.60	148.21	170.07
10	92.75	106.31	115.37	123.81	134.32	158.66	181.12

DIRICHLET 1, N-VALUES FOR INVERSE PROBLEM (LI)   B= 2

P*-VALUES

R	.7500	.9000	.9500	.9750	.9900	.9990	.9999
				P=1/ 8			
1	15.21	22.33	27.59	32.80	39.68	56.93	74.18
2	27.06	35.90	42.18	48.28	56.15	75.40	94.18
3	37.91	48.01	55.06	61.82	70.47	91.28	111.30
4	48.27	59.44	67.12	74.44	83.73	105.88	126.97
5	58.32	70.42	78.67	86.48	96.35	119.69	141.75
6	68.17	81.09	89.86	98.11	108.51	132.93	155.87
7	77.86	91.55	100.78	109.45	120.32	145.75	169.51
8	87.43	101.83	111.50	120.55	131.86	158.24	182.76
9	96.89	111.96	122.04	131.46	143.19	170.45	195.70
10	106.28	121.98	132.45	142.20	154.34	182.44	208.38
				P=1/ 9			
1	17.23	25.30	31.26	37.18	44.97	64.55	84.09
2	30.59	40.62	47.75	54.66	63.59	85.41	106.70
3	42.81	54.29	62.28	69.94	79.75	103.34	126.03
4	54.48	67.15	75.88	84.18	94.72	119.83	143.75
5	65.80	79.54	88.90	97.76	108.94	135.42	160.42
6	76.89	91.57	101.52	110.88	122.67	150.36	176.37
7	87.80	103.36	113.83	123.67	135.99	164.82	191.76
8	98.58	114.94	125.91	136.18	149.01	178.91	206.72
9	109.24	126.36	137.79	148.47	161.78	192.69	221.32
10	119.81	137.65	149.52	160.58	174.35	206.20	235.62
				P=1/10			
1	19.24	28.27	34.94	41.57	50.29	72.15	94.00
2	34.11	45.34	53.31	61.03	71.01	95.42	119.22
3	47.71	60.55	69.50	78.06	89.02	115.40	140.78
4	60.68	74.87	84.64	93.92	105.70	133.78	160.51
5	73.28	88.65	99.13	109.03	121.55	151.13	179.09
6	85.61	102.04	113.17	123.65	136.82	167.78	196.86
7	97.75	115.16	126.88	137.87	151.66	183.89	214.00
8	109.73	128.04	140.32	151.80	166.15	199.58	230.67
9	121.59	140.75	153.55	165.49	180.37	214.92	246.92
10	133.35	153.31	166.59	178.96	194.36	229.97	262.85
				P=1/11			
1	21.25	31.24	38.62	45.94	55.59	79.76	103.91
2	37.63	50.05	58.87	67.42	78.46	105.42	131.73
3	52.60	66.81	76.71	86.19	98.30	127.46	155.51
4	66.89	82.60	93.39	103.66	116.68	147.72	177.27
5	80.76	97.77	109.36	120.32	134.15	166.85	197.76
6	94.34	112.52	124.83	136.41	150.97	185.21	217.35
7	107.70	126.96	139.92	152.08	167.32	202.96	236.25
8	120.88	141.15	154.73	167.43	183.29	220.25	254.61
9	133.94	155.14	169.29	182.50	198.95	237.15	272.53
10	146.87	168.96	183.66	197.34	214.36	253.73	290.07
				P=1/12			
1	23.26	34.21	42.30	50.32	60.88	87.36	113.82
2	41.15	54.77	64.43	73.79	85.88	115.43	144.25
3	57.50	73.08	83.92	94.31	107.58	139.52	170.25
4	73.09	90.32	102.14	113.39	127.66	161.66	194.03
5	88.24	106.88	119.59	131.59	146.75	182.58	216.43
6	103.05	122.99	136.48	149.17	165.13	202.63	237.83
7	117.64	138.76	152.96	166.29	182.98	222.02	258.49
8	132.03	154.26	169.13	183.04	200.43	240.91	278.55
9	146.28	169.53	185.04	199.51	217.54	259.38	298.13
10	160.40	184.63	200.73	215.72	234.37	277.50	317.31

DIRICHLET 1, N-VALUES FOR INVERSE PROBLEM (LI)   B= 2
P*-VALUES

R	.7500	.9000	.9500	.9750	.9900	.9990	.9999
			P=1/13				
1	25.27	37.18	45.97	54.70	66.18	94.96	123.74
2	44.67	59.49	69.98	80.17	93.32	125.43	156.76
3	62.39	79.34	91.13	102.43	116.85	151.58	184.97
4	79.30	98.03	110.90	123.12	138.64	175.60	210.79
5	95.71	116.00	129.81	142.86	159.34	198.29	235.09
6	111.78	133.47	148.13	161.93	179.28	220.04	258.31
7	127.58	150.56	166.01	180.50	198.65	241.09	280.72
8	143.19	167.36	183.54	198.67	217.57	261.58	302.49
9	158.63	183.92	200.79	216.52	236.12	281.61	323.73
10	173.93	200.28	217.80	234.09	254.37	301.25	344.53
			P=1/14				
1	27.28	40.15	49.66	59.07	71.48	102.57	133.64
2	48.19	64.20	75.55	86.55	100.75	135.43	169.27
3	67.29	85.61	98.34	110.54	126.13	163.63	199.71
4	85.51	105.75	119.65	132.86	149.62	189.54	227.55
5	103.19	125.11	140.03	154.14	171.93	214.00	253.75
6	120.50	143.94	159.78	174.69	193.43	237.46	278.80
7	137.53	162.36	179.05	194.70	214.31	260.15	302.96
8	154.34	180.46	197.94	214.29	234.70	282.24	326.43
9	170.97	198.31	216.53	233.52	254.70	303.83	349.33
10	187.46	215.94	234.86	252.47	274.37	325.01	371.75
			P=1/15				
1	29.29	43.12	53.34	63.46	76.78	110.17	143.55
2	51.71	68.91	81.10	92.92	108.18	145.43	181.78
3	72.18	91.87	105.55	118.66	135.40	175.69	214.44
4	91.71	113.47	128.40	142.60	160.60	203.48	244.31
5	110.66	134.23	150.26	165.41	184.53	229.72	272.42
6	129.22	154.41	171.44	187.45	207.58	254.87	299.28
7	147.47	174.16	192.09	208.91	229.97	279.21	325.19
8	165.49	193.57	212.35	229.90	251.84	302.89	350.37
9	183.31	212.70	232.27	250.53	273.28	326.05	374.92
10	200.98	231.60	251.92	270.84	294.37	348.77	398.97
			P=1/16				
1	31.31	46.09	57.01	67.83	82.07	117.78	153.46
2	55.23	73.63	86.66	99.30	115.61	155.43	194.30
3	77.07	98.13	112.76	126.78	144.68	187.74	229.17
4	97.91	121.18	137.15	152.33	171.58	217.42	261.06
5	118.14	143.35	160.49	176.69	197.12	245.44	291.07
6	137.93	164.88	183.09	200.21	221.73	272.29	319.76
7	157.41	185.96	205.13	223.11	245.63	298.27	347.43
8	176.63	206.67	226.75	245.53	268.97	323.56	374.30
9	195.66	227.08	248.02	267.54	291.86	348.27	400.52
10	214.51	247.25	268.99	289.22	314.37	372.52	426.19
			P=1/17				
1	33.32	49.06	60.69	72.21	87.37	125.38	163.36
2	58.75	78.34	92.22	105.67	123.03	165.43	206.80
3	81.97	104.39	119.97	134.89	153.94	199.80	243.89
4	104.12	128.90	145.90	162.06	182.55	231.35	277.82
5	125.62	152.46	170.71	187.96	209.72	261.15	309.74
6	146.65	175.35	194.74	212.97	235.88	289.71	340.24
7	167.35	197.76	218.18	237.32	261.29	317.33	369.66
8	187.78	219.77	241.15	261.14	286.10	344.21	398.24
9	208.00	241.47	263.76	284.54	310.44	370.50	426.12
10	228.03	262.90	286.05	307.59	334.37	396.28	453.42

C 8

DIRICHLET 1, N-VALUES FOR INVERSE PROBLEM (LI)   B= 2

P#-VALUES

R	.7500	.9000	.9500	.9750	.9900	.9990	.9999
				P=1/18			
1	35.33	52.03	64.37	76.59	92.67	132.98	173.27
2	62.27	83.05	97.78	112.04	130.47	175.44	219.32
3	86.86	110.65	127.18	143.00	163.22	211.85	258.63
4	110.32	136.62	154.66	171.79	193.53	245.29	294.58
5	133.09	161.57	180.94	199.23	222.31	276.86	328.40
6	155.37	185.82	206.39	225.73	250.03	307.12	360.71
7	177.29	209.55	231.22	251.52	276.95	336.39	391.89
8	198.93	232.87	255.56	276.76	303.24	364.87	422.17
9	220.34	255.85	279.50	301.55	329.01	392.72	451.71
10	241.56	278.56	303.12	325.96	354.36	420.03	480.63
				P=1/19			
1	37.34	55.00	68.04	80.96	97.96	140.59	183.17
2	65.79	87.77	103.34	118.42	137.89	185.44	231.82
3	91.76	116.91	134.39	151.12	172.49	223.90	273.35
4	116.53	144.33	163.41	181.53	204.51	259.23	311.33
5	140.57	170.69	191.16	210.50	234.90	292.58	347.05
6	164.09	196.30	218.04	238.49	264.18	324.54	381.19
7	187.23	221.35	244.26	265.72	292.61	355.45	414.12
8	210.08	245.97	269.96	292.38	320.37	385.53	446.10
9	232.68	270.24	295.25	318.55	347.59	414.94	477.31
10	255.08	294.21	320.18	344.33	374.36	443.78	507.85
				P=1/20			
1	39.35	57.98	71.72	85.34	103.26	148.19	193.08
2	69.31	92.48	108.89	124.80	145.32	195.44	244.33
3	96.65	123.18	141.60	159.24	181.77	235.95	288.08
4	122.73	152.05	172.16	191.26	215.48	273.16	328.08
5	148.04	179.80	201.39	221.78	247.50	308.29	365.71
6	172.81	206.76	229.69	251.24	278.33	341.94	401.67
7	197.18	233.15	257.30	279.93	308.27	374.50	436.36
8	221.23	259.07	284.36	307.99	337.50	406.19	470.03
9	245.02	284.62	310.99	335.56	366.17	437.16	502.90
10	268.61	309.87	337.24	362.70	394.36	467.54	535.07
				P=1/21			
1	41.36	60.95	75.40	89.72	108.56	155.79	202.98
2	72.82	97.19	114.45	131.17	152.75	205.44	256.84
3	101.54	129.44	148.81	167.36	191.03	248.01	302.81
4	128.94	159.76	180.91	200.99	226.46	287.10	344.83
5	155.52	188.91	211.61	233.04	260.09	324.00	384.37
6	181.53	217.23	241.34	264.00	292.48	359.36	422.15
7	207.12	244.94	270.34	294.13	323.92	393.56	458.59
8	232.38	272.17	298.76	323.61	354.64	426.84	493.97
9	257.36	299.01	326.73	352.56	384.75	459.38	528.49
10	282.13	325.52	354.31	381.07	414.36	491.29	562.29
				P=1/ 3, B = 3			
1	6.11	8.43	10.11	11.84	14.08	19.78	25.47
2	10.68	13.46	15.48	17.45	19.98	26.32	32.49
3	14.81	17.93	20.19	22.38	25.17	31.95	38.55
4	18.76	22.19	24.64	26.97	29.97	37.20	44.09
5	22.59	26.29	28.89	31.40	34.59	42.16	49.38
6	26.32	30.27	33.01	35.67	38.99	46.93	54.44
7	29.99	34.16	37.05	39.83	43.33	51.58	59.33
8	33.63	37.98	41.01	43.90	47.55	56.07	64.06
9	37.22	41.78	44.92	47.91	51.70	60.52	68.74
10	40.78	45.52	48.79	51.88	55.77	64.86	73.31

DIRICHLET 1, N-VALUES FOR INVERSE PROBLEM (LI)   B= 3

P*-VALUES

R	.7500	.9000	.9500	.9750	.9900	.9990	.9999
			P=1/ 4				
1	8.56	11.82	14.25	16.67	19.85	27.85	35.85
2	14.73	18.70	21.56	24.34	27.94	36.84	45.54
3	20.33	24.81	27.98	31.06	35.01	44.61	53.84
4	25.66	30.58	34.01	37.35	41.59	51.76	61.49
5	30.81	36.11	39.81	43.34	47.83	58.54	68.69
6	35.86	41.50	45.41	49.13	53.85	65.02	75.60
7	40.81	46.77	50.87	54.78	59.71	71.35	82.27
8	45.70	51.94	56.24	60.31	65.44	77.49	88.76
9	50.53	57.04	61.52	65.75	71.05	83.51	95.08
10	55.31	62.09	66.73	71.09	76.59	89.41	101.31
			P=1/ 5				
1	10.95	15.20	18.35	21.47	25.59	35.89	46.22
2	18.75	23.90	27.61	31.21	35.87	47.34	58.54
3	25.81	31.65	35.76	39.74	44.83	57.18	69.10
4	32.52	38.91	43.39	47.68	53.14	66.26	78.80
5	39.00	45.90	50.69	55.24	61.03	74.83	87.93
6	45.35	52.69	57.75	62.57	68.66	83.06	96.69
7	51.58	59.33	64.65	69.69	76.04	91.03	105.12
8	57.71	65.84	71.41	76.66	83.28	98.81	113.35
9	63.78	72.27	78.05	83.52	90.37	106.42	121.37
10	69.79	78.62	84.62	90.26	97.35	113.87	129.22
			P=1/ 6				
1	13.36	18.59	22.44	26.26	31.30	43.92	56.57
2	22.77	29.11	33.66	38.06	43.79	57.81	71.53
3	31.28	38.47	43.52	48.39	54.64	69.75	84.34
4	39.37	47.25	52.73	57.97	64.68	80.74	96.08
5	47.18	55.68	61.55	67.13	74.23	91.11	107.15
6	54.82	63.86	70.07	75.96	83.43	101.07	117.74
7	62.32	71.86	78.40	84.57	92.37	110.72	127.95
8	69.70	79.72	86.55	92.98	101.09	120.11	137.89
9	77.00	87.47	94.57	101.26	109.66	129.30	147.60
10	84.23	95.11	102.48	109.40	118.07	138.31	157.09
			P=1/ 7				
1	15.76	21.95	26.53	31.04	36.99	51.94	66.88
2	26.78	34.32	39.69	44.91	51.69	68.27	84.50
3	36.75	45.28	51.26	57.02	64.43	82.30	99.56
4	46.21	55.57	62.06	68.28	76.21	95.20	113.36
5	55.35	65.44	72.40	79.00	87.41	107.39	126.35
6	64.28	75.02	82.39	89.37	98.19	119.07	138.78
7	73.04	84.39	92.13	99.45	108.67	130.39	150.77
8	81.68	93.59	101.68	109.31	118.89	141.40	162.43
9	90.22	102.65	111.06	118.98	128.92	152.16	173.80
10	98.67	111.60	120.33	128.53	138.79	162.72	184.94
			P=1/ 8				
1	18.15	25.33	30.61	35.83	42.72	59.96	77.21
2	30.79	39.52	45.73	51.77	59.59	78.73	97.46
3	42.21	52.08	59.00	65.67	74.20	94.84	114.77
4	53.04	63.88	71.40	78.58	87.73	109.66	130.62
5	63.51	75.20	83.24	90.88	100.58	123.65	145.54
6	73.73	86.18	94.69	102.76	112.94	137.05	159.80
7	83.77	96.91	105.86	114.31	124.96	150.03	173.58
8	93.66	107.45	116.80	125.62	136.69	162.67	186.94
9	103.43	117.83	127.56	136.71	148.19	175.01	199.99
10	113.09	128.07	138.17	147.64	159.50	187.13	212.79

DIRICHLET 1, N-VALUES FOR INVERSE PROBLEM (LI)   B= 3

R	.7500	.9000	.9500	P*-VALUES .9750	.9900	.9990	.9999

P=1/ 9

R	.7500	.9000	.9500	.9750	.9900	.9990	.9999
1	20.55	28.70	34.69	40.62	48.43	67.98	87.54
2	34.80	44.71	51.76	58.62	67.48	89.18	110.42
3	47.67	58.89	66.75	74.30	83.98	107.39	129.96
4	59.87	72.19	80.72	88.87	99.25	124.11	147.87
5	71.67	84.95	94.07	102.75	113.75	139.90	164.72
6	83.18	97.33	106.99	116.14	127.71	155.03	180.82
7	94.49	109.43	119.59	129.17	141.25	169.68	196.37
8	105.63	121.31	131.92	141.91	154.48	183.93	211.46
9	116.63	133.00	144.05	154.43	167.45	197.86	226.19
10	127.52	144.55	156.00	166.75	180.20	211.53	240.62

P=1/10

R	.7500	.9000	.9500	.9750	.9900	.9990	.9999
1	22.94	32.07	38.77	45.41	54.13	75.99	97.85
2	38.80	49.90	57.80	65.46	75.37	99.64	123.37
3	53.12	65.70	74.49	82.93	93.76	119.92	145.17
4	66.70	80.51	90.04	99.16	110.77	138.56	165.12
5	79.83	94.70	104.91	114.62	126.91	156.15	183.89
6	92.64	108.48	119.29	129.53	142.46	173.01	201.84
7	105.20	121.94	133.31	144.03	157.54	189.32	219.15
8	117.59	135.16	147.03	158.22	172.27	205.19	235.96
9	129.83	148.17	160.54	172.14	186.70	220.71	252.38
10	141.93	161.01	173.84	185.85	200.89	235.92	268.44

P=1/11

R	.7500	.9000	.9500	.9750	.9900	.9990	.9999
1	25.34	35.44	42.85	50.18	59.83	84.00	108.17
2	42.81	55.10	63.83	72.30	83.26	110.08	136.32
3	58.58	72.50	82.22	91.57	103.54	132.47	160.37
4	73.53	88.81	99.37	109.45	122.29	153.00	182.37
5	87.98	104.46	115.75	126.48	140.08	172.40	203.07
6	102.08	119.63	131.59	142.91	157.20	190.98	222.85
7	115.92	134.46	147.03	158.89	173.82	208.95	241.94
8	129.56	149.00	162.15	174.52	190.04	226.45	260.48
9	143.02	163.34	177.01	189.86	205.95	243.55	278.56
10	156.35	177.48	191.67	204.96	221.59	260.31	296.26

P=1/12

R	.7500	.9000	.9500	.9750	.9900	.9990	.9999
1	27.73	38.81	46.93	54.96	65.54	92.01	118.49
2	46.81	60.30	69.86	79.14	91.15	120.54	149.27
3	64.03	79.30	89.95	100.20	113.32	144.99	175.57
4	80.36	97.12	108.69	119.74	133.80	167.46	199.62
5	96.13	114.20	126.59	138.34	153.24	188.65	222.25
6	111.53	130.77	143.89	156.29	171.95	208.95	243.86
7	126.63	146.96	160.75	173.75	190.10	228.60	264.72
8	141.52	162.85	177.26	190.81	207.83	247.70	284.97
9	156.22	178.50	193.50	207.57	225.20	266.38	304.73
10	170.77	193.94	209.50	224.06	242.28	284.70	324.07

P=1/13

R	.7500	.9000	.9500	.9750	.9900	.9990	.9999
1	30.12	42.18	51.00	59.75	71.24	100.03	128.80
2	50.82	65.49	75.89	85.98	99.03	130.98	162.22
3	69.49	86.10	97.69	108.83	123.08	157.53	190.77
4	87.19	105.43	118.01	130.02	145.31	181.90	216.86
5	104.29	123.95	137.42	150.20	166.40	204.89	241.42
6	120.97	141.92	156.18	169.67	186.70	226.92	264.87
7	137.35	159.47	174.47	188.60	206.39	248.23	287.50
8	153.48	176.70	192.38	207.11	225.61	268.95	309.47
9	169.41	193.66	209.97	225.28	244.45	289.22	330.90
10	185.18	210.41	227.32	243.16	262.96	309.08	351.89

C11

DIRICHLET 1, N-VALUES FOR INVERSE PROBLEM (LI)　　B= 3

P*-VALUES

R	.7500	.9000	.9500	.9750	.9900	.9990	.9999
			P=1/14				
1	32.52	45.55	55.09	64.53	76.94	108.04	139.11
2	54.82	70.68	81.92	92.83	106.92	141.44	175.17
3	74.94	92.90	105.43	117.46	132.86	170.06	205.96
4	94.01	113.73	127.34	140.31	156.83	196.34	234.11
5	112.44	133.70	148.25	162.06	179.56	221.14	260.59
6	130.42	153.06	168.48	183.04	201.44	244.89	285.88
7	148.06	171.98	188.18	203.45	222.67	267.86	310.28
8	165.44	190.55	207.49	223.41	243.39	290.21	333.97
9	182.61	208.83	226.46	242.98	263.70	312.05	357.08
10	199.60	226.87	245.15	262.26	283.66	333.47	379.70
			P=1/15				
1	34.91	48.91	59.17	69.31	82.65	116.05	149.43
2	58.83	75.87	87.95	99.67	114.81	151.88	188.12
3	80.40	99.70	113.16	126.08	142.64	182.60	221.15
4	100.84	122.03	136.66	150.60	168.34	210.78	251.35
5	120.59	143.45	159.08	173.92	192.72	237.38	279.75
6	139.86	164.21	180.77	196.43	216.18	262.86	306.89
7	158.77	184.49	201.90	218.31	238.95	287.50	333.05
8	177.40	204.39	222.59	239.70	261.17	311.46	358.47
9	195.80	223.99	242.93	260.70	282.94	334.88	383.25
10	214.01	243.33	262.97	281.36	304.35	357.85	407.52
			P=1/16				
1	37.31	52.28	63.25	74.09	88.35	124.05	159.74
2	62.83	81.06	93.97	106.51	122.70	162.33	201.07
3	85.85	106.50	120.89	134.71	152.41	195.13	236.35
4	107.67	130.34	145.97	160.88	179.85	225.23	268.59
5	128.74	153.19	169.91	185.78	205.88	253.63	298.92
6	149.30	175.35	193.06	209.80	230.92	280.83	327.89
7	169.48	196.99	215.62	233.16	255.23	307.12	355.83
8	189.36	218.24	237.70	255.99	278.94	332.71	382.96
9	208.99	239.15	259.41	278.40	302.19	357.72	409.42
10	228.42	259.79	280.80	300.46	325.03	382.23	435.32
			P=1/17				
1	39.70	55.65	67.33	78.87	94.05	132.06	170.05
2	66.83	86.25	100.00	113.35	130.59	172.77	214.01
3	91.30	113.30	128.63	143.34	162.18	207.66	251.54
4	114.49	138.65	155.29	171.16	191.36	239.67	285.83
5	136.89	162.93	180.75	197.64	219.04	269.87	318.09
6	158.75	186.49	205.36	223.18	245.67	298.79	348.89
7	180.19	209.50	229.33	248.01	271.51	326.76	378.61
8	201.32	232.08	252.81	272.28	296.72	353.96	407.46
9	222.18	254.31	275.88	296.11	321.43	380.55	435.59
10	242.83	276.25	298.62	319.56	345.72	406.61	463.13
			P=1/18				
1	42.09	59.01	71.41	83.65	99.75	140.07	180.36
2	70.84	91.45	106.03	120.19	138.48	183.22	226.96
3	96.75	120.10	136.36	151.97	171.95	220.19	266.73
4	121.32	146.95	164.61	181.45	202.87	254.11	303.07
5	145.04	172.68	191.58	209.50	232.20	286.11	337.26
6	168.19	197.64	217.65	236.56	260.41	316.76	369.90
7	190.90	222.00	243.04	262.86	287.78	346.39	401.39
8	213.28	245.92	267.92	288.58	314.50	375.21	431.95
9	235.37	269.47	292.36	313.81	340.68	403.38	461.76
10	257.24	292.71	316.45	338.65	366.41	430.99	490.94

DIRICHLET 1, N-VALUES FOR INVERSE PROBLEM (LI)   B= 3

P*-VALUES

R	.7500	.9000	.9500	.9750	.9900	.9990	.9999
			P=1/19				
1	44.48	62.39	75.49	88.44	105.46	148.08	190.67
2	74.84	96.64	112.06	127.02	146.36	193.67	239.91
3	102.21	126.90	144.09	160.60	181.72	232.72	281.92
4	128.14	155.25	173.93	191.73	214.38	268.55	320.32
5	153.19	182.43	202.41	221.36	245.36	302.35	356.42
6	177.63	208.78	229.94	249.93	275.15	334.72	390.90
7	201.61	234.51	256.76	277.71	304.06	366.01	424.16
8	225.23	259.77	283.02	304.87	332.27	396.46	456.44
9	248.57	284.63	308.83	331.52	359.92	426.21	487.93
10	271.65	309.16	334.27	357.75	387.09	455.37	518.75
			P=1/20				
1	46.87	65.75	79.56	93.21	111.15	156.09	200.98
2	78.84	101.82	118.09	133.87	154.25	204.11	252.86
3	107.66	133.70	151.83	169.22	191.49	245.25	297.11
4	134.97	163.56	183.25	202.01	225.89	282.99	337.56
5	161.34	192.17	213.24	233.22	258.51	318.59	375.59
6	187.07	219.92	242.23	263.31	289.89	352.69	411.90
7	212.32	247.02	270.47	292.56	320.34	385.65	446.93
8	237.19	273.61	298.13	321.16	350.05	417.71	480.93
9	261.76	299.79	325.31	349.22	379.16	449.03	514.09
10	286.06	325.62	352.09	376.85	407.78	479.75	546.56
			P=1/21				
1	49.27	69.12	83.64	97.99	116.86	164.10	211.30
2	82.84	107.01	124.12	140.71	162.13	214.56	265.81
3	113.11	140.50	159.56	177.85	201.26	257.78	312.31
4	141.79	171.86	192.57	212.30	237.40	297.43	354.80
5	169.49	201.91	224.07	245.07	271.67	334.83	394.75
6	196.52	231.06	254.52	276.68	304.64	370.65	432.91
7	223.03	259.52	284.19	307.41	336.62	405.28	469.71
8	249.15	287.45	313.24	337.45	367.83	438.95	505.43
9	274.95	314.95	341.78	366.93	398.41	471.86	540.26
10	300.47	342.08	369.91	395.94	428.46	504.13	574.36
			P=1/22				
1	51.66	72.49	87.72	102.77	122.56	172.10	221.61
2	86.85	112.21	130.15	147.55	170.02	225.00	278.75
3	118.56	147.30	167.29	186.48	211.03	270.31	327.50
4	148.62	180.16	201.88	222.58	248.91	311.87	372.03
5	177.64	211.66	234.90	256.93	284.83	351.07	413.91
6	205.96	242.20	266.81	290.05	319.38	388.62	453.91
7	233.74	272.03	297.90	322.26	352.89	424.90	492.48
8	261.11	301.29	328.34	353.74	385.60	460.20	529.92
9	288.14	330.10	358.26	384.63	417.65	494.70	566.43
10	314.88	358.54	387.74	415.04	449.14	528.51	602.17
			P=1/ 4) B = 4				
1	9.56	12.82	15.25	17.67	20.85	28.85	36.85
2	15.95	19.87	22.71	25.49	29.07	37.92	46.62
3	21.74	26.13	29.28	32.33	36.25	45.77	54.97
4	27.21	32.00	35.42	38.70	42.88	52.97	62.68
5	32.51	37.66	41.29	44.77	49.21	59.82	69.92
6	37.66	43.13	46.96	50.64	55.30	66.39	76.87
7	42.72	48.50	52.52	56.36	61.22	72.74	83.60
8	47.70	53.75	57.94	61.94	66.99	78.92	90.11
9	52.63	58.92	63.30	67.45	72.68	84.97	96.51
10	57.50	64.03	68.57	72.85	78.26	90.92	102.75

DIRICHLET 1, N-VALUES FOR INVERSE PROBLEM (LI)   B= 4

P*-VALUES

R	.7500	.9000	.9500	.9750	.9900	.9990	.9999
				P=1/ 5			
1	12.24	16.50	19.64	22.76	26.86	37.19	47.52
2	20.34	25.44	29.09	32.69	37.33	48.75	59.92
3	27.63	33.36	37.42	41.35	46.42	58.70	70.58
4	34.52	40.78	45.17	49.41	54.82	67.85	80.34
5	41.17	47.89	52.59	57.08	62.81	76.51	89.54
6	47.66	54.80	59.76	64.50	70.52	84.81	98.35
7	54.02	61.55	66.76	71.71	77.98	92.84	106.84
8	60.29	68.17	73.61	78.77	85.30	100.68	115.10
9	66.48	74.69	80.35	85.70	92.46	108.33	123.17
10	72.59	81.12	86.98	92.53	99.50	115.84	131.07
				P=1/ 6			
1	14.91	20.15	24.00	27.83	32.87	45.51	58.13
2	24.71	30.97	35.49	39.86	45.57	59.54	73.22
3	33.50	40.56	45.55	50.36	56.57	71.60	86.13
4	41.81	49.52	54.91	60.10	66.74	82.69	97.96
5	49.82	58.11	63.87	69.39	76.41	93.16	109.11
6	57.64	66.45	72.54	78.34	85.71	103.20	119.77
7	65.30	74.59	80.97	87.04	94.74	112.92	130.05
8	72.84	82.57	89.24	95.57	103.56	122.39	140.05
9	80.29	90.43	97.37	103.93	112.21	131.65	149.81
10	87.66	98.19	105.38	112.17	120.71	140.72	159.38
				P=1/ 7			
1	17.60	23.81	28.39	32.91	38.87	53.82	68.76
2	29.06	36.53	41.85	47.04	53.78	70.32	86.50
3	39.36	47.75	53.66	59.37	66.70	84.49	101.68
4	49.08	58.25	64.66	70.79	78.65	97.51	115.59
5	58.47	68.32	75.15	81.68	89.97	109.80	128.67
6	67.61	78.07	85.30	92.17	100.88	121.59	141.18
7	76.57	87.61	95.18	102.38	111.48	132.98	153.25
8	85.39	96.95	104.86	112.36	121.81	144.08	164.97
9	94.09	106.15	114.38	122.15	131.94	154.93	176.43
10	102.70	115.23	123.76	131.80	141.90	165.58	187.64
				P=1/ 8			
1	20.27	27.48	32.76	37.98	44.87	62.12	79.37
2	33.42	42.05	48.23	54.23	61.99	81.08	99.77
3	45.21	54.93	61.77	68.37	76.83	97.37	117.22
4	56.35	66.97	74.38	81.48	90.55	112.32	133.18
5	67.09	78.52	86.42	93.95	103.55	126.44	148.21
6	77.57	89.70	98.05	105.99	116.06	139.95	162.58
7	87.82	100.62	109.39	117.70	128.20	153.04	176.44
8	97.92	111.33	120.48	129.14	140.05	165.77	189.89
9	107.88	121.86	131.38	140.37	151.68	178.22	203.02
10	117.74	132.27	142.12	151.42	163.09	190.42	215.89
				P=1/ 9			
1	22.94	31.12	37.12	43.05	50.86	70.43	89.97
2	37.77	47.60	54.60	61.40	70.22	91.85	113.03
3	51.06	62.11	69.87	77.36	86.96	110.25	132.75
4	63.62	75.70	84.10	92.15	102.44	127.12	150.79
5	75.73	88.71	97.68	106.24	117.11	143.06	167.76
6	87.52	101.32	110.80	119.81	131.23	158.33	183.96
7	99.07	113.63	123.58	133.01	144.93	173.09	199.62
8	110.45	125.70	136.09	145.91	158.30	187.46	214.80
9	121.67	137.58	148.38	158.58	171.40	201.50	229.63
10	132.77	149.29	160.49	171.03	184.27	215.26	244.13

DIRICHLET 1, N-VALUES FOR INVERSE PROBLEM (LI)   B= 4
P#-VALUES

R	.7500	.9000	.9500	.9750	.9900	.9990	.9999
				P=1/10			
1	25.62	34.78	41.50	48.13	56.86	78.73	100.59
2	42.11	53.12	60.95	68.58	78.44	102.63	126.31
3	56.91	69.30	77.98	86.35	97.09	123.12	148.28
4	70.88	84.43	93.83	102.83	114.33	141.93	168.38
5	84.35	98.90	108.94	118.51	130.68	159.69	187.29
6	97.47	112.93	123.55	133.63	146.40	176.69	205.36
7	110.32	126.63	137.77	148.32	161.65	193.13	222.78
8	122.97	140.06	151.70	162.69	176.54	209.13	239.71
9	135.46	153.28	165.38	176.78	191.12	224.77	256.21
10	147.80	166.32	178.85	190.64	205.45	240.09	272.38
				P=1/11			
1	28.29	38.44	45.86	53.20	62.85	87.02	111.19
2	46.47	58.66	67.32	75.75	86.65	113.39	139.57
3	62.76	76.48	86.08	95.34	107.22	135.99	163.81
4	78.14	93.14	103.55	113.51	126.22	156.74	185.97
5	92.97	109.09	120.20	130.79	144.24	176.31	206.82
6	107.42	124.55	136.29	147.44	161.56	195.05	226.74
7	121.57	139.64	151.96	163.63	178.37	213.17	245.95
8	135.50	154.43	167.30	179.46	194.77	230.80	264.61
9	149.24	168.98	182.37	194.97	210.83	248.03	282.80
10	162.83	183.34	197.21	210.25	226.63	264.92	300.61
				P=1/12			
1	30.95	42.08	50.22	58.27	68.84	95.33	121.79
2	50.81	64.19	73.69	82.91	94.86	124.15	152.83
3	68.60	83.66	94.19	104.33	117.35	148.86	179.34
4	85.41	101.86	113.27	124.18	138.11	171.54	203.57
5	101.60	119.28	131.46	143.06	157.79	192.92	226.36
6	117.36	136.16	149.03	161.25	176.72	213.41	248.12
7	132.81	152.64	166.15	178.93	195.08	233.21	269.11
8	148.01	168.79	182.90	196.22	212.99	252.48	289.50
9	163.02	184.68	199.36	213.17	230.55	271.30	309.38
10	177.85	200.36	215.56	229.86	247.80	289.75	328.84
				P=1/13			
1	33.63	45.74	54.59	63.34	74.83	103.63	132.40
2	55.16	69.72	80.04	90.08	103.07	134.92	166.09
3	74.45	90.83	102.29	113.32	127.48	161.74	194.86
4	92.67	110.58	122.98	134.85	149.99	186.34	221.15
5	110.22	129.47	142.72	155.33	171.35	209.55	245.88
6	127.31	147.77	161.78	175.06	191.88	231.77	269.50
7	144.06	165.64	180.34	194.24	211.80	253.25	292.28
8	160.54	183.15	198.50	212.98	231.23	274.14	314.39
9	176.80	200.38	216.34	231.37	250.26	294.56	335.96
10	192.87	217.38	233.91	249.46	268.96	314.58	357.07
				P=1/14			
1	36.30	49.39	58.95	68.41	80.82	111.92	142.99
2	59.51	75.24	86.41	97.25	111.29	145.68	179.35
3	80.30	98.01	110.40	122.31	137.60	174.61	210.38
4	99.92	119.29	132.70	145.53	161.88	201.13	238.74
5	118.84	139.66	153.97	167.60	184.90	226.16	265.41
6	137.25	159.39	174.52	188.87	207.04	250.12	290.88
7	155.30	178.63	194.52	209.55	228.51	273.29	315.44
8	173.05	197.51	214.10	229.75	249.46	295.81	339.29
9	190.57	216.08	233.33	249.57	269.97	317.82	362.54
10	207.89	234.39	252.27	269.06	290.13	339.40	385.30

C15

DIRICHLET 1, N-VALUES FOR INVERSE PROBLEM (LI)   B= 4
P*-VALUES

R	.7500	.9000	.9500	.9750	.9900	.9990	.9999
				P=1/15			
1	38.96	53.04	63.32	73.48	86.81	120.22	153.60
2	63.85	80.77	92.77	104.42	119.50	156.45	192.61
3	86.14	105.19	118.50	131.30	147.72	187.48	225.90
4	107.18	128.00	142.42	156.20	173.77	215.93	256.33
5	127.46	149.84	165.23	179.87	198.46	242.78	284.93
6	147.20	170.99	187.26	202.68	222.20	268.48	312.26
7	166.54	191.63	208.71	224.85	245.23	293.32	338.60
8	185.57	211.87	229.70	246.52	267.69	317.48	364.17
9	204.35	231.77	250.31	267.76	289.69	341.08	389.11
10	222.91	251.41	270.62	288.67	311.30	364.23	413.53
				P=1/16			
1	41.64	56.70	67.69	78.54	92.80	128.52	164.20
2	68.20	86.30	99.13	111.59	127.71	167.21	205.87
3	91.98	112.36	126.60	140.29	157.85	200.34	241.42
4	114.44	136.72	152.14	166.87	185.66	230.73	273.91
5	136.07	160.03	176.49	192.14	212.01	259.39	304.46
6	157.14	182.60	200.00	216.49	237.36	286.83	333.63
7	177.78	204.63	222.89	240.15	261.93	313.35	361.76
8	198.09	226.22	245.30	263.28	285.91	339.14	389.06
9	218.12	247.47	267.30	285.95	309.40	364.35	415.69
10	237.93	268.42	288.97	308.27	332.47	389.04	441.75
				P=1/17			
1	44.30	60.35	72.05	83.61	98.79	136.81	174.80
2	72.54	91.83	105.50	118.76	135.92	177.96	219.12
3	97.83	119.54	134.70	149.27	167.97	213.21	256.94
4	121.70	145.44	161.85	177.54	197.54	245.53	291.50
5	144.70	170.21	187.74	204.41	225.57	276.00	323.98
6	167.08	194.21	212.74	230.30	252.52	305.19	355.00
7	189.02	217.63	237.07	255.46	278.65	333.39	384.92
8	210.60	240.58	260.89	280.04	304.14	360.81	413.95
9	231.90	263.16	284.28	304.15	329.10	387.61	442.26
10	252.95	285.44	307.32	327.87	353.64	413.87	469.97
				P=1/18			
1	46.97	63.99	76.42	88.68	104.78	145.11	185.40
2	76.89	97.36	111.85	125.92	144.13	188.73	232.38
3	103.67	126.71	142.80	158.26	178.09	226.08	272.46
4	128.95	154.15	171.57	188.21	209.43	260.33	309.08
5	153.31	180.40	198.99	216.68	239.12	292.62	343.51
6	177.03	205.82	225.48	244.11	267.68	323.54	376.38
7	200.26	230.62	251.26	270.76	295.36	353.42	408.07
8	223.12	254.94	276.49	296.80	322.37	382.48	438.84
9	245.67	278.86	301.27	322.34	348.81	410.86	468.84
10	267.97	302.45	325.67	347.47	374.80	438.69	498.20
				P=1/19			
1	49.64	67.65	80.78	93.74	110.77	153.40	195.99
2	81.23	102.88	118.21	133.09	152.34	199.49	245.64
3	109.52	133.89	150.90	167.25	188.21	238.94	287.98
4	136.21	162.86	181.28	198.88	221.31	275.12	326.67
5	161.93	190.58	210.25	228.94	252.68	309.23	363.03
6	186.97	217.43	238.22	257.91	282.83	341.89	397.75
7	211.50	243.62	265.44	286.06	312.07	373.45	431.23
8	235.63	269.29	292.08	313.56	340.59	404.14	463.73
9	259.45	294.55	318.25	340.54	368.53	434.12	495.41
10	282.99	319.47	344.02	367.07	395.97	463.51	526.42

DIRICHLET 1, N-VALUES FOR INVERSE PROBLEM (LI)   B= 4

P*-VALUES

R	.7500	.9000	.9500	.9750	.9900	.9990	.9999
				P=1/20			
1	52.31	71.30	85.14	98.81	116.76	161.70	206.59
2	85.58	108.41	124.58	140.26	160.55	210.25	258.89
3	115.36	141.06	158.99	176.23	198.34	251.81	303.50
4	143.47	171.58	190.99	209.55	233.20	289.92	344.25
5	170.55	200.77	221.50	241.21	266.23	325.85	382.56
6	196.91	229.03	250.95	271.72	297.99	360.24	419.13
7	222.74	256.61	279.63	301.36	328.78	393.48	454.39
8	248.15	283.65	307.68	330.32	358.81	425.80	488.61
9	273.22	310.25	335.23	358.73	388.23	457.38	521.98
10	298.00	336.48	362.37	386.67	417.13	488.33	554.64
				P=1/21			
1	54.98	74.95	89.51	103.87	122.75	169.99	217.19
2	89.92	113.93	130.93	147.43	168.76	221.01	272.15
3	121.20	148.24	167.09	185.22	208.46	264.68	319.02
4	150.73	180.29	200.71	220.22	245.08	304.71	361.83
5	179.17	210.95	232.75	253.48	279.79	342.46	402.07
6	206.85	240.64	263.69	285.53	313.14	378.60	440.50
7	233.98	269.61	293.81	316.66	345.49	413.51	477.55
8	260.66	298.00	323.28	347.08	377.04	447.47	513.50
9	286.99	325.94	352.22	376.92	407.94	480.64	548.56
10	313.02	353.49	380.72	406.27	438.30	513.15	582.86
				P=1/22			
1	57.65	78.60	93.87	108.94	128.74	178.29	227.79
2	94.27	119.46	137.30	154.60	176.96	231.77	285.41
3	127.05	155.41	175.19	194.21	218.58	277.55	334.54
4	157.98	189.00	210.43	230.89	256.96	319.51	379.42
5	187.78	221.13	244.00	265.75	293.34	359.07	421.60
6	216.80	252.25	276.43	299.34	328.30	396.94	461.87
7	245.22	282.60	307.99	331.96	362.20	433.54	500.70
8	273.18	312.36	338.87	363.84	395.26	469.13	538.38
9	300.76	341.63	369.20	395.11	427.65	503.89	575.13
10	328.04	370.50	399.06	425.87	459.46	537.96	611.08
				P=1/23			
1	60.32	82.25	98.23	114.00	134.73	186.59	238.39
2	98.61	124.99	143.66	161.76	185.17	242.53	298.66
3	132.89	162.59	183.29	203.19	228.70	290.41	350.05
4	165.24	197.72	220.14	241.56	268.85	334.30	396.99
5	196.40	231.32	255.26	278.01	306.89	375.69	441.12
6	226.74	263.86	289.17	313.14	343.46	415.30	483.25
7	256.46	295.60	322.17	347.27	378.91	453.57	523.86
8	285.69	326.72	354.47	380.60	413.49	490.79	563.27
9	314.54	357.33	386.18	413.30	447.36	527.15	601.70
10	343.05	387.51	417.41	445.47	480.63	562.78	639.31
			P=1/ 5)	B = 5			
1	13.24	17.50	20.64	23.76	27.86	38.19	48.52
2	21.57	26.62	30.26	33.82	38.46	49.83	60.98
3	29.00	34.67	38.69	42.60	47.63	59.85	71.71
4	36.03	42.20	46.55	50.74	56.11	69.07	81.53
5	42.82	49.42	54.04	58.50	64.18	77.79	90.78
6	49.42	56.42	61.30	65.97	71.94	86.14	99.63
7	55.87	63.24	68.37	73.27	79.49	94.23	108.16
8	62.23	69.93	75.29	80.39	86.84	102.10	116.48
9	68.51	76.54	82.08	87.38	94.04	109.81	124.58
10	74.70	83.03	88.79	94.25	101.14	117.36	132.52

DIRICHLET 1, N-VALUES FOR INVERSE PROBLEM (LI)   B= 5

P*-VALUES

R	.7500	.9000	.9500	.9750	.9900	.9990	.9999
			P=1/ 6				
1	16.13	21.38	25.23	29.05	34.09	46.73	59.37
2	26.19	32.43	36.89	41.26	46.92	60.86	74.54
3	35.18	42.15	47.09	51.88	58.04	73.01	87.53
4	43.67	51.26	56.60	61.74	68.32	84.19	99.43
5	51.83	59.96	65.66	71.11	78.07	94.74	110.63
6	59.78	68.42	74.42	80.15	87.46	104.84	121.34
7	67.57	76.65	82.94	88.94	96.57	114.63	131.68
8	75.22	84.73	91.30	97.55	105.46	124.14	141.73
9	82.77	92.68	99.51	105.97	114.16	133.45	151.53
10	90.24	100.52	107.59	114.28	122.73	142.58	161.12
			P=1/ 7				
1	19.02	25.26	29.83	34.36	40.32	55.27	70.20
2	30.82	38.22	43.53	48.70	55.41	71.89	88.05
3	41.35	49.64	55.50	61.16	68.46	86.17	103.33
4	51.28	60.31	66.64	72.72	80.52	99.29	117.30
5	60.83	70.52	77.26	83.72	91.95	111.67	130.46
6	70.13	80.41	87.53	94.32	102.95	123.53	143.03
7	79.23	90.04	97.52	104.63	113.64	134.99	155.17
8	88.19	99.51	107.30	114.70	124.05	146.16	166.95
9	97.02	108.81	116.90	124.58	134.26	157.07	178.46
10	105.75	117.98	126.37	134.30	144.29	167.77	189.72
			P=1/ 8				
1	21.92	29.13	34.43	39.66	46.55	63.79	81.03
2	35.44	44.01	50.15	56.12	63.87	82.91	101.57
3	47.51	57.11	63.89	70.44	78.86	99.32	119.11
4	58.88	69.35	76.67	83.71	92.71	114.38	135.17
5	69.82	81.05	88.85	96.32	105.83	128.59	150.28
6	80.47	92.39	100.62	108.48	118.46	142.20	164.72
7	90.89	103.44	112.07	120.29	130.70	155.36	178.65
8	101.15	114.28	123.29	131.83	142.65	168.17	192.17
9	111.26	124.93	134.30	143.16	154.35	180.69	205.37
10	121.25	135.45	145.14	154.31	165.85	192.95	218.29
			P=1/ 9				
1	24.81	33.01	39.01	44.94	52.76	72.32	91.87
2	40.05	49.81	56.78	63.56	72.35	93.92	115.07
3	53.66	64.59	72.28	79.71	89.27	112.46	134.90
4	66.48	78.39	86.70	94.68	104.89	129.46	153.03
5	78.81	91.59	100.45	108.91	119.70	145.50	170.10
6	90.81	104.36	113.72	122.63	133.94	160.86	186.40
7	102.55	116.82	126.63	135.95	147.75	175.72	202.12
8	114.10	129.03	139.27	148.97	161.23	190.17	217.39
9	125.49	141.05	151.68	161.74	174.43	204.29	232.28
10	136.75	152.90	163.91	174.31	187.40	218.13	246.85
			P=1/10				
1	27.70	36.89	43.61	50.25	58.97	80.84	102.70
2	44.67	55.61	63.40	70.97	80.81	104.93	128.59
3	59.81	72.06	80.67	88.98	99.67	125.59	150.69
4	74.07	87.43	96.73	105.66	117.07	144.54	170.90
5	87.79	102.11	112.02	121.51	133.57	162.41	189.91
6	101.14	116.34	126.80	136.77	149.43	179.53	208.07
7	114.21	130.20	141.18	151.61	164.81	196.07	225.59
8	127.05	143.80	155.25	166.10	179.81	212.17	242.59
9	139.72	157.16	169.06	180.32	194.51	227.89	259.18
10	152.23	170.35	182.67	194.30	208.94	243.31	275.42

DIRICHLET 1, N-VALUES FOR INVERSE PROBLEM (LI)   B= 5

P*-VALUES

R	.7500	.9000	.9500	.9750	.9900	.9990	.9999
				P=1/11			
1	30.59	40.77	48.19	55.54	65.19	89.37	113.53
2	49.29	61.40	70.01	78.41	89.27	115.94	142.08
3	65.96	79.53	89.05	98.25	110.06	138.72	166.47
4	81.67	96.46	106.75	116.63	129.25	159.62	188.76
5	96.78	112.64	123.62	134.09	147.43	179.32	209.72
6	111.47	128.31	139.89	150.91	164.91	198.19	229.74
7	125.85	143.58	155.73	167.26	181.86	216.43	249.05
8	139.99	158.55	171.23	183.23	198.39	234.16	267.79
9	153.94	173.28	186.45	198.89	214.58	251.49	286.07
10	167.72	187.79	201.43	214.30	230.49	268.48	303.96
				P=1/12			
1	33.48	44.64	52.78	60.83	71.41	97.89	124.36
2	53.90	67.19	76.64	85.83	97.74	126.95	155.59
3	72.11	86.99	97.45	107.52	120.46	151.85	182.25
4	89.26	105.49	116.78	127.60	141.43	174.69	206.62
5	105.76	123.16	135.20	146.68	161.30	196.23	229.53
6	121.80	140.27	152.97	165.06	180.39	216.84	251.41
7	137.50	156.95	170.28	182.91	198.90	236.77	272.51
8	152.94	173.31	187.20	200.36	216.96	256.15	292.99
9	168.16	189.38	203.82	217.46	234.65	275.08	312.97
10	183.20	205.23	220.19	234.29	252.02	293.65	332.52
				P=1/13			
1	36.37	48.51	57.37	66.12	77.62	106.42	135.18
2	58.51	72.97	83.26	93.25	106.20	137.96	169.09
3	78.25	94.47	105.83	116.78	130.86	164.98	198.02
4	96.85	114.52	126.80	138.57	153.61	189.77	224.47
5	114.73	133.69	146.78	159.27	175.16	213.13	249.33
6	132.12	152.24	166.06	179.20	195.87	235.50	273.07
7	149.15	170.33	184.82	198.57	215.95	257.12	295.97
8	165.88	188.05	203.18	217.48	235.54	278.14	318.19
9	182.38	205.49	221.20	236.03	254.72	298.68	339.86
10	198.69	222.67	238.94	254.28	273.57	318.81	361.07
				P=1/14			
1	39.25	52.39	61.95	71.41	83.83	114.93	146.00
2	63.12	78.77	89.87	100.68	114.66	148.97	182.59
3	84.40	101.93	114.21	126.05	141.25	178.11	213.80
4	104.44	123.55	136.82	149.54	165.78	204.84	242.32
5	123.71	144.21	158.36	171.85	189.01	230.03	269.13
6	142.45	164.21	179.14	193.34	211.35	254.15	294.74
7	160.79	183.70	199.37	214.22	233.00	277.47	319.43
8	178.82	202.81	219.15	234.61	254.11	300.13	343.39
9	196.60	221.59	238.57	254.60	274.79	322.27	366.75
10	214.17	240.11	257.70	274.27	295.10	343.97	389.62
				P=1/15			
1	42.13	56.26	66.55	76.70	90.04	123.46	156.83
2	67.73	84.56	96.50	108.09	123.12	159.97	196.08
3	90.55	109.40	122.60	135.32	151.65	191.25	229.58
4	112.03	132.58	146.84	160.51	177.95	219.91	260.17
5	132.69	154.74	169.94	184.44	202.88	246.94	288.94
6	152.78	176.17	192.23	207.48	226.83	272.81	316.40
7	172.44	197.07	213.91	229.87	250.04	297.81	342.88
8	191.76	217.56	235.12	251.73	272.69	322.11	368.58
9	210.82	237.70	255.94	273.17	294.86	345.86	393.64
10	229.65	257.55	276.45	294.25	316.64	369.13	418.16

DIRICHLET 1, N-VALUES FOR INVERSE PROBLEM (LI)   B= 5

P*-VALUES

R	.7500	.9000	.9500	.9750	.9900	.9990	.9999
			P=1/16				
1	45.02	60.13	71.13	81.99	96.26	131.97	167.66
2	72.34	90.34	103.11	115.52	131.58	170.98	209.59
3	96.69	116.86	130.98	144.58	162.04	204.38	245.35
4	119.62	141.61	156.86	171.47	190.13	234.98	278.02
5	141.66	165.26	181.52	197.02	216.74	263.84	308.74
6	163.10	188.14	205.31	221.62	242.31	291.46	338.06
7	184.08	210.45	228.46	245.52	267.09	318.15	366.34
8	204.70	232.31	251.09	268.85	291.26	344.10	393.77
9	225.03	253.80	273.32	291.74	314.92	369.45	420.53
10	245.12	274.99	295.20	314.24	338.17	394.29	446.71
			P=1/17				
1	47.90	64.00	75.72	87.28	102.47	140.49	178.48
2	76.95	96.13	109.73	122.93	140.03	181.98	223.08
3	102.84	124.33	139.37	153.84	172.44	217.50	261.12
4	127.20	150.64	166.88	182.44	202.31	250.05	295.87
5	150.64	175.78	193.09	209.60	230.60	280.74	328.54
6	173.42	200.10	218.39	235.76	257.78	310.11	359.73
7	195.72	223.82	242.99	261.16	284.13	338.50	389.79
8	217.64	247.05	267.06	285.97	309.83	366.08	418.96
9	239.25	269.90	290.69	310.30	334.99	393.04	447.42
10	260.60	292.42	313.95	334.23	359.71	419.45	475.25
			P=1/18				
1	50.79	67.87	80.30	92.57	108.68	149.01	189.30
2	81.56	101.91	116.35	130.36	148.50	192.99	236.58
3	108.98	131.79	147.75	163.10	182.83	230.63	276.90
4	134.79	159.67	176.90	193.41	214.48	265.12	313.73
5	159.62	186.30	204.67	222.19	244.46	297.65	348.34
6	183.75	212.07	231.47	249.90	273.26	328.77	381.39
7	207.37	237.19	257.54	276.81	301.18	358.83	413.24
8	230.58	261.80	283.03	303.09	328.40	388.07	444.16
9	253.47	286.00	308.06	328.87	355.06	416.63	474.30
10	276.08	309.86	332.71	354.21	381.24	444.61	503.80
			P=1/19				
1	53.68	71.74	84.89	97.86	114.89	157.53	200.12
2	86.17	107.70	122.96	137.78	156.95	203.99	250.07
3	115.12	139.26	156.13	172.37	193.22	243.76	292.67
4	142.38	168.69	186.92	204.37	226.65	280.20	331.58
5	168.59	196.82	216.25	234.77	258.31	314.55	368.14
6	194.07	224.03	244.55	264.03	288.73	347.42	403.04
7	219.00	250.56	272.08	292.46	318.22	379.18	436.70
8	243.52	276.55	299.00	320.22	346.97	410.05	469.35
9	267.68	302.10	325.44	347.43	375.12	440.21	501.19
10	291.56	327.29	351.46	374.20	402.78	469.77	532.34
			P=1/20				
1	56.56	75.62	89.48	103.15	121.10	166.05	210.94
2	90.78	113.49	129.58	145.20	165.42	215.00	263.57
3	121.27	146.72	164.51	181.63	203.62	256.88	308.44
4	149.97	177.72	196.94	215.34	238.82	295.27	349.43
5	177.56	207.34	227.83	247.35	272.17	331.45	387.94
6	204.39	235.99	257.63	278.17	304.21	366.07	424.71
7	230.65	263.93	286.62	308.11	335.26	399.52	460.15
8	256.45	291.30	314.97	337.34	365.54	432.04	494.54
9	281.89	318.20	342.80	366.00	395.19	463.80	528.07
10	307.03	344.73	370.21	394.18	424.31	494.93	560.88

DIRICHLET 1, N-VALUES FOR INVERSE PROBLEM (LI)   B= 5
P*-VALUES

R	.7500	.9000	.9500	.9750	.9900	.9990	.9999
				P=1/21			
1	59.45	79.49	94.06	108.44	127.32	174.57	221.77
2	95.39	119.28	136.19	152.62	173.87	226.00	277.06
3	127.41	154.19	172.89	190.89	214.00	270.01	324.21
4	157.56	186.75	206.96	226.31	251.00	310.34	367.27
5	186.54	217.86	239.41	259.93	286.03	348.35	407.74
6	214.71	247.95	270.71	292.31	319.69	384.72	446.37
7	242.29	277.30	301.16	323.75	352.31	419.86	483.60
8	269.39	306.04	330.94	354.46	384.11	454.02	519.73
9	296.11	334.30	360.18	384.56	415.25	487.39	554.95
10	322.51	362.16	388.95	414.17	445.84	520.09	589.42
				P=1/22			
1	62.33	83.36	98.65	113.73	133.53	183.08	232.59
2	100.00	125.06	142.81	160.04	182.34	237.00	290.56
3	133.56	161.65	181.28	200.16	224.40	283.14	339.98
4	165.14	195.78	216.98	237.27	263.17	325.41	385.12
5	195.51	228.38	250.98	272.52	299.89	365.25	427.54
6	225.04	259.92	283.79	306.45	335.16	403.37	468.02
7	253.93	290.67	315.70	339.40	369.35	440.20	507.05
8	282.33	320.79	346.91	371.58	402.68	476.00	544.92
9	310.32	350.40	377.55	403.13	435.32	510.97	581.84
10	337.98	379.60	407.71	434.15	467.37	545.24	617.96
				P=1/23			
1	65.22	87.23	103.23	119.02	139.74	191.60	243.41
2	104.61	130.85	149.43	167.46	190.79	248.01	304.06
3	139.70	169.11	189.66	209.42	234.79	296.26	355.76
4	172.73	204.80	227.00	248.24	275.34	340.48	402.97
5	204.49	238.90	262.56	285.10	313.74	382.14	447.33
6	235.36	271.88	296.87	320.59	350.64	422.02	489.68
7	265.57	304.03	330.24	355.04	386.39	460.54	530.50
8	295.26	335.54	362.88	388.70	421.25	497.98	570.11
9	324.54	366.50	394.91	421.69	455.38	534.56	608.72
10	353.46	397.03	426.46	454.13	488.90	570.40	646.51
				P=1/24			
1	68.10	91.10	107.82	124.31	145.95	200.12	254.23
2	109.22	136.63	156.04	174.88	199.25	259.01	317.56
3	145.84	176.58	198.04	218.69	245.18	309.39	371.53
4	180.32	213.83	237.02	259.20	287.52	355.55	420.82
5	213.46	249.42	274.14	297.68	327.60	399.04	467.13
6	245.68	283.84	309.95	334.72	366.11	440.67	511.34
7	277.21	317.40	344.78	370.69	403.43	480.88	553.95
8	308.20	350.28	378.85	405.82	439.82	519.97	595.30
9	338.75	382.60	412.29	440.25	475.45	558.15	635.61
10	368.93	414.47	445.20	474.12	510.44	595.56	675.04
				P=1/ 6, B = 6			
1	17.13	22.38	26.23	30.05	35.09	47.73	60.37
2	27.41	33.60	38.03	42.40	48.04	61.95	75.61
3	36.56	43.46	48.36	53.11	59.26	74.18	88.66
4	45.16	52.67	57.95	63.05	69.61	85.42	100.61
5	53.45	61.48	67.10	72.52	79.43	96.00	111.86
6	61.50	69.99	75.93	81.63	88.87	106.18	122.62
7	69.38	78.32	84.55	90.49	98.03	115.99	132.99
8	77.12	86.47	92.95	99.14	106.98	125.57	143.07
9	84.76	94.49	101.23	107.64	115.75	134.91	152.91
10	92.30	102.40	109.37	115.98	124.37	144.08	162.56

DIRICHLET 1, N-VALUES FOR INVERSE PROBLEM (LI)   B= 6
P#-VALUES

R	.7500	.9000	.9500	.9750	.9900	.9990	.9999
				P=1/ 7			
1	20.21	26.45	31.00	35.55	41.51	56.45	71.39
2	32.25	39.61	44.88	50.03	56.73	73.18	89.33
3	42.96	51.17	56.98	62.63	69.88	87.55	104.67
4	53.04	61.96	68.24	74.29	82.03	100.73	118.71
5	62.74	72.29	78.96	85.38	93.56	113.18	131.91
6	72.16	82.28	89.33	96.05	104.64	125.10	144.55
7	81.38	92.01	99.40	106.44	115.39	136.63	156.73
8	90.44	101.57	109.26	116.59	125.86	147.85	168.56
9	99.36	110.95	118.94	126.54	136.13	158.81	180.10
10	108.18	120.21	128.48	136.32	146.23	169.56	191.41
				P=1/ 8			
1	23.28	30.51	35.79	41.01	47.90	65.16	82.41
2	37.08	45.62	51.72	57.68	65.40	84.40	103.02
3	49.37	58.88	65.62	72.12	80.52	100.90	120.67
4	60.91	71.27	78.53	85.51	94.47	116.04	136.79
5	72.01	83.10	90.82	98.23	107.68	130.34	151.96
6	82.81	94.55	102.70	110.49	120.39	144.01	166.47
7	93.37	105.71	114.25	122.39	132.72	157.25	180.45
8	103.74	116.65	125.56	134.01	144.74	170.11	194.02
9	113.96	127.41	136.65	145.42	156.52	182.69	207.27
10	124.06	138.00	147.58	156.64	168.08	195.01	220.24
				P=1/ 9			
1	26.35	34.56	40.57	46.50	54.31	73.87	93.42
2	41.91	51.62	58.56	65.31	74.07	95.62	116.74
3	55.76	66.60	74.24	81.63	91.13	114.25	136.66
4	68.78	80.56	88.80	96.73	106.88	131.36	154.87
5	81.30	93.90	102.68	111.08	121.80	147.49	172.00
6	93.45	106.81	116.06	124.90	136.14	162.92	188.38
7	105.35	119.40	129.09	138.33	150.04	177.86	204.16
8	117.03	131.73	141.84	151.45	163.60	192.38	219.49
9	128.55	143.85	154.35	164.30	176.88	206.57	234.44
10	139.92	155.80	166.67	176.95	189.93	220.47	249.06
				P=1/10			
1	29.42	38.62	45.34	51.97	60.71	82.58	104.44
2	46.75	57.63	65.39	72.93	82.74	106.82	130.44
3	62.16	74.30	82.85	91.12	101.76	127.60	152.65
4	76.65	89.85	99.07	107.94	119.30	146.66	172.95
5	90.57	104.71	114.53	123.92	135.91	164.63	192.04
6	104.09	119.07	129.44	139.32	151.88	181.83	210.29
7	117.32	133.08	143.93	154.27	167.37	198.47	227.87
8	130.32	146.80	158.12	168.86	182.47	214.64	244.94
9	143.13	160.29	172.04	183.18	197.25	230.44	261.59
10	155.78	173.59	185.75	197.26	211.77	245.92	277.88
				P=1/11			
1	32.49	42.67	50.10	57.45	67.10	91.28	115.45
2	51.58	63.63	72.21	80.57	91.41	118.03	144.14
3	68.55	82.00	91.47	100.62	112.38	140.94	168.64
4	84.51	99.14	109.35	119.16	131.71	161.95	191.02
5	99.84	115.51	126.38	136.77	150.02	181.77	212.08
6	114.73	131.34	142.80	153.73	167.63	200.74	232.19
7	129.30	146.76	158.78	170.20	184.69	219.06	251.58
8	143.61	161.87	174.40	186.29	201.32	236.89	270.39
9	157.71	176.73	189.74	202.05	217.61	254.30	288.74
10	171.64	191.37	204.84	217.57	233.62	271.36	306.70

DIRICHLET 1, N-VALUES FOR INVERSE PROBLEM (LI)   B= 6

P*-VALUES

R	.7500	.9000	.9500	.9750	.9900	.9990	.9999
			P=1/12				
1	35.55	46.73	54.87	62.92	73.50	99.98	126.45
2	56.41	69.63	79.04	88.20	100.07	129.24	157.84
3	74.94	89.71	100.08	110.10	122.99	154.29	184.63
4	92.37	108.43	119.62	130.37	144.12	177.26	209.10
5	109.10	126.30	138.22	149.62	164.14	198.91	232.11
6	125.37	143.59	156.16	168.14	183.37	219.64	254.09
7	141.27	160.44	173.62	186.13	202.00	239.67	275.28
8	156.89	176.94	190.68	203.71	220.18	259.14	295.84
9	172.29	193.17	207.43	220.93	237.97	278.16	315.89
10	187.49	209.15	223.92	237.87	255.45	296.80	335.52
			P=1/13				
1	38.62	50.78	59.65	68.40	79.89	108.69	137.46
2	61.24	75.63	85.87	95.83	108.74	140.45	171.54
3	81.34	97.41	108.70	119.60	133.61	167.64	200.61
4	100.23	117.71	129.89	141.58	156.54	192.56	227.17
5	118.37	137.10	150.06	162.46	178.25	216.05	252.14
6	136.00	155.85	169.53	182.55	199.10	238.54	275.98
7	153.24	174.12	188.45	202.06	219.32	260.27	298.97
8	170.17	192.01	206.95	221.12	239.03	281.39	321.28
9	186.86	209.60	225.12	239.80	258.33	302.02	343.03
10	203.34	226.93	242.99	258.17	277.29	322.24	364.32
			P=1/14				
1	41.68	54.83	64.42	73.87	86.29	117.40	148.47
2	66.06	81.63	92.70	103.46	117.41	151.65	185.23
3	87.73	105.11	117.32	129.08	144.23	180.97	216.59
4	108.08	126.99	140.16	152.79	168.94	207.85	245.24
5	127.64	147.89	161.91	175.30	192.36	233.19	272.17
6	146.63	168.10	182.88	196.96	214.84	257.44	297.88
7	165.21	187.79	203.28	217.99	236.64	280.86	322.67
8	183.45	207.08	223.23	238.54	257.89	303.64	346.73
9	201.44	226.03	242.81	258.67	278.69	325.88	370.18
10	219.19	244.72	262.07	278.47	299.12	347.68	393.13
			P=1/15				
1	44.74	58.88	69.18	79.34	92.69	126.09	159.48
2	70.89	87.63	99.52	111.08	126.07	162.85	198.92
3	94.11	112.81	125.93	138.58	154.84	194.32	232.57
4	115.94	136.28	150.43	164.00	181.36	223.15	263.31
5	136.90	158.69	173.75	188.13	206.46	250.33	292.20
6	157.26	180.36	196.24	211.37	230.58	276.33	319.78
7	177.17	201.47	218.11	233.92	253.95	301.46	346.37
8	196.73	222.14	239.50	255.95	276.74	325.88	372.17
9	216.00	242.47	260.49	277.54	299.05	349.74	397.32
10	235.04	262.49	281.15	298.77	320.96	373.12	421.94
			P=1/16				
1	47.80	62.94	73.95	84.81	99.08	134.80	170.48
2	75.72	93.63	106.35	118.71	134.73	174.06	212.62
3	100.51	120.51	134.54	148.06	165.46	207.66	248.56
4	123.80	145.57	160.70	175.21	193.77	238.45	281.38
5	146.17	169.48	185.59	200.97	220.57	267.46	312.23
6	167.89	192.61	209.60	225.78	246.32	295.23	341.68
7	189.14	215.14	232.95	249.85	271.27	322.05	370.06
8	210.01	237.21	255.78	273.36	295.59	348.13	397.61
9	230.58	258.89	278.18	296.41	319.41	373.60	424.47
10	250.89	280.27	300.23	319.07	342.79	398.56	450.74

DIRICHLET 1, N-VALUES FOR INVERSE PROBLEM (LI)   B= 6

P*-VALUES

R	.7500	.9000	.9500	.9750	.9900	.9990	.9999
				P=1/17			
1	50.87	66.99	78.72	90.29	105.48	143.50	181.49
2	80.54	99.63	113.17	126.34	143.40	185.26	226.31
3	106.89	128.21	143.15	157.56	176.07	220.99	264.54
4	131.65	154.85	170.96	186.42	206.17	253.74	299.45
5	155.43	180.27	197.43	213.81	234.68	284.60	332.25
6	178.53	204.86	222.96	240.18	262.05	314.12	363.57
7	201.11	228.81	247.78	265.78	288.58	342.65	393.76
8	223.29	252.27	272.05	290.78	314.44	370.38	423.05
9	245.15	275.33	295.86	315.28	339.76	397.46	451.61
10	266.73	298.04	319.31	339.37	364.63	423.99	479.55
				P=1/18			
1	53.93	71.04	83.49	95.76	111.87	152.20	192.49
2	85.37	105.63	120.00	133.96	152.06	196.47	240.00
3	113.28	135.91	151.77	167.04	186.68	234.34	280.52
4	139.51	164.13	181.23	197.63	218.58	269.03	317.51
5	164.69	191.06	209.27	226.65	248.78	301.73	352.28
6	189.15	217.11	236.32	254.59	277.79	333.02	385.47
7	213.07	242.48	262.61	281.71	305.89	363.24	417.45
8	236.57	267.33	288.32	308.19	333.29	392.62	448.49
9	259.72	291.75	313.55	334.14	360.11	421.31	478.75
10	282.58	315.82	338.38	359.66	386.46	449.43	508.35
				P=1/19			
1	56.99	75.09	88.25	101.23	118.27	160.90	203.50
2	90.19	111.63	126.82	141.60	160.72	207.67	253.70
3	119.67	143.61	160.38	176.53	197.30	247.68	296.50
4	147.36	173.41	191.50	208.83	230.99	284.33	335.58
5	173.95	201.86	221.11	239.49	262.89	318.87	372.30
6	199.78	229.36	249.67	268.99	293.52	351.91	407.36
7	225.04	256.15	277.44	297.64	323.21	383.83	441.14
8	249.84	282.40	304.59	325.60	352.14	414.86	473.92
9	274.29	308.18	331.23	353.01	380.47	445.16	505.88
10	298.43	333.60	357.45	379.96	408.29	474.86	537.16
				P=1/20			
1	60.05	79.15	93.02	106.70	124.66	169.60	214.50
2	95.01	117.62	133.65	149.22	169.39	218.87	267.39
3	126.06	151.31	168.99	186.02	207.91	261.01	312.47
4	155.22	182.69	201.77	220.04	243.40	299.62	353.64
5	183.22	212.65	232.95	252.33	276.99	336.00	392.33
6	210.41	241.61	263.03	283.40	309.26	370.81	429.25
7	237.00	269.83	292.27	313.57	340.52	404.43	464.84
8	263.12	297.46	320.86	343.01	370.99	437.11	499.36
9	288.85	324.61	348.92	371.88	400.82	469.02	533.02
10	314.27	351.37	376.53	400.25	430.12	500.29	565.96
				P=1/21			
1	63.12	83.20	97.79	112.17	131.05	178.30	225.50
2	99.84	123.62	140.48	156.85	178.04	230.07	281.08
3	132.45	159.01	177.60	195.51	218.52	274.36	328.45
4	163.07	191.97	212.03	231.25	255.81	314.91	371.71
5	192.48	223.44	244.79	265.16	291.10	353.13	412.35
6	221.04	253.86	276.39	297.81	324.99	389.70	451.14
7	248.97	283.50	307.10	329.49	357.83	425.02	488.53
8	276.40	312.52	337.13	360.42	389.84	459.35	524.80
9	303.42	341.04	366.60	390.74	421.18	492.87	560.16
10	330.11	369.14	395.60	420.55	451.95	525.72	594.76

DIRICHLET 1, N-VALUES FOR INVERSE PROBLEM (LI)   B= 6

P*-VALUES

R	.7500	.9000	.9500	.9750	.9900	.9990	.9999
			P=1/22				
1	66.18	87.25	102.56	117.64	137.45	187.00	236.51
2	104.67	129.62	147.30	164.48	186.71	241.28	294.77
3	138.83	166.71	186.21	204.99	229.13	287.69	344.43
4	170.93	201.26	222.30	242.46	268.21	330.21	389.77
5	201.74	234.23	256.63	278.00	305.20	370.27	432.37
6	231.67	266.11	289.74	312.21	340.72	408.60	473.03
7	260.93	297.17	321.93	345.42	375.14	445.61	512.22
8	289.67	327.58	353.40	377.83	408.69	481.59	550.24
9	317.99	357.47	384.28	409.61	441.53	516.73	587.30
10	345.96	386.92	414.68	440.84	473.78	551.16	623.56
			P=1/23				
1	69.24	91.31	107.32	123.11	143.84	195.71	247.51
2	109.49	135.62	154.12	172.10	195.37	252.48	308.46
3	145.22	174.41	194.82	214.49	239.75	301.03	360.41
4	178.78	210.54	232.57	253.66	280.62	345.50	407.84
5	211.00	245.02	268.47	290.84	319.31	387.40	452.40
6	242.30	278.36	303.10	326.62	356.46	427.49	494.92
7	272.89	310.84	336.76	361.34	392.46	466.20	535.91
8	302.95	342.64	369.67	395.25	427.54	503.83	575.68
9	332.56	373.89	401.96	428.47	461.88	540.58	614.44
10	361.80	404.69	433.75	461.14	495.61	576.59	652.36
			P=1/24				
1	72.31	95.36	112.09	128.59	150.23	204.41	258.52
2	114.32	141.61	160.95	179.73	204.03	263.68	322.15
3	151.61	182.10	203.43	223.97	250.36	314.37	376.39
4	186.63	219.82	242.83	264.87	293.02	360.79	425.90
5	220.27	255.81	280.31	303.68	333.41	404.54	472.42
6	252.92	290.61	316.46	341.02	372.19	446.38	516.82
7	284.86	324.51	351.59	377.27	409.77	486.80	559.60
8	316.22	357.70	385.94	412.66	446.39	526.08	601.11
9	347.13	390.32	419.65	447.34	482.24	564.44	641.58
10	377.65	422.46	452.82	481.44	517.44	602.02	681.16
			P=1/25				
1	75.37	99.41	116.86	134.05	156.63	213.10	269.52
2	119.14	147.61	167.77	187.35	212.70	274.88	335.84
3	157.99	189.80	212.04	233.46	260.97	327.71	392.37
4	194.49	229.10	253.09	276.07	305.44	376.09	443.96
5	229.53	266.61	292.15	316.51	347.52	421.67	492.44
6	263.55	302.86	329.81	355.43	387.92	465.27	538.71
7	296.82	338.18	366.42	393.20	427.08	507.39	583.29
8	329.50	372.76	402.21	430.06	465.24	548.32	626.55
9	361.69	406.75	437.33	466.20	502.59	588.29	668.71
10	393.49	440.24	471.89	501.73	539.27	627.45	709.96
			P=1/ 7,	B = 7			
1	21.21	27.45	32.00	36.55	42.51	57.45	72.39
2	33.45	40.77	46.02	51.16	57.84	74.27	90.40
3	44.32	52.46	58.25	63.85	71.09	88.71	105.80
4	54.53	63.37	69.60	75.61	83.32	101.95	119.89
5	64.33	73.78	80.41	86.77	94.90	114.46	133.15
6	73.85	83.85	90.83	97.53	106.04	126.44	145.82
7	83.16	93.67	100.97	107.96	116.85	138.00	158.04
8	92.30	103.29	110.90	118.17	127.39	149.27	169.91
9	101.31	112.74	120.65	128.17	137.71	160.27	181.50
10	110.21	122.05	130.24	138.01	147.85	171.05	192.83

DIRICHLET 1, N-VALUES FOR INVERSE PROBLEM (LI)   B= 7

P*-VALUES

R	.7500	.9000	.9500	.9750	.9900	.9990	.9999
			P=1/ 8				
1	24.44	31.66	36.94	42.17	49.05	66.32	83.56
2	38.47	46.95	53.03	58.97	66.69	85.66	104.27
3	50.92	60.37	67.06	73.55	81.90	102.24	121.97
4	62.62	72.87	80.08	87.02	95.94	117.46	138.16
5	73.85	84.81	92.48	99.84	109.24	131.81	153.39
6	84.76	96.36	104.44	112.17	122.01	145.56	167.93
7	95.42	107.62	116.07	124.15	134.42	158.83	181.96
8	105.89	118.64	127.45	135.84	146.50	171.76	195.59
9	116.21	129.47	138.62	147.32	158.33	184.38	208.87
10	126.40	140.14	149.61	158.60	169.95	196.75	221.88
			P=1/ 9				
1	27.66	35.86	41.87	47.81	55.62	75.18	94.73
2	43.49	53.14	60.04	66.79	75.53	97.03	118.14
3	57.53	68.28	75.88	83.23	92.71	115.77	138.14
4	70.72	82.39	90.57	98.45	108.56	132.95	156.43
5	83.37	95.85	104.56	112.90	123.57	149.15	173.63
6	95.66	108.87	118.04	126.81	137.98	164.67	190.04
7	107.67	121.56	131.16	140.33	151.96	179.66	205.88
8	119.47	133.97	143.98	153.52	165.60	194.24	221.26
9	131.09	146.19	156.58	166.45	178.94	208.48	236.25
10	142.57	158.22	168.97	179.17	192.05	222.44	250.92
			P=1/10				
1	30.87	40.07	46.80	53.44	62.17	84.03	105.89
2	48.50	59.33	67.05	74.59	84.37	108.42	132.00
3	64.13	76.18	84.69	92.92	103.52	129.30	154.31
4	78.80	91.89	101.05	109.87	121.17	148.45	174.69
5	92.89	106.88	116.63	125.96	137.89	166.50	193.85
6	106.56	121.37	131.64	141.46	153.95	183.78	212.15
7	119.92	135.49	146.25	156.50	169.52	200.48	229.79
8	133.04	149.32	160.53	171.18	184.69	216.71	246.92
9	145.97	162.90	174.54	185.58	199.56	232.57	263.62
10	158.74	176.30	188.33	199.74	214.15	248.11	279.96
			P=1/11				
1	34.09	44.29	51.72	59.06	68.72	92.90	117.05
2	53.51	65.51	74.06	82.40	93.21	119.79	145.87
3	70.73	84.08	93.50	102.60	114.33	142.82	170.47
4	86.89	101.39	111.54	121.29	133.78	163.93	192.94
5	102.40	117.90	128.69	139.02	152.21	183.84	214.07
6	117.45	133.87	145.24	156.09	169.91	202.88	234.25
7	132.17	149.43	161.33	172.67	187.06	221.29	253.70
8	146.62	164.65	177.06	188.85	203.79	239.18	272.58
9	160.85	179.62	192.50	204.71	220.16	256.66	290.98
10	174.90	194.37	207.69	220.31	236.24	273.79	309.00
			P=1/12				
1	37.31	48.50	56.65	64.70	75.27	101.76	128.22
2	58.52	71.69	81.06	90.20	102.04	131.17	159.74
3	77.33	91.98	102.31	112.29	125.12	156.35	186.63
4	94.97	110.89	122.01	132.70	146.40	179.43	211.20
5	111.91	128.93	140.76	152.08	166.53	201.17	234.30
6	128.35	146.37	158.83	170.73	185.87	221.99	256.35
7	144.41	163.36	176.41	188.84	204.61	242.10	277.61
8	160.18	179.98	193.59	206.51	222.88	261.66	298.24
9	175.72	196.33	210.45	223.84	240.76	280.75	318.35
10	191.06	212.43	227.04	240.87	258.33	299.47	338.03

DIRICHLET 1, N-VALUES FOR INVERSE PROBLEM (LI)   B= 7

P*-VALUES

R	.7500	.9000	.9500	.9750	.9900	.9990	.9999
			P=1/13				
1	40.52	52.70	61.57	70.32	81.82	110.62	139.39
2	63.53	77.86	88.07	98.00	110.88	142.55	173.60
3	83.92	99.86	111.12	121.96	135.93	169.87	202.79
4	103.06	120.39	132.49	144.11	159.00	194.91	229.46
5	121.42	139.95	152.82	165.13	180.84	218.51	254.52
6	139.24	158.86	172.43	185.37	201.82	241.09	278.44
7	156.65	177.28	191.49	205.00	222.15	262.91	301.51
8	173.75	195.32	210.12	224.17	241.96	284.12	323.89
9	190.59	213.03	228.40	242.96	261.37	304.84	345.71
10	207.22	230.50	246.39	261.43	280.42	325.13	367.06
			P=1/14				
1	43.74	56.90	66.49	75.94	88.37	119.48	150.55
2	68.54	84.04	95.07	105.81	119.72	153.91	187.46
3	90.52	107.79	119.92	131.65	146.73	183.39	218.94
4	111.14	129.89	142.96	155.53	171.61	210.40	247.71
5	130.92	150.98	164.88	178.19	195.16	235.84	274.74
6	150.12	171.36	186.02	200.00	217.78	260.20	300.54
7	168.89	191.21	206.57	221.17	239.70	283.72	325.41
8	187.31	210.65	226.65	241.83	261.05	306.59	349.54
9	205.46	229.74	246.35	262.08	281.96	328.92	373.07
10	223.38	248.56	265.74	281.99	302.50	350.81	396.09
			P=1/15				
1	46.95	61.11	71.41	81.58	94.92	128.33	161.71
2	73.55	90.22	102.07	113.61	128.56	165.29	201.32
3	97.11	115.68	128.73	141.33	157.53	196.91	235.10
4	119.22	139.39	153.44	166.94	184.22	225.88	265.95
5	140.43	162.00	176.95	191.25	209.48	253.18	294.95
6	161.01	183.85	199.61	214.63	233.74	279.30	322.63
7	181.13	205.14	221.64	237.33	257.24	304.53	349.31
8	200.88	225.97	243.17	259.49	280.14	329.05	375.19
9	220.33	246.45	264.30	281.21	302.57	353.00	400.43
10	239.53	266.62	285.09	302.56	324.59	376.48	425.12
			P=1/16				
1	50.17	65.32	76.34	87.20	101.47	137.19	172.87
2	78.56	96.40	109.08	121.41	137.40	176.66	215.18
3	103.71	123.58	137.54	151.00	168.33	210.43	251.26
4	127.30	148.86	163.92	178.35	196.82	241.37	284.21
5	149.93	173.02	189.01	204.30	223.79	270.52	315.17
6	171.90	196.35	213.20	229.27	249.69	298.40	344.72
7	193.37	219.06	236.72	253.50	274.78	325.34	373.20
8	214.44	241.30	259.70	277.15	299.23	351.52	400.84
9	235.20	263.15	282.25	300.33	323.17	377.09	427.78
10	255.69	284.68	304.44	323.11	346.67	402.14	454.14
			P=1/17				
1	53.38	69.53	81.26	92.83	108.01	146.04	184.02
2	83.57	102.58	116.08	129.21	146.23	188.03	229.03
3	110.30	131.48	146.34	160.68	179.13	223.94	267.42
4	135.38	158.38	174.39	189.76	209.43	256.85	302.46
5	159.44	184.04	201.07	217.35	238.11	287.84	335.39
6	182.79	208.84	226.79	243.90	265.64	317.50	366.81
7	205.60	232.99	251.79	269.66	292.32	346.14	397.10
8	228.00	256.63	276.22	294.80	318.32	373.98	426.49
9	250.06	279.86	300.19	319.45	343.76	401.17	455.14
10	271.84	302.74	323.78	343.67	368.75	427.81	483.17

C27

DIRICHLET 1, N-VALUES FOR INVERSE PROBLEM (LI)   B= 7

P*-VALUES

R	.7500	.9000	.9500	.9750	.9900	.9990	.9999
			P=1/18				
1	56.60	73.73	86.18	98.45	114.57	154.90	195.19
2	88.57	108.75	123.08	137.01	155.06	199.41	242.89
3	116.89	139.38	155.15	170.36	189.93	237.47	283.57
4	143.46	167.88	184.87	201.17	222.03	272.33	320.71
5	168.94	195.06	213.13	230.41	252.42	305.18	355.60
6	193.67	221.33	240.38	258.53	281.60	336.60	388.90
7	217.84	246.91	266.86	285.82	309.86	366.95	420.99
8	241.56	271.96	292.75	312.46	337.40	396.45	452.14
9	264.93	296.56	318.14	338.57	364.36	425.25	482.49
10	287.99	320.80	343.13	364.23	390.83	453.48	512.19
			P=1/19				
1	59.81	77.93	91.10	104.08	121.11	163.75	206.35
2	93.58	114.93	130.08	144.81	163.90	210.77	256.75
3	123.49	147.27	163.95	180.04	200.73	250.98	299.72
4	151.54	177.37	195.34	212.58	234.64	287.81	338.95
5	178.45	206.08	225.19	243.46	266.74	322.51	375.81
6	204.56	233.82	253.97	273.16	297.55	355.70	410.99
7	230.07	260.84	281.94	301.98	327.40	387.76	444.89
8	255.12	287.28	309.27	330.12	356.49	418.91	477.78
9	279.79	313.26	336.09	357.69	384.96	449.33	509.85
10	304.14	338.86	362.48	384.79	412.91	479.14	541.21
			P=1/20				
1	63.02	82.14	96.02	109.70	127.66	172.61	217.51
2	98.59	121.10	137.08	152.61	172.73	222.14	270.61
3	130.08	155.17	172.76	189.72	211.53	264.50	315.87
4	159.62	186.87	205.82	223.99	247.25	303.30	357.20
5	187.95	217.10	237.25	256.51	281.05	339.84	396.02
6	215.44	246.32	267.56	287.79	313.50	374.60	433.08
7	242.31	274.76	297.01	318.15	344.93	408.56	468.79
8	268.68	302.61	325.79	347.77	375.57	441.38	503.43
9	294.66	329.96	354.03	376.81	405.56	473.41	537.20
10	320.30	356.92	381.82	405.34	434.99	504.80	570.24
			P=1/21				
1	66.24	86.35	100.94	115.33	134.21	181.46	228.66
2	103.59	127.28	144.08	160.41	181.57	233.52	284.47
3	136.67	163.06	181.56	199.39	222.33	278.02	332.03
4	167.69	196.36	216.29	235.40	259.85	318.78	375.45
5	197.45	228.12	249.32	269.56	295.36	357.17	416.24
6	226.33	258.81	281.15	302.42	329.45	393.90	455.17
7	254.54	288.69	312.08	334.31	362.48	429.37	492.68
8	282.24	317.93	342.32	365.43	394.66	463.84	529.07
9	309.52	346.67	371.98	395.92	426.15	497.49	564.55
10	336.45	374.97	401.16	425.90	457.08	530.47	599.26
			P=1/22				
1	69.45	90.55	105.86	120.95	140.76	190.32	239.82
2	108.60	133.46	151.08	168.21	190.40	244.88	298.33
3	143.26	170.96	190.37	209.07	233.12	291.54	348.18
4	175.77	205.86	226.76	246.81	272.46	334.26	393.70
5	206.95	239.14	261.38	282.61	309.67	374.50	436.46
6	237.21	271.30	294.73	317.05	345.40	412.99	477.26
7	266.78	302.61	327.16	350.47	380.01	450.17	516.58
8	295.80	333.26	358.84	383.08	413.74	486.30	554.72
9	324.38	363.37	389.92	415.04	446.75	521.57	591.90
10	352.60	393.03	420.51	446.46	479.16	556.13	628.28

DIRICHLET 1, N-VALUES FOR INVERSE PROBLEM (LI)   B= 7

P*-VALUES

R	.7500	.9000	.9500	.9750	.9900	.9990	.9999
				P=1/23			
1	72.66	94.76	110.78	126.58	147.31	199.17	250.97
2	113.61	139.63	158.08	176.01	199.23	256.26	312.18
3	149.85	178.86	199.17	218.75	243.92	305.05	364.34
4	183.85	215.35	237.24	258.22	285.06	349.74	411.94
5	216.46	250.16	273.44	295.66	323.98	391.83	456.67
6	248.09	283.79	308.32	331.68	361.36	432.09	499.35
7	279.01	316.53	342.23	366.63	397.55	470.97	540.47
8	309.36	348.58	375.36	400.74	432.82	508.76	580.37
9	339.24	380.07	407.87	434.16	467.35	545.65	619.26
10	368.75	411.09	439.85	467.01	501.24	581.80	657.30
				P=1/24			
1	75.87	98.96	115.70	132.20	153.85	208.02	262.13
2	118.61	145.81	165.08	183.81	208.07	267.63	326.04
3	156.45	186.75	207.97	228.43	254.72	318.57	380.49
4	191.93	224.85	247.71	269.63	297.67	365.22	430.19
5	225.96	261.18	285.49	308.71	338.30	409.16	476.88
6	258.98	296.28	321.91	346.31	377.31	451.19	521.43
7	291.24	330.45	357.30	382.79	415.09	491.78	564.37
8	322.91	363.91	391.88	418.39	451.91	531.22	606.01
9	354.11	396.77	425.81	453.28	487.94	569.73	646.61
10	384.90	429.15	459.20	487.57	523.32	607.46	686.32
				P=1/25			
1	79.09	103.16	120.62	137.83	160.40	216.88	273.29
2	123.62	151.98	172.08	191.61	216.90	278.99	339.89
3	163.04	194.65	216.78	238.10	265.52	332.09	396.64
4	200.00	234.34	258.18	281.04	310.27	380.70	448.44
5	235.46	272.19	297.55	321.76	352.61	426.49	497.09
6	269.86	308.77	335.50	360.94	393.26	470.29	543.52
7	303.48	344.38	372.37	398.95	432.63	512.58	588.26
8	336.47	379.23	408.41	436.04	470.99	553.68	631.66
9	368.97	413.47	443.76	472.40	508.54	593.81	673.96
10	401.05	447.20	478.54	508.12	545.40	633.12	715.34
				P=1/26			
1	82.30	107.37	125.55	143.45	166.95	225.74	284.45
2	128.62	158.16	179.08	199.41	225.74	290.37	353.75
3	169.63	202.54	225.58	247.78	276.31	345.61	412.79
4	208.08	243.83	268.66	292.45	322.87	396.18	466.68
5	244.96	283.21	309.61	334.81	366.92	443.82	517.31
6	280.74	321.26	349.08	375.57	409.21	489.39	565.61
7	315.71	358.30	387.45	415.11	450.16	533.39	612.15
8	350.03	394.56	424.93	453.70	490.07	576.14	657.30
9	383.83	430.17	461.70	491.51	529.13	617.89	701.31
10	417.20	465.26	497.88	528.68	567.48	658.79	744.36
				P=1/ 8, B = 8			
1	25.44	32.66	37.94	43.17	50.05	67.32	84.56
2	39.67	48.11	54.18	60.10	67.80	86.75	105.35
3	52.27	61.66	68.32	74.77	83.10	103.40	123.10
4	64.09	74.26	81.43	88.34	97.22	118.68	139.35
5	75.43	86.29	93.90	101.22	110.59	133.08	154.62
6	86.43	97.92	105.93	113.63	123.43	146.87	169.21
7	97.18	109.25	117.64	125.67	135.87	160.21	183.28
8	107.74	120.34	129.08	137.42	148.01	173.17	196.93
9	118.13	131.24	140.31	148.94	159.89	185.83	210.26
10	128.39	141.97	151.36	160.28	171.57	198.25	223.30

DIRICHLET 1, N-VALUES FOR INVERSE PROBLEM (LI)   B= 8
P*-VALUES

R	.7500	.9000	.9500	.9750	.9900	.9990	.9999
			P=1/ 9				
1	28.79	36.99	42.99	48.93	56.75	76.31	95.86
2	44.84	54.46	61.34	68.06	76.79	98.27	119.36
3	59.05	69.73	77.30	84.63	94.07	117.08	139.43
4	72.38	83.95	92.09	99.93	110.00	134.34	157.77
5	85.15	97.52	106.17	114.48	125.09	150.60	175.01
6	97.56	110.64	119.74	128.47	139.58	166.17	191.49
7	109.67	123.41	132.94	142.04	153.63	181.21	207.37
8	121.56	135.91	145.84	155.30	167.32	195.84	222.79
9	133.27	148.19	158.50	168.30	180.72	210.13	237.82
10	144.83	160.30	170.95	181.07	193.88	224.13	252.54
			P=1/10				
1	32.13	41.35	48.06	54.71	63.44	85.31	107.16
2	50.01	60.79	68.51	76.01	85.78	109.79	133.37
3	65.83	77.81	86.28	94.48	105.03	130.77	155.74
4	80.66	93.64	102.75	111.53	122.79	149.98	176.18
5	94.88	108.75	118.43	127.72	139.59	168.11	195.41
6	108.67	123.35	133.54	143.30	155.73	185.46	213.76
7	122.15	137.56	148.23	158.42	171.37	202.21	231.46
8	135.38	151.48	162.59	173.18	186.62	218.51	248.63
9	148.41	165.15	176.68	187.65	201.54	234.42	265.38
10	161.26	178.62	190.55	201.87	216.19	250.01	281.77
			P=1/11				
1	35.49	45.69	53.12	60.47	70.12	94.30	118.46
2	55.18	67.13	75.66	83.97	94.77	121.32	147.38
3	72.61	85.88	95.25	104.32	116.00	144.45	172.05
4	88.93	103.33	113.42	123.12	135.57	165.65	194.60
5	104.60	119.97	130.69	140.96	154.09	185.62	215.80
6	119.79	136.05	147.34	158.13	171.88	204.74	236.03
7	134.63	151.71	163.53	174.79	189.11	223.21	255.54
8	149.19	167.03	179.34	191.05	205.91	241.17	274.47
9	163.54	182.09	194.86	206.99	222.36	258.70	292.93
10	177.69	196.93	210.13	222.66	238.50	275.88	310.99
			P=1/12				
1	38.83	50.02	58.18	66.23	76.80	103.29	129.76
2	60.35	73.47	82.82	91.93	103.76	132.84	161.39
3	79.38	93.95	104.23	114.16	126.96	158.12	188.37
4	97.21	113.01	124.07	134.71	148.35	181.29	213.01
5	114.31	131.19	142.94	154.20	168.59	203.13	236.18
6	130.90	148.76	161.13	172.96	188.02	224.02	258.30
7	147.10	165.86	178.81	191.16	206.85	244.21	279.62
8	163.00	182.60	196.09	208.92	225.20	263.83	300.31
9	178.66	199.04	213.04	226.34	243.17	282.98	320.48
10	194.11	215.24	229.72	243.45	260.80	301.76	340.21
			P=1/13				
1	42.18	54.37	63.23	71.98	83.49	112.29	141.05
2	65.52	79.80	89.97	99.88	112.74	144.36	175.39
3	86.15	102.02	113.20	124.00	137.93	171.80	204.68
4	105.49	122.70	134.73	146.30	161.13	196.94	231.43
5	124.03	142.41	155.20	167.45	183.08	220.64	256.57
6	142.01	161.46	174.92	187.79	204.17	243.31	280.57
7	159.58	180.00	194.10	207.53	224.59	265.20	303.70
8	176.81	198.15	212.84	226.80	244.50	286.49	326.14
9	193.78	215.98	231.22	245.68	263.97	307.27	348.02
10	210.54	233.55	249.31	264.24	283.10	327.63	369.44

DIRICHLET 1, N-VALUES FOR INVERSE PROBLEM (LI)   B= 8

			P*-VALUES				
R	.7500	.9000	.9500	.9750	.9900	.9990	.9999

P=1/14

R	.7500	.9000	.9500	.9750	.9900	.9990	.9999
1	45.53	58.71	68.29	77.75	90.17	121.28	152.35
2	70.68	86.13	97.13	107.83	121.73	155.87	189.39
3	92.92	110.09	122.17	133.85	148.89	185.48	220.98
4	113.76	132.38	145.39	157.89	173.91	212.59	249.83
5	133.74	153.63	167.45	180.69	197.59	238.14	276.95
6	153.12	174.16	188.72	202.62	220.31	262.59	302.83
7	172.04	194.15	209.39	223.90	242.33	286.19	327.77
8	190.62	213.71	229.58	244.67	263.79	309.14	351.97
9	208.91	232.93	249.40	265.02	284.79	331.55	375.57
10	226.95	251.85	268.89	285.02	305.41	353.50	398.65

P=1/15

R	.7500	.9000	.9500	.9750	.9900	.9990	.9999
1	48.87	63.04	73.35	83.51	96.85	130.27	163.64
2	75.85	92.46	104.28	115.79	130.71	167.40	203.40
3	99.69	118.16	131.15	143.70	159.85	199.15	237.29
4	122.04	142.06	156.04	169.48	186.69	228.24	268.25
5	143.46	164.85	179.70	193.92	212.08	255.65	297.33
6	164.23	186.86	202.51	217.45	236.46	281.86	325.09
7	184.52	208.29	224.67	240.27	260.06	307.18	351.85
8	204.43	229.27	246.33	262.54	283.08	331.79	377.80
9	224.03	249.87	267.57	284.36	305.60	355.82	403.11
10	243.38	270.16	288.47	305.81	327.71	379.37	427.87

P=1/16

R	.7500	.9000	.9500	.9750	.9900	.9990	.9999
1	52.22	67.38	78.40	89.27	103.54	139.26	174.93
2	81.01	98.79	111.44	123.74	139.70	178.91	217.40
3	106.47	126.23	140.12	153.54	170.81	212.82	253.60
4	130.31	151.74	166.69	181.06	199.47	243.88	286.65
5	153.17	176.06	191.95	207.16	226.57	273.15	317.71
6	175.33	199.56	216.30	232.27	252.60	301.14	347.35
7	196.98	222.44	239.95	256.63	277.80	328.17	375.92
8	218.23	244.82	263.07	280.41	302.37	354.45	403.64
9	239.15	266.81	285.74	303.70	326.41	380.10	430.65
10	259.79	288.47	308.05	326.59	350.00	405.24	457.08

P=1/17

R	.7500	.9000	.9500	.9750	.9900	.9990	.9999
1	55.56	71.72	83.46	95.02	110.22	148.24	186.23
2	86.18	105.12	118.59	131.69	148.68	190.43	231.40
3	113.23	134.29	149.09	163.38	181.77	226.50	269.90
4	138.58	161.42	177.35	192.65	212.24	259.53	305.06
5	162.88	187.28	204.21	220.40	241.06	290.65	338.09
6	186.44	212.26	230.09	247.09	268.74	320.42	369.62
7	209.45	236.58	255.24	272.99	295.54	349.16	399.99
8	232.03	260.38	279.81	298.27	321.66	377.10	429.46
9	254.27	283.75	303.91	323.03	347.21	404.38	458.19
10	276.21	306.77	327.63	347.38	372.31	431.10	486.30

P=1/18

R	.7500	.9000	.9500	.9750	.9900	.9990	.9999
1	58.91	76.06	88.51	100.79	116.90	157.23	197.52
2	91.34	111.46	125.74	139.64	157.66	201.94	245.40
3	120.00	142.36	158.06	173.22	192.73	240.17	286.21
4	146.85	171.10	188.00	204.23	225.02	275.18	323.47
5	172.59	198.50	216.46	233.64	255.56	308.15	358.47
6	197.55	224.96	243.88	261.92	284.88	339.70	391.87
7	221.92	250.72	270.52	289.36	313.27	370.15	424.06
8	245.84	275.93	296.55	316.14	340.94	399.75	455.29
9	269.38	300.69	322.09	342.37	368.02	428.66	485.73
10	292.62	325.07	347.21	368.16	394.60	456.97	515.51

DIRICHLET 1, N-VALUES FOR INVERSE PROBLEM (LI)   B= 8

P*-VALUES

R	.7500	.9000	.9500	.9750	.9900	.9990	.9999
				P=1/19			
1	62.25	80.40	93.57	106.55	123.59	166.22	208.81
2	96.51	117.78	132.89	147.60	166.65	213.46	259.40
3	126.77	150.43	167.03	183.06	203.69	253.84	302.52
4	155.13	180.78	198.65	215.82	237.79	290.82	341.87
5	182.30	209.72	228.71	246.87	270.05	325.66	378.85
6	208.65	237.60	257.67	276.75	301.02	358.97	414.13
7	234.39	264.86	285.80	305.73	331.01	391.14	448.13
8	259.64	291.48	313.29	334.00	360.23	422.41	481.12
9	284.50	317.62	340.26	361.71	388.83	452.93	513.27
10	309.04	343.37	366.79	388.94	416.90	482.83	544.72
				P=1/20			
1	65.60	84.73	98.62	112.31	130.27	175.21	220.10
2	101.67	124.11	140.05	155.55	175.63	224.97	273.40
3	133.54	158.50	176.00	192.90	214.65	267.52	318.82
4	163.40	190.46	209.31	227.40	250.57	306.47	360.28
5	192.01	220.93	240.96	260.11	284.55	343.16	399.23
6	219.75	250.36	271.46	291.57	317.16	378.25	436.39
7	246.85	279.00	301.08	322.09	348.74	412.13	472.20
8	273.44	307.03	330.03	351.87	379.52	445.06	506.94
9	299.62	334.56	358.43	381.04	409.63	477.20	540.81
10	325.45	361.68	386.37	409.72	439.20	508.70	573.93
				P=1/21			
1	68.94	89.07	103.67	118.06	136.94	184.20	231.40
2	106.83	130.45	147.20	163.50	184.61	236.49	287.40
3	140.31	166.56	184.97	202.75	225.61	281.19	335.12
4	171.67	200.14	219.96	238.98	263.35	322.11	378.69
5	201.72	232.14	253.21	273.35	299.04	360.66	419.61
6	230.86	263.05	285.24	306.40	333.30	397.52	458.65
7	259.32	293.14	316.37	338.46	366.48	433.12	496.28
8	287.24	322.58	346.78	369.74	398.80	467.71	532.77
9	314.73	351.50	376.60	400.38	430.44	501.48	568.35
10	341.86	379.97	405.94	430.50	461.49	534.56	603.14
				P=1/22			
1	72.29	93.41	108.73	123.82	143.63	193.19	242.69
2	111.99	136.77	154.36	171.45	193.59	248.00	301.40
3	147.08	174.63	193.94	212.59	236.57	294.86	351.43
4	179.94	209.82	230.61	250.57	276.12	337.76	397.09
5	211.43	243.36	265.46	286.58	313.53	378.16	439.98
6	241.96	275.75	299.03	321.22	349.44	416.80	480.91
7	271.78	307.28	331.65	354.82	384.21	454.10	520.35
8	301.04	338.13	363.52	387.60	418.09	490.36	558.60
9	329.85	368.44	394.77	419.72	451.24	525.75	595.88
10	358.28	398.28	425.52	451.28	483.79	560.43	632.36
				P=1/23			
1	75.63	97.75	113.78	129.58	150.31	202.18	253.98
2	117.16	143.10	161.51	179.40	202.58	259.52	315.40
3	153.84	182.69	202.92	222.43	247.53	308.53	367.73
4	188.21	219.50	241.26	262.15	288.89	353.40	415.50
5	221.14	254.58	277.71	299.82	328.02	395.66	460.36
6	253.06	288.45	312.82	336.04	365.58	436.07	503.17
7	284.25	321.42	346.93	371.18	401.94	475.09	544.42
8	314.84	353.68	380.26	405.47	437.38	513.01	584.42
9	344.96	385.37	412.94	439.05	472.05	550.02	623.42
10	374.69	416.58	445.10	472.06	506.08	586.29	661.57

DIRICHLET 1, N-VALUES FOR INVERSE PROBLEM (LI)   B= 8
P*-VALUES

R	.7500	.9000	.9500	.9750	.9900	.9990	.9999
			P=1/24				
1	78.97	102.08	118.83	135.34	156.99	211.16	265.27
2	122.32	149.43	168.66	187.35	211.56	271.03	329.40
3	160.61	190.76	211.89	232.27	258.49	322.20	384.03
4	196.48	229.17	251.91	273.74	301.67	369.04	433.90
5	230.84	265.79	289.95	313.05	342.52	413.16	480.74
6	264.17	301.15	326.61	350.87	381.72	455.35	525.43
7	296.71	335.56	362.21	387.54	419.68	496.08	568.48
8	328.65	369.23	396.99	423.33	456.66	535.66	610.25
9	360.08	402.31	431.11	458.39	492.85	574.30	650.95
10	391.10	434.88	464.67	492.84	528.38	612.15	690.77
			P=1/25				
1	82.32	106.42	123.89	141.09	163.67	220.15	276.57
2	127.49	155.76	175.81	195.30	220.54	282.55	343.40
3	167.38	198.82	220.86	242.11	269.44	335.88	400.34
4	204.75	238.85	262.57	285.32	314.45	384.69	452.31
5	240.55	277.00	302.20	326.29	357.00	430.66	501.11
6	275.27	313.84	340.39	365.69	397.85	474.62	547.68
7	309.18	349.70	377.49	403.91	437.41	517.06	592.55
8	342.45	384.78	413.73	441.19	475.95	558.31	636.07
9	375.20	419.24	449.28	477.72	513.66	598.57	678.49
10	407.52	453.18	484.25	513.62	550.68	638.01	719.98
			P=1/26				
1	85.66	110.76	128.94	146.85	170.35	229.14	287.85
2	132.65	162.09	182.96	203.25	229.52	294.07	357.40
3	174.15	206.89	229.83	251.95	280.40	349.55	416.64
4	213.02	248.53	273.22	296.90	327.22	400.33	470.71
5	250.26	288.22	314.45	339.53	371.50	448.16	521.49
6	286.38	326.54	354.18	380.51	413.99	493.90	569.94
7	321.64	363.84	392.77	420.27	455.14	538.05	616.62
8	356.25	400.34	430.47	459.06	495.23	580.96	661.90
9	390.31	436.18	467.45	497.06	534.46	622.85	706.03
10	423.93	471.48	503.83	534.40	572.97	663.88	749.19
			P=1/27				
1	89.00	115.09	133.99	152.61	177.03	238.13	299.15
2	137.81	168.42	190.11	211.19	238.50	305.58	371.40
3	180.91	214.95	238.80	261.79	291.36	363.22	432.94
4	221.29	258.21	283.87	308.49	339.99	415.97	489.11
5	259.97	299.43	326.70	352.76	385.99	465.66	541.87
6	297.48	339.24	367.96	395.34	430.13	513.17	592.20
7	334.11	377.98	408.05	436.63	472.87	559.04	640.69
8	370.04	415.88	447.21	476.92	514.52	603.61	687.72
9	405.43	453.11	485.62	516.39	555.27	647.12	733.56
10	440.34	489.78	523.40	555.18	595.27	689.74	778.40
			P=1/ 9, B = 9				
1	29.79	37.99	43.99	49.93	57.75	77.31	96.86
2	46.02	55.62	62.48	69.19	77.90	99.36	120.43
3	60.39	71.00	78.54	85.84	95.27	118.24	140.56
4	73.83	85.33	93.43	101.24	111.28	135.55	158.94
5	86.71	98.98	107.59	115.85	126.43	151.87	176.24
6	99.21	112.18	121.23	129.91	140.97	167.49	192.76
7	111.41	125.02	134.50	143.55	155.08	182.58	208.68
8	123.38	137.60	147.46	156.87	168.82	197.25	224.13
9	135.17	149.95	160.18	169.92	182.28	211.58	239.21
10	146.80	162.11	172.69	182.75	195.49	225.62	253.95

DIRICHLET 1, N-VALUES FOR INVERSE PROBLEM (LI)   B= 9

P*-VALUES

R	.7500	.9000	.9500	.9750	.9900	.9990	.9999
			P=1/10				
1	33.25	42.46	49.18	55.82	64.56	86.43	108.28
2	51.34	62.08	69.77	77.28	87.01	111.00	134.57
3	67.32	79.23	87.67	95.84	106.38	132.05	157.00
4	82.28	95.18	104.25	112.98	124.21	151.35	177.51
5	96.62	110.39	120.01	129.26	141.09	169.54	196.78
6	110.52	125.07	135.21	144.91	157.30	186.93	215.18
7	124.09	139.37	149.97	160.10	172.99	203.74	232.92
8	137.42	153.37	164.41	174.93	188.30	220.08	250.14
9	150.53	167.11	178.56	189.46	203.28	236.04	266.92
10	163.47	180.65	192.49	203.74	217.98	251.68	283.35
			P=1/11				
1	36.72	46.91	54.36	61.70	71.36	95.54	119.70
2	56.65	68.56	77.06	85.37	96.14	122.66	148.70
3	74.25	87.46	96.79	105.83	117.49	145.87	173.45
4	90.73	105.02	115.06	124.73	137.14	167.14	196.06
5	106.52	121.78	132.44	142.67	155.75	187.19	217.31
6	121.83	137.96	149.18	159.91	173.61	206.38	237.60
7	136.78	153.71	165.45	176.66	190.91	224.90	257.16
8	151.44	169.12	181.35	192.99	207.78	242.91	276.13
9	165.88	184.26	196.93	208.99	224.28	260.50	294.64
10	180.12	199.17	212.28	224.73	240.49	277.73	312.74
			P=1/12				
1	40.18	51.38	59.53	67.58	78.16	104.65	131.11
2	61.95	75.02	84.36	93.45	105.26	134.31	162.83
3	81.18	95.68	105.91	115.82	128.59	159.69	189.89
4	99.18	114.87	125.88	136.48	150.07	182.94	214.62
5	116.42	133.17	144.86	156.07	170.41	204.85	237.84
6	133.13	150.85	163.15	174.91	189.92	225.81	260.01
7	149.45	168.05	180.92	193.20	208.82	246.05	281.39
8	165.47	184.88	198.28	211.04	227.25	265.74	302.13
9	181.23	201.42	215.32	228.53	245.28	284.95	322.35
10	196.78	217.70	232.07	245.71	262.97	303.78	342.13
			P=1/13				
1	43.65	55.83	64.71	73.46	84.96	113.76	142.53
2	67.26	81.50	91.65	101.54	114.38	145.95	176.96
3	88.10	103.89	115.03	125.81	139.70	173.50	206.34
4	107.62	124.72	136.69	148.22	163.00	198.73	233.16
5	126.32	144.57	157.28	169.48	185.05	222.51	258.37
6	144.44	163.73	177.12	189.91	206.23	245.25	282.44
7	162.13	182.39	196.39	209.75	226.74	267.21	305.63
8	179.49	200.64	215.22	229.10	246.72	288.56	328.12
9	196.57	218.57	233.69	248.06	266.27	309.41	350.05
10	213.43	236.22	251.86	266.70	285.47	329.82	371.52
			P=1/14				
1	47.11	60.29	69.88	79.34	91.76	122.86	153.93
2	72.57	87.96	98.93	109.62	123.49	157.60	191.09
3	95.03	112.11	124.15	135.79	150.80	187.31	222.78
4	116.06	134.56	147.51	159.96	175.93	214.53	251.71
5	136.21	155.95	169.70	182.87	199.71	240.16	278.90
6	155.74	176.62	191.08	204.91	222.54	264.69	304.85
7	174.80	196.72	211.86	226.30	244.65	288.37	329.85
8	193.51	216.39	232.15	247.15	266.19	311.38	354.11
9	211.91	235.71	252.06	267.60	287.27	333.85	377.76
10	230.08	254.74	271.65	287.68	307.96	355.87	400.90

DIRICHLET 1, N-VALUES FOR INVERSE PROBLEM (LI)   B= 9

P*-VALUES

R	.7500	.9000	.9500	.9750	.9900	.9990	.9999
			P=1/15				
1	50.57	64.75	75.05	85.22	98.57	131.97	165.35
2	77.87	94.43	106.22	117.71	132.61	169.25	205.22
3	101.96	120.33	133.27	145.78	161.90	201.12	239.22
4	124.51	144.41	158.32	171.70	188.86	230.32	270.26
5	146.10	167.35	182.12	196.28	214.37	257.82	299.43
6	167.04	189.50	205.05	219.91	238.84	284.12	327.26
7	187.48	211.05	227.33	242.84	262.56	309.52	354.08
8	207.52	232.15	249.09	265.21	285.65	334.20	380.10
9	227.26	252.86	270.44	287.12	308.26	358.30	405.47
10	246.73	273.26	291.43	308.66	330.45	381.91	430.29
			P=1/16				
1	54.03	69.20	80.22	91.09	105.36	141.08	176.76
2	83.17	100.90	113.51	125.79	141.72	180.89	219.35
3	108.88	128.55	142.39	155.77	173.00	214.93	255.66
4	132.95	154.25	169.13	183.44	201.79	246.10	288.81
5	155.99	178.74	194.54	209.68	229.01	275.47	319.95
6	178.34	202.38	219.01	234.91	255.15	303.55	349.67
7	200.15	225.39	242.79	259.38	280.46	330.67	378.31
8	221.54	247.90	266.02	283.26	305.12	357.02	406.09
9	242.60	270.00	288.81	306.66	329.25	382.75	433.17
10	263.37	291.78	311.21	329.64	352.93	407.95	459.67
			P=1/17				
1	57.49	73.66	85.40	96.97	112.16	150.19	188.17
2	88.48	107.37	120.80	133.87	150.83	192.54	233.48
3	115.81	136.77	151.51	165.75	184.10	228.74	272.10
4	141.39	164.09	179.94	195.18	214.72	261.89	307.36
5	165.89	190.12	206.95	223.08	243.67	293.12	340.48
6	189.64	215.26	232.98	249.90	271.46	322.98	372.08
7	212.82	239.72	258.26	275.92	298.37	351.82	402.54
8	235.56	263.65	282.95	301.31	324.59	379.84	432.08
9	257.94	287.15	307.18	326.18	350.24	407.20	460.88
10	280.02	310.29	331.00	350.62	375.43	434.00	489.04
			P=1/18				
1	60.95	78.11	90.57	102.85	118.96	159.30	199.58
2	93.78	113.83	128.08	141.96	159.95	204.18	247.61
3	122.73	144.98	160.63	175.74	195.20	242.56	288.54
4	149.83	173.93	190.75	206.92	227.64	277.69	325.90
5	175.78	201.51	219.37	236.48	258.32	310.77	361.00
6	200.93	228.14	246.94	264.90	287.76	342.42	394.49
7	225.49	254.05	273.73	292.47	316.28	372.97	426.76
8	249.57	279.40	299.89	319.36	344.05	402.66	458.06
9	273.28	304.30	325.55	345.71	371.23	431.65	488.58
10	296.66	328.81	350.78	371.60	397.91	460.04	518.43
			P=1/19				
1	64.42	82.57	95.74	108.72	125.76	168.40	210.99
2	99.08	120.30	135.37	150.04	169.06	215.83	261.73
3	129.66	153.20	169.75	185.73	206.30	256.36	304.97
4	158.27	183.77	201.56	218.66	240.57	293.48	344.45
5	185.67	212.90	231.79	249.88	272.97	328.43	381.52
6	212.23	241.02	260.91	279.89	304.07	361.85	416.90
7	238.16	268.38	289.19	309.01	334.18	394.12	450.98
8	263.59	295.15	316.82	337.41	363.52	425.48	484.05
9	288.61	321.44	343.91	365.24	392.22	456.09	516.28
10	313.31	347.32	370.56	392.58	420.40	486.08	547.81

DIRICHLET 1, N-VALUES FOR INVERSE PROBLEM (LI)   B= 9
P#-VALUES

R	.7500	.9000	.9500	.9750	.9900	.9990	.9999
				P=1/20			
1	67.87	87.02	100.91	114.60	132.56	177.51	222.40
2	104.38	126.76	142.66	158.12	178.18	227.47	275.86
3	136.58	161.42	178.86	195.71	217.40	270.17	321.42
4	166.71	193.61	212.37	230.40	253.49	309.26	362.99
5	195.56	224.29	244.21	263.28	287.62	346.07	402.04
6	223.53	253.90	274.87	294.89	320.37	381.28	439.31
7	250.82	282.71	304.66	325.55	352.08	415.27	475.21
8	277.60	310.90	333.75	355.46	382.98	448.30	510.04
9	303.95	338.58	362.28	384.77	413.21	480.54	543.98
10	329.95	365.84	390.35	413.56	442.88	512.12	577.18
				P=1/21			
1	71.34	91.48	106.08	120.48	139.36	186.62	233.81
2	109.68	133.23	149.94	166.21	187.29	239.11	289.98
3	143.50	169.63	187.98	205.70	228.50	283.98	337.85
4	175.15	203.45	223.18	242.13	266.42	325.05	381.54
5	205.45	235.67	256.62	276.68	302.27	363.73	422.57
6	234.82	266.78	288.83	309.88	336.68	400.71	461.72
7	263.49	297.04	320.12	342.09	369.99	436.42	499.44
8	291.61	326.65	350.68	373.51	402.45	471.11	536.02
9	319.29	355.72	380.65	404.29	434.20	504.98	571.68
10	346.59	384.35	410.13	434.54	465.37	538.16	606.56
				P=1/22			
1	74.79	95.93	111.25	126.35	146.16	195.72	245.22
2	114.98	139.69	157.23	174.29	196.40	250.76	304.11
3	150.43	177.85	197.09	215.68	239.60	297.79	354.29
4	183.59	213.29	233.99	253.87	279.34	340.84	400.08
5	215.34	247.06	269.04	290.07	316.92	381.38	443.09
6	246.12	279.66	302.80	324.88	352.98	420.14	484.12
7	276.16	311.37	335.59	358.63	387.90	457.57	523.66
8	305.63	342.40	367.61	391.56	421.91	493.93	562.00
9	334.63	372.87	399.02	423.82	455.19	529.43	599.38
10	363.24	402.86	429.91	455.51	487.85	564.20	635.94
				P=1/23			
1	78.26	100.39	116.43	132.23	152.96	204.83	256.63
2	120.29	146.16	164.52	182.37	205.52	262.40	318.24
3	157.35	186.06	206.21	225.66	250.70	311.60	370.73
4	192.03	223.13	244.80	265.61	292.27	356.63	418.63
5	225.23	258.45	281.46	303.47	331.57	399.03	463.61
6	257.41	292.54	316.76	339.87	369.29	439.57	506.53
7	288.83	325.70	351.05	375.17	405.80	478.72	547.88
8	319.64	358.15	384.54	409.61	441.38	516.75	587.99
9	349.96	390.01	417.39	443.34	476.18	553.87	627.08
10	379.88	421.38	449.69	476.49	510.34	590.24	665.32
				P=1/24			
1	81.72	104.84	121.60	138.10	159.76	213.93	268.04
2	125.59	152.62	171.80	190.46	214.63	274.04	332.37
3	164.27	194.28	215.33	235.65	261.80	325.41	387.16
4	200.47	232.97	255.61	277.35	305.19	372.41	437.17
5	235.12	269.83	293.87	316.87	346.22	416.68	484.13
6	268.71	305.41	330.72	354.86	385.59	459.00	528.94
7	301.50	340.03	366.52	391.71	423.71	499.86	572.11
8	333.65	373.90	401.47	427.66	460.84	539.57	613.97
9	365.30	407.15	435.75	462.87	497.17	578.32	654.78
10	396.52	439.89	469.47	497.47	532.82	616.28	694.69

DIRICHLET 1, N-VALUES FOR INVERSE PROBLEM (LI)   B= 9

P*-VALUES

R	.7500	.9000	.9500	.9750	.9900	.9990	.9999
				P=1/25			
1	85.17	109.30	126.77	143.98	166.56	223.03	279.45
2	130.89	159.09	179.09	198.54	223.74	285.69	346.49
3	171.19	202.49	224.45	245.63	272.90	339.21	403.60
4	208.90	242.81	266.42	289.08	318.11	388.20	455.71
5	245.01	281.22	306.29	330.27	360.87	434.33	504.66
6	280.00	318.29	344.69	369.86	401.89	478.43	551.35
7	314.16	354.36	381.98	408.25	441.61	521.01	596.33
8	347.66	389.65	418.40	445.71	480.30	562.38	639.96
9	380.63	424.29	454.12	482.40	518.15	602.76	682.48
10	413.16	458.40	489.25	518.44	555.31	642.32	724.07
				P=1/26			
1	88.64	113.75	131.94	149.85	173.36	232.14	290.86
2	136.19	165.55	186.38	206.62	232.85	297.33	360.62
3	178.11	210.71	233.56	255.62	283.99	353.02	420.03
4	217.34	252.65	277.23	300.82	331.04	403.99	474.26
5	254.90	292.60	318.70	343.67	375.52	451.98	525.18
6	291.30	331.17	358.65	384.85	418.20	497.86	573.75
7	326.83	368.69	397.44	424.79	459.52	542.16	620.55
8	361.68	405.39	435.33	463.76	499.76	585.20	665.94
9	395.97	441.43	472.49	501.92	539.14	627.21	710.18
10	429.80	476.92	509.03	539.42	577.79	668.36	753.45
				P=1/27			
1	92.09	118.21	137.11	155.73	180.15	241.25	302.27
2	141.49	172.01	193.66	214.70	241.97	308.97	374.74
3	185.04	218.92	242.68	265.60	295.09	366.83	436.47
4	225.78	262.49	288.03	312.56	343.96	419.77	492.80
5	264.79	303.99	331.12	357.06	390.17	469.63	545.70
6	302.60	344.04	372.61	399.84	434.50	517.29	596.16
7	339.50	383.02	412.90	441.33	477.42	563.31	644.78
8	375.69	421.14	452.26	481.80	519.23	608.01	691.93
9	411.30	458.57	490.85	521.45	560.13	651.65	737.88
10	446.44	495.43	528.81	560.40	600.28	694.39	782.82
				P=1/28			
1	95.56	122.66	142.28	161.61	186.95	250.35	313.68
2	146.79	178.48	200.95	222.78	251.08	320.62	388.87
3	191.96	227.13	251.79	275.58	306.19	380.64	452.91
4	234.22	272.33	298.84	324.30	356.89	435.56	511.34
5	274.68	315.38	343.53	370.46	404.82	487.28	566.22
6	313.89	356.92	386.57	414.84	450.80	536.72	618.57
7	352.16	397.35	428.37	457.87	495.32	584.45	669.00
8	389.70	436.89	469.19	499.85	538.69	630.83	717.91
9	426.64	475.71	509.22	540.97	581.12	676.09	765.58
10	463.08	513.94	548.59	581.37	622.76	720.43	812.20
				P=1/10, B = 10			
1	34.25	43.46	50.18	56.82	65.56	87.43	109.28
2	52.53	63.24	70.91	78.40	88.13	112.09	135.64
3	68.64	80.50	88.90	97.05	107.58	133.21	158.13
4	83.72	96.55	105.58	114.29	125.48	152.57	178.69
5	98.16	111.84	121.42	130.64	142.43	170.80	198.00
6	112.16	126.61	136.69	146.36	158.69	188.25	216.45
7	125.82	140.98	151.52	161.61	174.45	205.10	234.23
8	139.22	155.04	166.01	176.49	189.80	221.49	251.49
9	152.41	168.85	180.23	191.07	204.83	237.49	268.30
10	165.42	182.45	194.21	205.40	219.59	253.16	284.76

DIRICHLET 1, N-VALUES FOR INVERSE PROBLEM (LI)   B=10
P#-VALUES

R	.7500	.9000	.9500	.9750	.9900	.9990	.9999
			P=1/11				
1	37.82	48.02	55.46	62.81	72.47	96.64	120.80
2	57.95	69.83	78.32	86.61	97.37	123.86	149.88
3	75.71	88.86	98.16	107.18	118.81	147.15	174.70
4	92.32	106.54	116.54	126.17	138.55	168.49	197.37
5	108.23	123.39	133.99	144.18	157.22	188.60	218.67
6	123.64	139.66	150.82	161.51	175.15	207.83	239.00
7	138.68	155.49	167.16	178.32	192.52	226.41	258.61
8	153.44	170.98	183.12	194.71	209.44	244.47	277.62
9	167.95	186.19	198.78	210.78	225.99	262.09	296.16
10	182.28	201.16	214.18	226.57	242.25	279.37	314.31
			P=1/12				
1	41.39	52.59	60.74	68.79	79.37	105.86	132.32
2	63.39	76.42	85.73	94.81	106.60	135.62	164.13
3	82.78	97.21	107.42	117.30	130.03	161.08	191.26
4	100.92	116.53	127.49	138.05	151.61	184.41	216.04
5	118.29	134.94	146.57	157.73	172.02	206.39	239.33
6	135.11	152.71	164.94	176.66	191.61	227.41	261.56
7	151.54	169.99	182.79	195.02	210.58	247.71	282.97
8	167.65	186.91	200.23	212.93	229.06	267.44	303.76
9	183.50	203.53	217.34	230.48	247.15	286.71	324.01
10	199.14	219.88	234.16	247.73	264.91	305.58	343.84
			P=1/13				
1	44.95	57.15	66.01	74.78	86.28	115.07	143.84
2	68.82	83.01	93.14	103.01	115.83	147.39	178.38
3	89.85	105.57	116.67	127.41	141.27	175.02	207.82
4	109.52	126.52	138.45	149.93	164.68	200.33	234.72
5	128.35	146.49	159.13	171.28	186.82	224.18	259.98
6	146.59	165.75	179.06	191.81	208.06	246.98	284.10
7	164.40	184.51	198.43	211.73	228.65	269.00	307.35
8	181.86	202.85	217.34	231.15	248.70	290.41	329.89
9	199.04	220.86	235.88	250.18	268.32	311.31	351.87
10	215.99	238.59	254.13	268.89	287.58	331.78	373.38
			P=1/14				
1	48.53	61.71	71.30	80.76	93.18	124.29	155.36
2	74.25	89.60	100.55	111.22	125.06	159.14	192.62
3	96.91	113.92	125.91	137.53	152.50	188.95	224.39
4	118.11	136.51	149.40	161.81	177.74	216.25	253.39
5	138.40	158.02	171.71	184.83	201.61	241.96	280.64
6	158.06	178.80	193.19	206.96	224.52	266.56	306.65
7	177.25	199.00	214.06	228.43	246.71	290.31	331.71
8	196.06	218.78	234.45	249.37	268.32	313.38	356.02
9	214.58	238.19	254.44	269.88	289.47	335.91	379.73
10	232.85	257.30	274.09	290.04	310.23	357.98	402.91
			P=1/15				
1	52.09	66.27	76.58	86.74	100.09	133.50	166.87
2	79.67	96.19	107.95	119.42	134.30	170.90	206.86
3	103.97	122.27	135.17	147.64	163.72	202.89	240.94
4	126.70	146.50	160.35	173.69	190.80	232.17	272.06
5	148.46	169.57	184.27	198.38	216.41	259.75	301.30
6	169.53	191.84	207.31	222.11	240.97	286.13	329.20
7	190.10	213.51	229.69	245.13	264.77	311.60	356.08
8	210.27	234.71	251.55	267.59	287.95	336.35	382.16
9	230.12	255.52	272.98	289.59	310.63	360.52	407.58
10	249.70	276.00	294.06	311.20	332.89	384.18	432.44

DIRICHLET 1, N-VALUES FOR INVERSE PROBLEM (LI)   B=10

P*-VALUES

R	.7500	.9000	.9500	.9750	.9900	.9990	.9999
			P=1/16				
1	55.66	70.83	81.85	92.73	106.99	142.71	178.40
2	85.10	102.78	115.37	127.63	143.53	182.67	221.10
3	111.04	130.62	144.42	157.76	174.95	216.82	257.50
4	135.30	156.48	171.30	185.57	203.86	248.09	290.73
5	158.51	181.11	196.84	211.92	231.20	277.54	321.95
6	181.00	204.88	221.44	237.26	257.43	305.71	351.74
7	202.95	228.01	245.32	261.83	282.83	332.90	380.45
8	224.48	250.64	268.65	285.80	307.58	359.32	408.29
9	245.66	272.85	291.53	309.29	331.78	385.11	435.43
10	266.55	294.71	314.03	332.36	355.55	410.39	461.97
			P=1/17				
1	59.22	75.40	87.13	98.71	113.90	151.92	189.91
2	90.53	109.37	122.77	135.83	152.76	194.43	235.34
3	118.10	138.97	153.67	167.87	186.18	230.75	274.06
4	143.89	166.47	182.25	197.44	216.92	264.00	309.41
5	168.57	192.65	209.41	225.47	245.99	295.33	342.61
6	192.47	217.92	235.56	252.41	273.88	325.28	374.29
7	215.80	242.51	260.95	278.53	300.89	354.19	404.81
8	238.68	266.56	285.75	304.02	327.20	382.29	434.42
9	261.19	290.17	310.07	328.98	352.94	409.72	463.27
10	283.40	313.42	333.99	353.51	378.20	436.58	491.50
			P=1/18				
1	62.79	79.95	92.41	104.69	120.81	161.14	201.43
2	95.95	115.95	130.18	144.02	161.99	206.18	249.58
3	125.16	147.32	162.91	177.98	197.41	244.68	290.62
4	152.48	176.45	193.20	209.32	229.98	279.92	328.07
5	178.62	204.19	221.97	239.01	260.78	313.11	363.26
6	203.94	230.97	249.68	267.55	290.34	344.85	396.83
7	228.65	257.01	276.58	295.23	318.95	375.49	429.17
8	252.88	282.49	302.85	322.23	346.82	405.26	460.55
9	276.73	307.50	328.62	348.68	374.09	434.32	491.12
10	300.24	332.12	353.96	374.67	400.86	462.78	521.03
			P=1/19				
1	66.35	84.52	97.69	110.67	127.71	170.35	212.94
2	101.38	122.54	137.59	152.23	171.22	217.94	263.82
3	132.23	155.67	172.16	188.10	208.63	258.61	307.17
4	161.07	186.44	204.15	221.19	243.04	295.84	346.75
5	188.67	215.73	234.54	252.55	275.57	330.90	383.91
6	215.41	244.01	263.79	282.70	306.79	364.42	419.37
7	241.50	271.52	292.21	311.93	337.00	396.78	453.54
8	267.09	298.42	319.95	340.45	366.45	428.22	486.67
9	292.26	324.83	347.16	368.38	395.25	458.91	518.97
10	317.09	350.83	373.92	395.82	423.51	488.98	550.56
			P=1/20				
1	69.91	89.07	102.96	116.65	134.62	179.56	224.46
2	106.80	129.12	144.99	160.43	180.46	229.70	278.06
3	139.29	164.02	181.41	198.21	219.85	272.55	323.73
4	169.66	196.42	215.10	233.06	256.10	311.76	365.42
5	198.72	227.27	247.10	266.09	290.36	348.68	404.56
6	226.88	257.05	277.91	297.84	323.24	383.99	441.91
7	254.35	286.01	307.84	328.63	355.06	418.07	477.90
8	281.29	314.34	337.06	358.66	386.07	451.19	512.80
9	307.79	342.15	365.71	388.07	416.40	483.51	546.81
10	333.94	369.53	393.88	416.97	446.17	515.17	580.09

DIRICHLET 1, N-VALUES FOR INVERSE PROBLEM (LI)   B=10

P*-VALUES

R	.7500	.9000	.9500	.9750	.9900	.9990	.9999
				P=1/21			
1	73.48	93.63	108.24	122.64	141.52	188.77	235.97
2	112.23	135.71	152.40	168.63	189.69	241.46	292.30
3	146.35	172.37	190.66	208.33	231.08	286.48	340.29
4	178.25	206.40	226.05	244.94	269.16	327.67	384.08
5	208.77	238.81	259.67	279.64	305.15	366.47	425.21
6	238.34	270.09	292.03	312.99	339.69	403.56	464.46
7	267.20	300.51	323.46	345.33	373.12	439.37	502.26
8	295.49	330.27	354.16	376.88	405.69	474.15	538.92
9	323.33	359.48	384.25	407.77	437.55	508.11	574.66
10	350.78	388.24	413.85	438.12	468.82	541.37	609.62
				P=1/22			
1	77.04	98.19	113.52	128.62	148.43	197.98	247.49
2	117.65	142.30	159.80	176.83	198.91	253.22	306.54
3	153.41	180.72	199.90	218.44	242.31	300.40	356.84
4	186.84	216.39	237.00	256.82	282.22	343.59	402.75
5	218.82	250.35	272.23	293.18	319.94	384.25	445.87
6	249.81	283.13	306.15	328.14	356.14	423.13	486.99
7	280.04	315.01	339.09	362.03	391.18	460.66	526.62
8	309.69	346.19	371.26	395.09	425.31	497.12	565.05
9	338.86	376.80	402.80	427.47	458.70	532.71	602.51
10	367.63	406.94	433.81	459.28	491.48	567.57	639.14
				P=1/23			
1	80.61	102.75	118.79	134.60	155.33	207.20	259.00
2	123.07	148.88	167.21	185.03	208.14	264.97	320.78
3	160.47	189.06	209.15	228.55	253.53	314.33	373.40
4	195.43	226.37	247.95	268.69	295.27	359.51	421.42
5	228.87	261.89	284.79	306.73	334.74	402.03	466.52
6	261.28	296.17	320.27	343.28	372.59	442.70	509.54
7	292.89	329.51	354.72	378.73	409.24	481.95	550.98
8	323.89	362.12	388.36	413.30	444.93	520.08	591.18
9	354.39	394.13	421.34	447.16	479.86	557.31	630.35
10	384.47	425.64	453.77	480.43	514.13	593.76	668.67
				P=1/24			
1	84.17	107.31	124.07	140.58	162.23	216.41	270.52
2	128.50	155.47	174.62	193.23	217.38	276.73	335.01
3	167.53	197.41	218.40	238.66	264.76	328.26	389.95
4	204.01	236.35	258.90	280.57	308.33	375.42	440.09
5	238.92	273.43	297.36	320.27	349.53	419.82	487.17
6	272.74	309.21	334.39	358.43	389.04	462.27	532.08
7	305.74	344.01	370.35	395.43	427.30	503.24	575.34
8	338.09	378.04	405.46	431.52	464.56	543.05	617.31
9	369.92	411.45	439.88	466.86	501.01	581.90	658.20
10	401.32	444.34	473.73	501.58	536.78	619.96	698.19
				P=1/25			
1	87.73	111.87	129.35	146.56	169.14	225.62	282.03
2	133.92	162.05	182.02	201.44	226.60	288.49	349.25
3	174.59	205.76	227.65	248.78	275.98	342.19	406.51
4	212.60	246.34	269.85	292.44	321.39	391.34	458.76
5	248.97	284.97	309.92	333.81	364.32	437.60	507.82
6	284.21	322.25	348.51	373.57	405.49	481.83	554.62
7	318.59	358.51	385.97	412.13	445.36	524.53	599.71
8	352.29	393.96	422.56	449.73	484.18	566.01	643.43
9	385.45	428.78	458.43	486.55	522.16	606.50	686.04
10	418.16	463.04	493.70	522.73	559.43	646.15	727.72

DIRICHLET 1, N-VALUES FOR INVERSE PROBLEM (LI)   B=10
P*-VALUES

R	.7500	.9000	.9500	.9750	.9900	.9990	.9999
				P=1/26			
1	91.30	116.43	134.62	152.54	176.04	234.83	293.55
2	139.35	168.64	189.43	209.64	235.83	300.25	363.49
3	181.65	214.11	236.89	258.89	287.21	356.12	423.07
4	221.19	256.32	280.80	304.31	334.45	407.25	477.43
5	259.02	296.51	322.49	347.35	379.10	455.39	528.47
6	295.67	335.29	362.62	388.72	421.94	501.40	577.16
7	331.43	373.00	401.60	428.82	463.41	545.82	624.06
8	366.49	409.89	439.66	467.94	503.80	588.97	669.56
9	400.98	446.10	476.97	506.25	543.31	631.09	713.88
10	435.01	481.75	513.66	543.88	582.08	672.35	757.24
				P=1/27			
1	94.86	120.99	139.90	158.52	182.94	244.04	305.06
2	144.77	175.23	196.83	217.83	245.06	312.00	377.73
3	188.71	222.46	246.14	269.00	298.43	370.05	439.62
4	229.78	266.30	291.75	316.19	347.51	423.16	496.09
5	269.08	308.05	335.05	360.89	393.89	473.17	549.12
6	307.14	348.32	376.74	403.86	438.39	520.97	599.70
7	344.28	387.50	417.22	445.52	481.47	567.11	648.43
8	380.69	425.81	456.75	486.15	523.42	611.94	695.68
9	416.52	463.43	495.51	525.94	564.46	655.69	741.73
10	451.85	500.45	533.62	565.03	604.74	698.54	786.77
				P=1/28			
1	98.43	125.55	145.17	164.50	189.85	253.25	316.57
2	150.19	181.81	204.24	226.03	254.29	323.76	391.97
3	195.77	230.81	255.39	279.11	309.66	383.98	456.18
4	238.37	276.29	302.69	328.06	360.57	439.08	514.76
5	279.13	319.59	347.61	374.44	408.68	490.95	569.77
6	318.60	361.36	390.86	419.00	454.84	540.54	622.24
7	357.12	402.00	432.85	462.22	499.53	588.41	672.79
8	394.89	441.74	473.85	504.37	543.04	634.90	721.80
9	432.05	480.75	514.05	545.64	585.61	680.29	769.57
10	468.70	519.15	553.58	586.19	627.39	724.74	816.29
				P=1/29			
1	101.99	130.11	150.45	170.48	196.75	262.46	328.09
2	155.62	188.40	211.64	234.24	263.52	335.52	406.21
3	202.83	239.15	264.63	289.22	320.88	397.91	472.73
4	246.95	286.27	313.64	339.93	373.62	454.99	533.43
5	289.18	331.12	360.18	387.98	423.47	508.73	590.42
6	330.07	374.40	404.97	434.15	471.29	560.11	644.78
7	369.97	416.50	448.48	478.92	517.58	609.70	697.15
8	409.09	457.66	490.95	522.58	562.67	657.87	747.93
9	447.58	498.07	532.59	565.33	606.76	704.88	797.42
10	485.54	537.85	573.54	607.34	650.04	750.93	845.82

TABLE D

This table gives the mean and variance of the minimum frequency in a multinomial with b blue cells each having a common cell probability p

        for $p = 1/j$, $j = b(1)b + 9$,

        for $b = 2(1)10$,

        for $n = b(1)b + 54$.

The formulas on which these are based are (4.41) and (4.42).

DIRICHLET 1 (MEANS AND VARIANCES),   B= 1

	P=1/ 2		P=1/ 3	
N	MEAN	VARIANCE	MEAN	VARIANCE
1	.5000000000	.2500000000	.3333333333	.2222222222
2	1.0000000000	.5000000000	.6666666667	.4444444444
3	1.5000000000	.7500000000	1.0000000000	.6666666667
4	2.0000000000	1.0000000000	1.3333333333	.8888888889
5	2.5000000000	1.2500000000	1.6666666667	1.1111111111
6	3.0000000000	1.5000000000	2.0000000000	1.3333333333
7	3.5000000000	1.7500000000	2.3333333333	1.5555555556
8	4.0000000000	2.0000000000	2.6666666667	1.7777777778
9	4.5000000000	2.2500000000	3.0000000000	2.0000000000
10	5.0000000000	2.5000000000	3.3333333333	2.2222222222
11	5.5000000000	2.7500000000	3.6666666667	2.4444444444
12	6.0000000000	3.0000000000	4.0000000000	2.6666666667
13	6.5000000000	3.2500000000	4.3333333333	2.8888888889
14	7.0000000000	3.5000000000	4.6666666667	3.1111111111
15	7.5000000000	3.7500000000	5.0000000000	3.3333333333
16	8.0000000000	4.0000000000	5.3333333333	3.5555555556
17	8.5000000000	4.2500000000	5.6666666667	3.7777777778
18	9.0000000000	4.5000000000	6.0000000000	4.0000000000
19	9.5000000000	4.7500000000	6.3333333333	4.2222222222
20	10.0000000000	5.0000000000	6.6666666667	4.4444444444
21	10.5000000000	5.2500000000	7.0000000000	4.6666666667
22	11.0000000000	5.5000000000	7.3333333333	4.8888888889
23	11.5000000000	5.7500000000	7.6666666667	5.1111111111
24	12.0000000000	6.0000000000	8.0000000000	5.3333333333
25	12.5000000000	6.2500000000	8.3333333333	5.5555555556
26	13.0000000000	6.5000000000	8.6666666667	5.7777777778
27	13.5000000000	6.7500000000	9.0000000000	6.0000000000
28	14.0000000000	7.0000000000	9.3333333333	6.2222222222
29	14.5000000000	7.2500000000	9.6666666667	6.4444444444
30	15.0000000000	7.5000000000	10.0000000000	6.6666666667
31	15.5000000000	7.7500000000	10.3333333333	6.8888888889
32	16.0000000000	8.0000000000	10.6666666667	7.1111111111
33	16.5000000000	8.2500000000	11.0000000000	7.3333333333
34	17.0000000000	8.5000000000	11.3333333333	7.5555555556
35	17.5000000000	8.7500000000	11.6666666667	7.7777777778
36	18.0000000000	9.0000000000	12.0000000000	8.0000000000
37	18.5000000000	9.2500000000	12.3333333333	8.2222222222
38	19.0000000000	9.5000000000	12.6666666667	8.4444444444
39	19.5000000000	9.7500000000	13.0000000000	8.6666666667
40	20.0000000000	10.0000000000	13.3333333333	8.8888888889
41	20.5000000000	10.2500000000	13.6666666667	9.1111111111
42	21.0000000000	10.5000000000	14.0000000000	9.3333333333
43	21.5000000000	10.7500000000	14.3333333333	9.5555555556
44	22.0000000000	11.0000000000	14.6666666667	9.7777777778
45	22.5000000000	11.2500000000	15.0000000000	10.0000000000
46	23.0000000000	11.5000000000	15.3333333333	10.2222222222
47	23.5000000000	11.7500000000	15.6666666667	10.4444444444
48	24.0000000000	12.0000000000	16.0000000000	10.6666666667
49	24.5000000000	12.2500000000	16.3333333333	10.8888888889
50	25.0000000000	12.5000000000	16.6666666667	11.1111111111
51	25.5000000000	12.7500000000	17.0000000000	11.3333333333
52	26.0000000000	13.0000000000	17.3333333333	11.5555555556
53	26.5000000000	13.2500000000	17.6666666667	11.7777777778
54	27.0000000000	13.5000000000	18.0000000000	12.0000000000
55	27.5000000000	13.7500000000	18.3333333333	12.2222222222

D 2

DIRICHLET 1 (MEANS AND VARIANCES),   B= 1

	P=1/ 4		P=1/ 5	
N	MEAN	VARIANCE	MEAN	VARIANCE
1	.2500000000	.1875000000	.2000000000	.1600000000
2	.5000000000	.3750000000	.4000000000	.3200000000
3	.7500000000	.5625000000	.6000000000	.4800000000
4	1.0000000000	.7500000000	.8000000000	.6400000000
5	1.2500000000	.9375000000	1.0000000000	.8000000000
6	1.5000000000	1.1250000000	1.2000000000	.9600000000
7	1.7500000000	1.3125000000	1.4000000000	1.1200000000
8	2.0000000000	1.5000000000	1.6000000000	1.2800000000
9	2.2500000000	1.6875000000	1.8000000000	1.4400000000
10	2.5000000000	1.8750000000	2.0000000000	1.6000000000
11	2.7500000000	2.0625000000	2.2000000000	1.7600000000
12	3.0000000000	2.2500000000	2.4000000000	1.9200000000
13	3.2500000000	2.4375000000	2.6000000000	2.0800000000
14	3.5000000000	2.6250000000	2.8000000000	2.2400000000
15	3.7500000000	2.8125000000	3.0000000000	2.4000000000
16	4.0000000000	3.0000000000	3.2000000000	2.5600000000
17	4.2500000000	3.1875000000	3.4000000000	2.7200000000
18	4.5000000000	3.3750000000	3.6000000000	2.8800000000
19	4.7500000000	3.5625000000	3.8000000000	3.0400000000
20	5.0000000000	3.7500000000	4.0000000000	3.2000000000
21	5.2500000000	3.9375000000	4.2000000000	3.3600000000
22	5.5000000000	4.1250000000	4.4000000000	3.5200000000
23	5.7500000000	4.3125000000	4.6000000000	3.6800000000
24	6.0000000000	4.5000000000	4.8000000000	3.8400000000
25	6.2500000000	4.6875000000	5.0000000000	4.0000000000
26	6.5000000000	4.8750000000	5.2000000000	4.1600000000
27	6.7500000000	5.0625000000	5.4000000000	4.3200000000
28	7.0000000000	5.2500000000	5.6000000000	4.4800000000
29	7.2500000000	5.4375000000	5.8000000000	4.6400000000
30	7.5000000000	5.6250000000	6.0000000000	4.8000000000
31	7.7500000000	5.8125000000	6.2000000000	4.9600000000
32	8.0000000000	6.0000000000	6.4000000000	5.1200000000
33	8.2500000000	6.1875000000	6.6000000000	5.2800000000
34	8.5000000000	6.3750000000	6.8000000000	5.4400000000
35	8.7500000000	6.5625000000	7.0000000000	5.6000000000
36	9.0000000000	6.7500000000	7.2000000000	5.7600000000
37	9.2500000000	6.9375000000	7.4000000000	5.9200000000
38	9.5000000000	7.1250000000	7.6000000000	6.0800000000
39	9.7500000000	7.3125000000	7.8000000000	6.2400000000
40	10.0000000000	7.5000000000	8.0000000000	6.4000000000
41	10.2500000000	7.6875000000	8.2000000000	6.5600000000
42	10.5000000000	7.8750000000	8.4000000000	6.7200000000
43	10.7500000000	8.0625000000	8.6000000000	6.8800000000
44	11.0000000000	8.2500000000	8.8000000000	7.0400000000
45	11.2500000000	8.4375000000	9.0000000000	7.2000000000
46	11.5000000000	8.6250000000	9.2000000000	7.3600000000
47	11.7500000000	8.8125000000	9.4000000000	7.5200000000
48	12.0000000000	9.0000000000	9.6000000000	7.6800000000
49	12.2500000000	9.1875000000	9.8000000000	7.8400000000
50	12.5000000000	9.3750000000	10.0000000000	8.0000000000
51	12.7500000000	9.5625000000	10.2000000000	8.1600000000
52	13.0000000000	9.7500000000	10.4000000000	8.3200000000
53	13.2500000000	9.9375000000	10.6000000000	8.4800000000
54	13.5000000000	10.1250000000	10.8000000000	8.6400000000
55	13.7500000000	10.3125000000	11.0000000000	8.8000000000

DIRICHLET 1 (MEANS AND VARIANCES),   B= 1

N	P=1/ 6 MEAN	VARIANCE	P=1/ 7 MEAN	VARIANCE
1	.1666666667	.1388888889	.1428571429	.1224489796
2	.3333333333	.2777777778	.2857142857	.2448979592
3	.5000000000	.4166666667	.4285714286	.3673469388
4	.6666666667	.5555555556	.5714285714	.4897959184
5	.8333333333	.6944444444	.7142857143	.6122448980
6	1.0000000000	.8333333333	.8571428571	.7346938776
7	1.1666666667	.9722222222	1.0000000000	.8571428571
8	1.3333333333	1.1111111111	1.1428571429	.9795918367
9	1.5000000000	1.2500000000	1.2857142857	1.1020408163
10	1.6666666667	1.3888888889	1.4285714286	1.2244897959
11	1.8333333333	1.5277777778	1.5714285714	1.3469387755
12	2.0000000000	1.6666666667	1.7142857143	1.4693877551
13	2.1666666667	1.8055555556	1.8571428571	1.5918367347
14	2.3333333333	1.9444444444	2.0000000000	1.7142857143
15	2.5000000000	2.0833333333	2.1428571429	1.8367346939
16	2.6666666667	2.2222222222	2.2857142857	1.9591836735
17	2.8333333333	2.3611111111	2.4285714286	2.0816326531
18	3.0000000000	2.5000000000	2.5714285714	2.2040816327
19	3.1666666667	2.6388888889	2.7142857143	2.3265306122
20	3.3333333333	2.7777777778	2.8571428571	2.4489795918
21	3.5000000000	2.9166666667	3.0000000000	2.5714285714
22	3.6666666667	3.0555555556	3.1428571429	2.6938775510
23	3.8333333333	3.1944444444	3.2857142857	2.8163265306
24	4.0000000000	3.3333333333	3.4285714286	2.9387755102
25	4.1666666667	3.4722222222	3.5714285714	3.0612244898
26	4.3333333333	3.6111111111	3.7142857143	3.1836734694
27	4.5000000000	3.7500000000	3.8571428571	3.3061224490
28	4.6666666667	3.8888888889	4.0000000000	3.4285714286
29	4.8333333333	4.0277777778	4.1428571429	3.5510204082
30	5.0000000000	4.1666666667	4.2857142857	3.6734693878
31	5.1666666667	4.3055555556	4.4285714286	3.7959183673
32	5.3333333333	4.4444444444	4.5714285714	3.9183673469
33	5.5000000000	4.5833333333	4.7142857143	4.0408163265
34	5.6666666667	4.7222222222	4.8571428571	4.1632653061
35	5.8333333333	4.8611111111	5.0000000000	4.2857142857
36	6.0000000000	5.0000000000	5.1428571429	4.4081632653
37	6.1666666667	5.1388888889	5.2857142857	4.5306122449
38	6.3333333333	5.2777777778	5.4285714286	4.6530612245
39	6.5000000000	5.4166666667	5.5714285714	4.7755102041
40	6.6666666667	5.5555555556	5.7142857143	4.8979591837
41	6.8333333333	5.6944444444	5.8571428571	5.0204081633
42	7.0000000000	5.8333333333	6.0000000000	5.1428571429
43	7.1666666667	5.9722222222	6.1428571429	5.2653061224
44	7.3333333333	6.1111111111	6.2857142857	5.3877551020
45	7.5000000000	6.2500000000	6.4285714286	5.5102040816
46	7.6666666667	6.3888888889	6.5714285714	5.6326530612
47	7.8333333333	6.5277777778	6.7142857143	5.7551020408
48	8.0000000000	6.6666666667	6.8571428571	5.8775510204
49	8.1666666667	6.8055555556	7.0000000000	6.0000000000
50	8.3333333333	6.9444444444	7.1428571429	6.1224489796
51	8.5000000000	7.0833333333	7.2857142857	6.2448979592
52	8.6666666667	7.2222222222	7.4285714286	6.3673469388
53	8.8333333333	7.3611111111	7.5714285714	6.4897959184
54	9.0000000000	7.5000000000	7.7142857143	6.6122448980
55	9.1666666667	7.6388888889	7.8571428571	6.7346938776

DIRICHLET 1 (MEANS AND VARIANCES),   B= 1

N	P=1/ 8 MEAN	VARIANCE	P=1/ 9 MEAN	VARIANCE
1	.1250000000	.1093750000	.1111111111	.0987654321
2	.2500000000	.2187500000	.2222222222	.1975308642
3	.3750000000	.3281250000	.3333333333	.2962962963
4	.5000000000	.4375000000	.4444444444	.3950617284
5	.6250000000	.5468750000	.5555555556	.4938271605
6	.7500000000	.6562500000	.6666666667	.5925925926
7	.8750000000	.7656250000	.7777777778	.6913580247
8	1.0000000000	.8750000000	.8888888889	.7901234568
9	1.1250000000	.9843750000	1.0000000000	.8888888889
10	1.2500000000	1.0937500000	1.1111111111	.9876543210
11	1.3750000000	1.2031250000	1.2222222222	1.0864197531
12	1.5000000000	1.3125000000	1.3333333333	1.1851851852
13	1.6250000000	1.4218750000	1.4444444444	1.2839506173
14	1.7500000000	1.5312500000	1.5555555556	1.3827160494
15	1.8750000000	1.6406250000	1.6666666667	1.4814814815
16	2.0000000000	1.7500000000	1.7777777778	1.5802469136
17	2.1250000000	1.8593750000	1.8888888889	1.6790123457
18	2.2500000000	1.9687500000	2.0000000000	1.7777777778
19	2.3750000000	2.0781250000	2.1111111111	1.8765432099
20	2.5000000000	2.1875000000	2.2222222222	1.9753086420
21	2.6250000000	2.2968750000	2.3333333333	2.0740740741
22	2.7500000000	2.4062500000	2.4444444444	2.1728395062
23	2.8750000000	2.5156250000	2.5555555556	2.2716049383
24	3.0000000000	2.6250000000	2.6666666667	2.3703703704
25	3.1250000000	2.7343750000	2.7777777778	2.4691358025
26	3.2500000000	2.8437500000	2.8888888889	2.5679012346
27	3.3750000000	2.9531250000	3.0000000000	2.6666666667
28	3.5000000000	3.0625000000	3.1111111111	2.7654320988
29	3.6250000000	3.1718750000	3.2222222222	2.8641975309
30	3.7500000000	3.2812500000	3.3333333333	2.9629629630
31	3.8750000000	3.3906250000	3.4444444444	3.0617283951
32	4.0000000000	3.5000000000	3.5555555556	3.1604938272
33	4.1250000000	3.6093750000	3.6666666667	3.2592592593
34	4.2500000000	3.7187500000	3.7777777778	3.3580246914
35	4.3750000000	3.8281250000	3.8888888889	3.4567901235
36	4.5000000000	3.9375000000	4.0000000000	3.5555555556
37	4.6250000000	4.0468750000	4.1111111111	3.6543209877
38	4.7500000000	4.1562500000	4.2222222222	3.7530864198
39	4.8750000000	4.2656250000	4.3333333333	3.8518518519
40	5.0000000000	4.3750000000	4.4444444444	3.9506172840
41	5.1250000000	4.4843750000	4.5555555556	4.0493827160
42	5.2500000000	4.5937500000	4.6666666667	4.1481481481
43	5.3750000000	4.7031250000	4.7777777778	4.2469135802
44	5.5000000000	4.8125000000	4.8888888889	4.3456790123
45	5.6250000000	4.9218750000	5.0000000000	4.4444444444
46	5.7500000000	5.0312500000	5.1111111111	4.5432098765
47	5.8750000000	5.1406250000	5.2222222222	4.6419753086
48	6.0000000000	5.2500000000	5.3333333333	4.7407407407
49	6.1250000000	5.3593750000	5.4444444444	4.8395061728
50	6.2500000000	5.4687500000	5.5555555556	4.9382716049
51	6.3750000000	5.5781250000	5.6666666667	5.0370370370
52	6.5000000000	5.6875000000	5.7777777778	5.1358024691
53	6.6250000000	5.7968750000	5.8888888889	5.2345679012
54	6.7500000000	5.9062500000	6.0000000000	5.3333333333
55	6.8750000000	6.0156250000	6.1111111111	5.4320987654

DIRICHLET 1 (MEANS AND VARIANCES),   B= 1

	P=1/10		P=1/11	
N	MEAN	VARIANCE	MEAN	VARIANCE
1	.1000000000	.0900000000	.0909090909	.0826446281
2	.2000000000	.1800000000	.1818181818	.1652892562
3	.3000000000	.2700000000	.2727272727	.2479338843
4	.4000000000	.3600000000	.3636363636	.3305785124
5	.5000000000	.4500000000	.4545454545	.4132231405
6	.6000000000	.5400000000	.5454545455	.4958677686
7	.7000000000	.6300000000	.6363636364	.5785123967
8	.8000000000	.7200000000	.7272727273	.6611570248
9	.9000000000	.8100000000	.8181818182	.7438016529
10	1.0000000000	.9000000000	.9090909091	.8264462810
11	1.1000000000	.9900000000	1.0000000000	.9090909091
12	1.2000000000	1.0800000000	1.0909090909	.9917355372
13	1.3000000000	1.1700000000	1.1818181818	1.0743801653
14	1.4000000000	1.2600000000	1.2727272727	1.1570247934
15	1.5000000000	1.3500000000	1.3636363636	1.2396694215
16	1.6000000000	1.4400000000	1.4545454545	1.3223140496
17	1.7000000000	1.5300000000	1.5454545455	1.4049586777
18	1.8000000000	1.6200000000	1.6363636364	1.4876033058
19	1.9000000000	1.7100000000	1.7272727273	1.5702479339
20	2.0000000000	1.8000000000	1.8181818182	1.6528925620
21	2.1000000000	1.8900000000	1.9090909091	1.7355371901
22	2.2000000000	1.9800000000	2.0000000000	1.8181818182
23	2.3000000000	2.0700000000	2.0909090909	1.9008264463
24	2.4000000000	2.1600000000	2.1818181818	1.9834710744
25	2.5000000000	2.2500000000	2.2727272727	2.0661157025
26	2.6000000000	2.3400000000	2.3636363636	2.1487603306
27	2.7000000000	2.4300000000	2.4545454545	2.2314049587
28	2.8000000000	2.5200000000	2.5454545455	2.3140495868
29	2.9000000000	2.6100000000	2.6363636364	2.3966942149
30	3.0000000000	2.7000000000	2.7272727273	2.4793388430
31	3.1000000000	2.7900000000	2.8181818182	2.5619834711
32	3.2000000000	2.8800000000	2.9090909091	2.6446280992
33	3.3000000000	2.9700000000	3.0000000000	2.7272727273
34	3.4000000000	3.0600000000	3.0909090909	2.8099173554
35	3.5000000000	3.1500000000	3.1818181818	2.8925619835
36	3.6000000000	3.2400000000	3.2727272727	2.9752066116
37	3.7000000000	3.3300000000	3.3636363636	3.0578512397
38	3.8000000000	3.4200000000	3.4545454545	3.1404958678
39	3.9000000000	3.5100000000	3.5454545455	3.2231404959
40	4.0000000000	3.6000000000	3.6363636364	3.3057851240
41	4.1000000000	3.6900000000	3.7272727273	3.3884297521
42	4.2000000000	3.7800000000	3.8181818182	3.4710743802
43	4.3000000000	3.8700000000	3.9090909091	3.5537190083
44	4.4000000000	3.9600000000	4.0000000000	3.6363636364
45	4.5000000000	4.0500000000	4.0909090909	3.7190082645
46	4.6000000000	4.1400000000	4.1818181818	3.8016528926
47	4.7000000000	4.2300000000	4.2727272727	3.8842975207
48	4.8000000000	4.3200000000	4.3636363636	3.9669421488
49	4.9000000000	4.4100000000	4.4545454545	4.0495867769
50	5.0000000000	4.5000000000	4.5454545455	4.1322314050
51	5.1000000000	4.5900000000	4.6363636364	4.2148760331
52	5.2000000000	4.6800000000	4.7272727273	4.2975206612
53	5.3000000000	4.7700000000	4.8181818182	4.3801652893
54	5.4000000000	4.8600000000	4.9090909091	4.4628099174
55	5.5000000000	4.9500000000	5.0000000000	4.5454545455

DIRICHLET 1 (MEANS AND VARIANCES),   B= 2

	P=1/ 2		P=1/ 3	
N	MEAN	VARIANCE	MEAN	VARIANCE
2	.5000000000	.2500000000	.2222222222	.1728395062
3	.7500000000	.1875000000	.4444444444	.2469135802
4	1.2500000000	.4375000000	.6913580247	.3615302545
5	1.5625000000	.3710937500	.9465020576	.4621585463
6	2.0625000000	.6210937500	1.2098765432	.5691205609
7	2.4062500000	.5537109375	1.4787379973	.6747879821
8	2.9062500000	.8037109375	1.7521719250	.7817453971
9	3.2695312500	.7359466553	2.0292638317	.8889470118
10	3.7695312500	.9859466553	2.3094379244	.9966629451
11	4.1464843750	.9179954529	2.5922313107	1.1047199712
12	4.6464843750	1.1679954529	2.8772939988	1.2131153203
13	5.0336914063	1.0999391079	3.1643449916	1.3217983368
14	5.5336914063	1.3499391079	3.4531576517	1.4307438398
15	5.9289550781	1.2818178535	3.7435450658	1.5399247360
16	6.4289550781	1.5318178535	4.0353512640	1.6493200493
17	6.8307647705	1.4636537486	4.3284443895	1.7589110703
18	7.3307647705	1.7136537486	4.6227118205	1.8686818564
19	7.7380294800	1.6454598866	4.9180564480	1.9786184229
20	8.2380294800	1.8954598866	5.2143938452	2.0887085222
21	8.6499309540	1.8272445250	5.5116500632	2.1989413248
22	9.1499309540	2.0772445250	5.8097598973	2.3093072128
23	9.5658369064	2.0090131275	6.1086655050	2.4197975952
24	10.0658369064	2.2590131275	6.4083152929	2.5304047631
25	10.4852467775	2.1907694526	6.7086630106	2.6411217680
26	10.9852467775	2.4407694526	7.0096670072	2.7519423214
27	11.4077562690	2.3725161700	7.3112896154	2.8628607103
28	11.9077562690	2.6225161700	7.6134966377	2.9738717269
29	12.3330332786	2.5542552283	7.9162569153	3.0849706078
30	12.8330332786	2.8042552283	8.2195419626	3.1961529833
31	13.2608010545	2.7359880826	8.5233256571	3.3074148332
32	13.7608010545	2.9859880826	8.8275839738	3.4187524489
33	14.1908260875	2.9177158418	9.1322947566	3.5301624010
34	14.6908260875	3.1677158418	9.4374375210	3.6416415106
35	15.1229092077	3.0994393652	9.7429932830	3.7531868243
36	15.6229092077	3.3494393652	10.0489444090	3.8647955930
37	16.0568789079	3.2811593295	10.3552744857	3.9764652520
38	16.5568789079	3.5311593295	10.6619682046	4.0881934048
39	16.9925862476	3.4628762743	10.9690112609	4.1999778076
40	17.4925862476	3.7128762743	11.2763902637	4.3118163564
41	17.9299009038	3.6445906357	11.5840926561	4.4237070749
42	18.4299009038	3.8945906357	11.8921066444	4.5356481039
43	18.8687080682	3.8263027695	12.2004211347	4.6476376920
44	19.3687080682	4.0763027695	12.5090256762	4.7596741869
45	19.8089059788	4.0080129691	12.8179104100	4.8717560279
46	20.3089059788	4.2580129691	13.1270660233	4.9838817388
47	20.7504039349	4.1897214787	13.4364837081	5.0960499217
48	21.2504039349	4.4397214787	13.7461551234	5.2082592513
49	21.6931206835	4.3714285027	14.0560723617	5.3205084698
50	22.1931206835	4.6214285027	14.3662279176	5.4327963820
51	22.6369830972	4.5531342142	14.6766146601	5.5451218513
52	23.1369830972	4.8031342142	14.9872258068	5.6574837955
53	23.5819250798	4.7348387603	15.2980549007	5.7698811835
54	24.0819250798	4.9848387603	15.6090957885	5.8823130317
55	24.5278866554	4.9165422667	15.9203426011	5.9947784011
56	25.0278866554	5.1665422667	16.2317897355	6.1072763946

DIRICHLET 1 (MEANS AND VARIANCES),   B= 2

	P=1/ 4		P=1/ 5	
N	MEAN	VARIANCE	MEAN	VARIANCE
2	.1250000000	.1093750000	.0800000000	.0736000000
3	.2812500000	.2021484375	.1920000000	.1551360000
4	.4531250000	.2946777344	.3200000000	.2368000000
5	.6347656250	.3880882263	.4576000000	.3186022400
6	.8232421875	.4824285507	.6017280000	.4009314140
7	1.0168457031	.5775970817	.7506688000	.4839203527
8	1.2144775391	.6734818071	.9033523200	.5675629060
9	1.4153823853	.7699852450	1.0590566400	.6518062653
10	1.6190147400	.8670267453	1.2172666880	.7365888503
11	1.8249654770	.9645401847	1.3775993651	.8218528647
12	2.0329184532	1.0624711392	1.5397604229	.9075474861
13	2.2426233888	1.1607744442	1.7035179819	.9936289513
14	2.4538781345	1.2594122626	1.8686854316	1.0800597522
15	2.6665166393	1.3583525931	2.0351099339	1.1668076807
16	2.8804005273	1.4575681245	2.2026644326	1.2538449534
17	3.0954130437	1.5570353505	2.3712419431	1.3411474680
18	3.3114546039	1.6567338823	2.5407513733	1.4286941928
19	3.5284394540	1.7566459084	2.7111144010	1.5164666676
20	3.7462931238	1.8567557670	2.8822630992	1.6044485987
21	3.9649504519	1.9570496039	3.0541381008	1.6926255283
22	4.1843540341	2.0575150964	3.2266871602	1.7809845638
23	4.4044529894	2.1581412285	3.3998640113	1.8695141546
24	4.6252019674	2.2589181060	3.5736274513	1.9582039077
25	4.8465603418	2.3598368047	3.7479405968	2.0470444337
26	5.0684915486	2.4608892424	3.9227702755	2.1360272178
27	5.2909625399	2.5620680729	4.0980865231	2.2251445109
28	5.5139433277	2.6633665956	4.2738621656	2.3143892376
29	5.7374066014	2.7647786793	4.4500724679	2.4037549179
30	5.9613274049	2.8662986971	4.6266948383	2.4932355992
31	6.1856828616	2.9679214702	4.8037085768	2.5828257992
32	6.4104519400	3.0696422197	4.9810946610	2.6725204555
33	6.6356152516	3.1714565249	5.1588355611	2.7623148821
34	6.8611548766	3.2733602870	5.3369150813	2.8522047318
35	7.0870542130	3.3753496970	5.5153182216	2.9421859627
36	7.3132978447	3.4774212082	5.6940310579	3.0322548095
37	7.5398714258	3.5795715116	5.8730406369	3.1224077575
38	7.7667615802	3.6817975139	6.0523348839	3.2126415203
39	7.9939558116	3.7840963189	6.2319025216	3.3029530192
40	8.2214424245	3.8864652101	6.4117329980	3.3933393657
41	8.4492104548	3.9889016353	6.5918164233	3.4837978454
42	8.6772496067	4.0914031934	6.7721435125	3.5743259035
43	8.9055501973	4.1939676222	6.9527055352	3.6649211323
44	9.1341031065	4.2965927869	7.1334942703	3.7555812596
45	9.3628997328	4.3992766711	7.3145019652	3.8463041380
46	9.5919319520	4.5020173670	7.4957212995	3.9370877357
47	9.8211920819	4.6048130681	7.6771453515	4.0279301283
48	10.0506728487	4.7076620613	7.8587675692	4.1188294905
49	10.2803673576	4.8105627206	8.0405817425	4.2097840893
50	10.5102690653	4.9135135008	8.2225819791	4.3007922780
51	10.7403717560	5.0165129325	8.4047626823	4.3918524897
52	10.9706695183	5.1195596161	8.5871185302	4.4829632323
53	11.2011567252	5.2226522183	8.7696444577	4.5741230837
54	11.4318280151	5.3257894669	8.9523356391	4.6653306872
55	11.6626782745	5.4289701475	9.1351874727	4.7565847472
56	11.8937026224	5.5321930997	9.3181955665	4.8478840257

DIRICHLET 1 (MEANS AND VARIANCES) .   B= 2

	P=1/ 6		P=1/ 7	
N	MEAN	VARIANCE	MEAN	VARIANCE
2	.0555555556	.0524691358	.0408163265	.0391503540
3	.1388888889	.1195987654	.1049562682	.0939404500
4	.2376543210	.1904340040	.1832569763	.1546717744
5	.3459362140	.2622725803	.2707205331	.2176605923
6	.4606481481	.3346139832	.3644909859	.2816164824
7	.5800325789	.4074322349	.4628513630	.3461303698
8	.7030035437	.4807517240	.5647140292	.4110882245
9	.8288412351	.5545720612	.6693541422	.4764600842
10	.9570405114	.6288703391	.7762655694	.5422317497
11	1.0872305094	.7036127838	.8850800952	.6083868397
12	1.2191291100	.7787627916	.9955203119	.6749046845
13	1.3525155633	.8542850203	1.1073710245	.7417620808
14	1.4872131562	.9301470683	1.2204611954	.8089352469
15	1.6230777464	1.0063199302	1.3346521094	.8764011510
16	1.7599898931	1.0827779038	1.4498293440	.9441382290
17	1.8978492870	1.1594982852	1.5658971522	1.0121266824
18	2.0365707049	1.2364610090	1.6827744243	1.0803485297
19	2.1760810054	1.3136482989	1.8003917154	1.1487875265
20	2.3163168543	1.3910443543	1.9186890121	1.2174290268
21	2.4572229737	1.4686350782	2.0376140231	1.2862598247
22	2.5987507749	1.5464078448	2.1571208512	1.3552679976
23	2.7408572760	1.6243513026	2.2771689463	1.4244427588
24	2.8835042345	1.7024552075	2.3977222695	1.4937743263
25	3.0266574461	1.7807102820	2.5187486187	1.5632538055
26	3.1702861715	1.8591080948	2.6402190790	1.6328730867
27	3.3143626633	1.9376409593	2.7621075708	1.7026247563
28	3.4588617737	2.0163018461	2.8843904739	1.7725020188
29	3.6037606253	2.0950843083	3.0070463148	1.8424986296
30	3.7490383340	2.1739824168	3.1300555023	1.9126088358
31	3.8946757739	2.2529907051	3.2534001045	1.9828273253
32	4.0406553769	2.3321041204	3.3770636578	2.0531491816
33	4.1869609603	2.4113179816	3.5010310041	2.1235698449
34	4.3335775786	2.4906279427	3.6252881501	2.1940850770
35	4.4804913950	2.5700299602	3.7498221455	2.2646909313
36	4.6276895692	2.6495202646	3.8746209773	2.3353837255
37	4.7751601604	2.7290953357	3.9996734777	2.4061600180
38	4.9228920406	2.8087518797	4.1249692424	2.4770165867
39	5.0708748197	2.8884868097	4.2504985601	2.5479504098
40	5.2190987787	2.9682972282	4.3762523493	2.6189586492
41	5.3675548104	3.0481804110	4.5022221022	2.6900386353
42	5.5162343669	3.1281337935	4.6283998358	2.7611878535
43	5.6651294124	3.2081549574	4.7547780470	2.8324039317
44	5.8142323817	3.2882416199	4.8813496737	2.9036846298
45	5.9635361420	3.3683916231	5.0081080589	2.9750278292
46	6.1130339595	3.4486029244	5.1350469191	3.0464315242
47	6.2627194684	3.5288735886	5.2621603157	3.1178938135
48	6.4125866437	3.6092017800	5.3894426287	3.1894128931
49	6.5626297761	3.6895857551	5.5168885336	3.2609870493
50	6.7128434491	3.7700238566	5.6444929799	3.3326146525
51	6.8632225186	3.8505145076	5.7722511719	3.4042941517
52	7.0137620942	3.9310562059	5.9001585508	3.4760240691
53	7.1644575217	4.0116475196	6.0282107791	3.5478029957
54	7.3153043677	4.0922870823	6.1564037251	3.6196295867
55	7.4662984051	4.1729735890	6.2847334504	3.6915025573
56	7.6174355998	4.2537057925	6.4131961965	3.7634206795

DIRICHLET 1 (MEANS AND VARIANCES),   B= 2

N	P=1/ 8 MEAN	VARIANCE	P=1/ 9 MEAN	VARIANCE
2	.0312500000	.0302734375	.02469135$8_0$	.0240816949
3	.0820312500	.0753021240	.0658436214	.0615082389
4	.1455078125	.1272649765	.1182746533	.1061147491
5	.2175903320	.1824518107	.1786651764	.1545340715
6	.2957611084	.2391855230	.2448474995	.2049935057
7	.3784198761	.2967417110	.3153752408	.2565880158
8	.4645142555	.3548212940	.3892605897	.3088680545
9	.5533275157	.4133050771	.4658121579	.3616154381
10	.6443543546	.4721446437	.5445336997	.4147254855
11	.7372266329	.5313164282	.6250599965	.4681469513
12	.8316679228	.5908038436	.7071156608	.5218524411
13	.9274649828	.6505912348	.7904882334	.5758244467
14	1.0244493779	.7106624428	.8750103067	.6300491353
15	1.1224853143	.7710009325	.9605474246	.6845138434
16	1.2214613710	.8315902846	1.0469897359	.7392062427
17	1.3212847334	.8924146397	1.1342461243	.7941141990
18	1.4218770724	.9534589932	1.2222400004	.8492258705
19	1.5231715345	1.0147093514	1.3109062279	.9045298519
20	1.6251104963	1.0761527834	1.4001888377	.9600152929
21	1.7276438584	1.1377774084	1.4900392994	1.0156719728
22	1.8307277283	1.1995723429	1.5804151923	1.0714903336
23	1.9343233874	1.2615276301	1.6712791695	1.1274614823
24	2.0383964712	1.3236341611	1.7625981396	1.1835771714
25	2.1429163096	1.3858835961	1.8543426120	1.2398297662
26	2.2478553919	1.4482682899	1.9464861668	1.2962122050
27	2.3531889273	1.5107812218	2.0390050225	1.3527179563
28	2.4588944813	1.5734159333	2.1318776791	1.4093409761
29	2.5649516733	1.6361664715	2.2250846216	1.4660756667
30	2.6713419217	1.6990273383	2.3186080713	1.5229168373
31	2.7780482285	1.7619934457	2.4124317759	1.5798596687
32	2.8850549967	1.8250600759	2.5065408320	1.6368996796
33	2.9923478730	1.8882228463	2.6009215326	1.6940326971
34	3.0999136132	1.9514776777	2.6955612362	1.7512548294
35	3.2077399657	2.0148207667	2.7904482547	1.8085624413
36	3.3158155703	2.0782485606	2.8855717548	1.8659521323
37	3.4241298696	2.1417577360	2.9809216730	1.9234207167
38	3.5326730319	2.2053451780	3.0764886413	1.9809652054
39	3.6414358830	2.2690079637	3.1722639212	2.0385827903
40	3.7504098463	2.3327433454	3.2682393459	2.0962708293
41	3.8595868894	2.3965487370	3.3644072696	2.1540268334
42	3.9689594774	2.4604217009	3.4607605216	2.2118484548
43	4.0785205302	2.5243599367	3.5572923664	2.2697334761
44	4.1882633852	2.5883612704	3.6539964669	2.3276798007
45	4.2981817638	2.6524236453	3.7508668528	2.3856854437
46	4.4082697407	2.7165451132	3.8478978912	2.4437485239
47	4.5185217168	2.7807238268	3.9450842604	2.5018672568
48	4.6289323949	2.8449580325	4.0424209263	2.5600399473
49	4.7394967568	2.9092460637	4.1399031216	2.6182649841
50	4.8502100437	2.9735863356	4.2375263256	2.6765408340
51	4.9610677373	3.0379773387	4.3352862474	2.7348660364
52	5.0720655437	3.1024176350	4.4331788091	2.7932391992
53	5.1831993773	3.1669058524	4.5312001323	2.8516589939
54	5.2944653478	3.2314406814	4.6293465237	2.9101241518
55	5.4058597467	3.2960208706	4.7276144634	2.9686334606
56	5.5173790359	3.3606452235	4.8260005939	3.0271857605

DIRICHLET 1 (MEANS AND VARIANCES),   B= 2

N	P=1/10 MEAN	VARIANCE	P=1/11 MEAN	VARIANCE
2	.0200000000	.0196000000	.0165289256	.0162557202
3	.0540000000	.0510840000	.0450788881	.0430467819
4	.0980000000	.0895960000	.0825080254	.0765200673
5	.1493000000	.1322095100	.1266058578	.1141781582
6	.2060520000	.1772145733	.1758008897	.1544458358
7	.2669716000	.2236361648	.2289630445	.1963501563
8	.3311511200	.2709296557	.2852695242	.2393003750
9	.3979376100	.3187960845	.3441132444	.2829429099
10	.4668525620	.3670734254	.4050399564	.3270685456
11	.5375385844	.4156741535	.4677047889	.3715543916
12	.6097236267	.4645502089	.5318419929	.4163282635
13	.6831967125	.5136737970	.5972437149	.4613472065
14	.7577912827	.5630271489	.6637449770	.5065847734
15	.8333736145	.6125972442	.7312129523	.5520236313
16	.9098346618	.6623732113	.7995392313	.5976513596
17	.9870842234	.7123451340	.8686341894	.6434581267
18	1.0650467160	.7625035632	.9384228391	.6894354504
19	1.1436580639	.8128393644	1.0088417421	.7355755695
20	1.2228633786	.8633437069	1.0798366849	.7818711479
21	1.3026152012	.9140081025	1.1513609076	.8283151542
22	1.3828721539	.9648244492	1.2233737404	.8749008248
23	1.4635978905	1.0157850648	1.2958395409	.9216216654
24	1.5447602706	1.0668827040	1.3687268577	.9684714632
25	1.6263307010	1.1181105627	1.4420077642	1.0154443002
26	1.7082836048	1.1694622698	1.5156573231	1.0625345604
27	1.7905959902	1.2209318716	1.5896531521	1.1097369327
28	1.8732470954	1.2725138112	1.6639750676	1.1570464068
29	1.9562180949	1.3242029050	1.7386047905	1.2044582654
30	2.0394918539	1.3759943174	1.8135257015	1.2519680724
31	2.1230527217	1.4278835359	1.8887226353	1.2995716599
32	2.2068863567	1.4798663472	1.9641817073	1.3472651134
33	2.2909795774	1.5319388139	2.0398901657	1.3950447558
34	2.3753202346	1.5840972527	2.1158362655	1.4429071330
35	2.4598971008	1.6363382143	2.1920091598	1.4908489982
36	2.5446997754	1.6886584649	2.2683988060	1.5388672980
37	2.6297186012	1.7410549693	2.3449958842	1.5869591582
38	2.7149445918	1.7935248749	2.4217917264	1.6351218718
39	2.8003693682	1.8460654977	2.4987782537	1.6833528860
40	2.8859851026	1.8986743092	2.5759479216	1.7316497921
41	2.9717844687	1.9513489249	2.6532936719	1.7800103146
42	3.0577605980	2.0040870930	2.7308088895	1.8284323018
43	3.1439070404	2.0568866849	2.8084873643	1.8769137173
44	3.2302177292	2.1097456862	2.8863232575	1.9254526319
45	3.3166869504	2.1626621885	2.9643110708	1.9740472158
46	3.4033093140	2.2156343821	3.0424456193	2.0226957324
47	3.4900797291	2.2686605486	3.1207220070	2.0713965317
48	3.5769933813	2.3217390554	3.1991356050	2.1201480446
49	3.6640457118	2.3748683495	3.2776820308	2.1689487776
50	3.7512323986	2.4280469524	3.3563571311	2.2177973082
51	3.8385493403	2.4812734552	3.4351569646	2.2666922796
52	3.9259926400	2.5345465141	3.5140777877	2.3156323977
53	4.0135585915	2.5878648466	3.5931160406	2.3646164262
54	4.1012436666	2.6412272275	3.6722683348	2.4136431839
55	4.1890445029	2.6946324851	3.7515314421	2.4627115407
56	4.2769578939	2.7480794988	3.8309022837	2.5118204152

DIRICHLET 1 (MEANS AND VARIANCES),   B= 3

N	P=1/ 3 MEAN	VARIANCE	P=1/ 4 MEAN	VARIANCE
3	.2222222222	.1728395062	.0937500000	.0849609375
4	.4444444444	.2469135802	.2343750000	.1794433594
5	.6172839506	.2362444749	.3808593750	.2358055115
6	.8641975309	.3642737388	.5346679688	.2927434444
7	1.1138545953	.4493142230	.6998291016	.3638769239
8	1.3315043439	.4548053730	.8715820313	.4323613644
9	1.5921353452	.5679857009	1.0472030640	.4965893146
10	1.8549001677	.6547576527	1.2271213531	.5636369145
11	2.0942268286	.6703876768	1.4109506607	.6324415378
12	2.3635963352	.7761958150	1.5977010727	.7002907450
13	2.6345840837	.8631105340	1.7870066464	.7682223611
14	2.8870950240	.8855553824	1.9787343219	.8370611805
15	3.1626276482	.9868832983	2.1725824568	.9061228327
16	3.4394268497	1.0735484648	2.3682836853	.9751875659
17	3.7009722065	1.1009332535	2.5656903995	1.0445627878
18	3.9811488081	1.1992081951	2.7646651180	1.1142334275
19	4.2623482570	1.2855209728	2.9650622334	1.1840442900
20	4.5305544660	1.3166873991	3.1667665148	1.2540154314
21	4.8143931008	1.4127253754	3.3696853946	1.3241858400
22	5.0990793212	1.4986705262	3.5737310817	1.3945172768
23	5.3724456913	1.5328486429	3.7788240838	1.4649847177
24	5.6592688537	1.6271638004	3.9848965358	1.5355947301
25	5.9468091117	1.7127551237	4.1918876758	1.6063418650
26	6.2243184696	1.7494038636	4.3997416683	1.6772114559
27	6.5136354820	1.8423431839	4.6084083574	1.7481975028
28	6.8035695056	1.9276036594	4.8178426621	1.8192968543
29	7.0844962037	1.9663261471	5.0280033294	1.8905028373
30	7.3759381750	2.0581360600	5.2388525617	1.9618092546
31	7.6679184590	2.1430905188	5.4503558389	2.0332120436
32	7.9517240955	2.1835856338	5.6624814543	2.1047070565
33	8.2450053575	2.2744481894	5.8752001348	2.1762898315
34	8.5387617421	2.3591205494	6.0884848371	2.2479565512
35	8.8250342337	2.4011535081	6.3023105333	2.3197038492
36	9.1199281823	2.4912074749	6.5166539813	2.3915283982
37	9.4152456085	2.5756197705	6.7314935515	2.4634270570
38	9.7036614978	2.6190033612	6.9468090842	2.5353969765
39	9.9999847657	2.7083572468	7.1625817511	2.6074354778
40	10.2966886578	2.7925293905	7.3787939320	2.6795400089
41	10.5869886963	2.8371115100	7.5954291104	2.7517081832
42	10.8845903967	2.9258519727	7.8124717802	2.8239377664
43	11.1825366947	3.0098018151	8.0299073604	2.8962266429
44	11.4745094132	3.0554569041	8.2477221193	2.9685728118
45	11.7732635992	3.1436543972	8.4659031072	3.0409743863
46	12.0723317482	3.2273978984	8.6844380953	3.1134295809
47	12.3658020461	3.2740208853	8.9033155210	3.1859367015
48	12.6656022832	3.3617335655	9.1225244374	3.2584941412
49	12.9566901719	3.4452849987	9.3420544695	3.3311003751
50	13.2605111737	3.4927869093	9.5618957721	3.4037539530
51	13.5612665443	3.5800634184	9.7820389933	3.4764534945
52	13.8622867671	3.6634355707	10.0024752404	3.5491976853
53	14.1583338750	3.7117402764	10.2231960491	3.6219852721
54	14.4599659779	3.7986217675	10.4441933548	3.6948150591
55	14.7618430204	3.8818261236	10.6654594670	3.7676859042
56	15.0590094801	3.9308678881	10.8869870452	3.8405967157
57	15.3614501382	4.0173895318	11.1087690772	3.9135464495

DIRICHLET 1 (MEANS AND VARIANCES),   B= 3

N	P=1/ 5 MEAN	VARIANCE	P=1/ 6 MEAN	VARIANCE
3	.0480000000	.0456960000	.0277777778	.0270061728
4	.1344000000	.1163366400	.0833333333	.0763888889
5	.2400000000	.1824000000	.1581790123	.1331584124
6	.3552000000	.2405529600	.2449845679	.1888251541
7	.4773888000	.2978727336	.3395276063	.2422527261
8	.6055526400	.3571306402	.4397005030	.2948750589
9	.7385794560	.4173235552	.5443583533	.3477551303
10	.8754923520	.4776270616	.6527280640	.4011515190
11	1.0156401869	.5381553129	.7642156170	.4549800994
12	1.1585844019	.5991277419	.8783572489	.5091519841
13	1.3039764062	.6605259418	.9947937702	.5636412256
14	1.4515226355	.7222566105	1.1132439625	.6184457360
15	1.6009822991	.7842777745	1.2334818676	.6735568022
16	1.7521592006	.8465800559	1.3553213181	.7289560334
17	1.9048895732	.9091489913	1.4786062833	.7846223465
18	2.0590336928	.9719627276	1.6032045067	.8405365767
19	2.2144713208	1.0350016442	1.7290028360	.8966821860
20	2.3710984571	1.0982506010	1.8559035947	.9530444968
21	2.5288244561	1.1616964953	1.9838217483	1.0096100937
22	2.6875696531	1.2253269506	2.1126827016	1.0663666410
23	2.8472635584	1.2891305404	2.2424205778	1.1233028563
24	3.0078434591	1.3530970263	2.3729768525	1.1804084591
25	3.1692532766	1.4172171955	2.5042992527	1.2376740737
26	3.3314426198	1.4814826388	2.6363408580	1.2950911246
27	3.4943660051	1.5458856427	2.7690593615	1.3526517437
28	3.6579822115	1.6104191362	2.9024164574	1.4103486957
29	3.8222537417	1.6750766296	3.0363773290	1.4681753138
30	3.9871463673	1.7398521465	3.1709102190	1.5261254420
31	4.1526287429	1.8047401666	3.3059860648	1.5841933850
32	4.3186720762	1.8697355799	3.4415781887	1.6423738620
33	4.4852498466	1.9348336471	3.5776620321	1.7006619676
34	4.6523375612	2.0000299637	3.7142149264	1.7590531359
35	4.8199125441	2.0653204277	3.8512158958	1.8175431095
36	4.9879537532	2.1307012120	3.9886454846	1.8761279118
37	5.1564416197	2.1961687393	4.1264856076	1.9348038219
38	5.3253579076	2.2617196603	4.2647194181	1.9935673531
39	5.4946855897	2.3273508341	4.4033311924	2.0524152327
40	5.6644087387	2.3930593106	4.5423062276	2.1113443844
41	5.8345124295	2.4588423145	4.6816307510	2.1703519125
42	6.0049826537	2.5246972320	4.8212918397	2.2294350875
43	6.1758062423	2.5906215972	4.9612773489	2.2885913333
44	6.3469707967	2.6566130814	5.1015758476	2.3478182155
45	6.5184646275	2.7226694821	5.2421765615	2.4071134308
46	6.6902766985	2.7887887143	5.3830693212	2.4664747973
47	6.8623965768	2.8549688016	5.5242445155	2.5259002461
48	7.0348143880	2.9212078685	5.6656930495	2.5853878132
49	7.2075207745	2.9875041336	5.8074063066	2.6449356320
50	7.3805068589	3.0538559027	5.9493761140	2.7045419272
51	7.5537642101	3.1202615635	6.0915947111	2.7642050085
52	7.7272848122	3.1867195800	6.2340547215	2.8239232648
53	7.9010610370	3.2532284873	6.3767491260	2.8836951595
54	8.0750856175	3.3197868874	6.5196712396	2.9435192257
55	8.2493516250	3.3863934450	6.6628146894	3.0033940617
56	8.4238524473	3.4530468834	6.8061733943	3.0633183273
57	8.5985817683	3.5197459812	6.9497415471	3.1232907398

DIRICHLET 1 (MEANS AND VARIANCES),   B= 3

	P=1/ 7		P=1/ 8	
N	MEAN	VARIANCE	MEAN	VARIANCE
3	.0174927114	.0171867164	.0117187500	.0115814209
4	.0549770929	.0519546121	.0380859375	.0366353989
5	.1088832034	.0970276514	.0778198242	.0717638992
6	.1746721179	.1456917439	.1282882690	.1125170346
7	.2487738952	.1945353181	.1869134903	.1555817264
8	.3288425741	.2427076297	.2516613007	.1991926251
9	.4134326332	.2904536264	.3210779876	.2427104433
10	.5016357734	.3381782891	.3941654507	.2860721439
11	.5928377746	.3861255714	.4702374148	.3294073727
12	.6865887363	.4343790701	.5488076170	.3728518512
13	.7825408270	.4829425842	.6295163400	.4164908438
14	.8804174922	.5317982717	.7120861660	.4603620991
15	.9799954096	.5809292143	.7962960921	.5044744573
16	1.0810918794	.6303222571	.8819659994	.5488235221
17	1.1835553688	.6799654998	.9689466988	.5934004258
18	1.2872584447	.7298465931	1.0571130450	.6381954071
19	1.3920925829	.7799526299	1.1463588963	.6831987744
20	1.4979643635	.8302707382	1.2365933083	.7284009820
21	1.6047926395	.8807886311	1.3277376241	.7737925972
22	1.7125063775	.9314948750	1.4197232324	.8193643579
23	1.8210429702	.9823789363	1.5124898284	.8651072916
24	1.9303468956	1.0334311245	1.6059840494	.9110128306
25	2.0403686382	1.0846425128	1.7001583927	.9570728865
26	2.1510638123	1.1360048655	1.7949703442	1.0032798799
27	2.2623924441	1.1875105766	1.8903816697	1.0496267371
28	2.3743183787	1.2391526152	1.9863578313	1.0961068666
29	2.4868087862	1.2909244743	2.0828675030	1.1427141280
30	2.5998337475	1.3428201245	2.1798821654	1.1894427988
31	2.7133659040	1.3948339694	2.2773757642	1.2362875414
32	2.8273801606	1.4469608069	2.3753244203	1.2832433734
33	2.9418534308	1.4991957927	2.4737061830	1.3303056386
34	3.0567644183	1.5515344090	2.5725008176	1.3774699810
35	3.1720934281	1.6039724356	2.6716896230	1.4247323211
36	3.2878222021	1.6565059239	2.7712552730	1.4720888331
37	3.4039337768	1.7091311745	2.8711816784	1.5195359254
38	3.5204123576	1.7618447158	2.9714538663	1.5670702216
39	3.6372432089	1.8146432857	3.0720578746	1.6146885440
40	3.7544125575	1.8675238149	3.1729806587	1.6623878981
41	3.8719075062	1.9204834110	3.2742100087	1.7101654591
42	3.9897159579	1.9735193458	3.3757344767	1.7580185587
43	4.1078265478	2.0266290419	3.4775433114	1.8059446739
44	4.2262285824	2.0798100620	3.5796264004	1.8539414163
45	4.3449119853	2.1330600986	3.6819742178	1.9020065228
46	4.4638672485	2.1863769645	3.7845777777	1.9501378464
47	4.5830853884	2.2397585843	3.8874285925	1.9983333488
48	4.7025579063	2.2932029870	3.9905186344	2.0465910922
49	4.8222767519	2.3467082983	4.0938403018	2.0949092333
50	4.9422342916	2.4002727350	4.1973863879	2.1432860170
51	5.0624232785	2.4538945980	4.3011500525	2.1917197702
52	5.1828368253	2.5075722676	4.4051247967	2.2402088973
53	5.3034683803	2.5613041984	4.5093044391	2.2887518748
54	5.4243117042	2.6150889143	4.6136830947	2.3373472471
55	5.5453608506	2.6689250046	4.7182551553	2.3859936226
56	5.6666101460	2.7228111198	4.8230152716	2.4346896693
57	5.7880541734	2.7767459681	4.9279583367	2.4834341120

DIRICHLET 1 (MEANS AND VARIANCES),   B= 3

N	P=1/ 9 MEAN	VARIANCE	P=1/10 MEAN	VARIANCE
3	.0082304527	.0081627123	.0060000000	.0059640000
4	.0274348422	.0266821717	.0204000000	.0199838400
5	.0574099477	.0541140456	.0435000000	.0416077500
6	.0966993514	.0876872886	.0745500000	.0691722975
7	.1435543488	.1247905406	.1124046000	.1007778059
8	.1963856899	.1635748867	.1558838400	.1348180684
9	.2539148568	.2029832963	.2039409180	.1701579960
10	.3151758192	.2425379907	.2557166130	.2061054888
11	.3794591946	.2820975922	.3105355546	.2422995723
12	.4462463307	.3216746023	.3678783029	.2785895678
13	.5151530254	.3613280532	.4273478101	.3149390151
14	.5858878287	.4011138359	.4886389316	.3513620541
15	.6582234488	.4410692132	.5515139764	.3878874488
16	.7319779058	.4812132221	.6157844392	.4245416849
17	.8070022336	.5215520849	.6812979031	.4613432422
18	.8831723273	.5620845740	.7479288410	.4983023626
19	.9603833574	.6028057489	.8155721801	.5354228734
20	1.0385457803	.6437090844	.8841387553	.5727043119
21	1.1175823796	.6847875057	.9535520326	.6101436504
22	1.1974260078	.7260338286	1.0237456885	.6477364749
23	1.2780178284	.7674409345	1.0946617743	.6854777031
24	1.3593059283	.8090018520	1.1662492931	.7233619866
25	1.4412442138	.8507098125	1.2384630754	.7613839250
26	1.5237915231	.8925582968	1.3112628786	.7995381778
27	1.6069109073	.9345410725	1.3846126571	.8378195266
28	1.6905690430	.9766522188	1.4584799660	.8762229127
29	1.7747357476	1.0188861373	1.5328354702	.9147434629
30	1.8593835759	1.0612375516	1.6076525364	.9533765064
31	1.9444874819	1.1037014986	1.6829068928	.9921175846
32	2.0300245334	1.1462733140	1.7585763424	1.0309624562
33	2.1159736682	1.1889486153	1.8346405203	1.0699070959
34	2.2023154868	1.2317232844	1.9110806868	1.1089476901
35	2.2890320727	1.2745934489	1.9878795497	1.1480806293
36	2.3761068381	1.3175554656	2.0650211115	1.1873024995
37	2.4635243890	1.3606059040	2.1424905366	1.2266100711
38	2.5512704076	1.4037415310	2.2202740357	1.2660002895
39	2.6393315498	1.4469592968	2.2983587648	1.3054702634
40	2.7276953545	1.4902563218	2.3767327363	1.3450172558
41	2.8163501633	1.5336298847	2.4553847411	1.3846386732
42	2.9052850501	1.5770774110	2.5343042787	1.4243320569
43	2.9944897579	1.6205964630	2.6134814961	1.4640950744
44	3.0839546425	1.6641847301	2.6929071330	1.5039255109
45	3.1736706225	1.7078400203	2.7725724728	1.5438212621
46	3.2636291344	1.7515602519	2.8524692988	1.5837803269
47	3.3538220914	1.7953434467	2.9325898548	1.6238008009
48	3.4442418478	1.8391877225	3.0129268098	1.6638808706
49	3.5348811653	1.8830912873	3.0934732254	1.7040188072
50	3.6257331838	1.9270524336	3.1742225278	1.7442129619
51	3.7167913934	1.9710695326	3.2551684806	1.7844617608
52	3.8080496108	2.0151410298	3.3363051616	1.8247637004
53	3.8995019560	2.0592654404	3.4176269404	1.8651173433
54	3.9911428322	2.1034413448	3.4991284594	1.9055213148
55	4.0829669068	2.1476673849	3.5808046148	1.9459742985
56	4.1749690946	2.1919422605	3.6626505407	1.9864750335
57	4.2671445416	2.2362647261	3.7446615932	2.0270223114

DIRICHLET 1 (MEANS AND VARIANCES),   B= 3
P=1/11   P=1/12

N	MEAN	VARIANCE	MEAN	VARIANCE
3	.0045078888	.0044875677	.0034722222	.0034601659
4	.0155727068	.0153301976	.0121527778	.0120050878
5	.0337160278	.0325792573	.0266444830	.0259345545
6	.0586093282	.0552758801	.0468689718	.0447325530
7	.0895253600	.0820924913	.0723833416	.0674956364
8	.1256134939	.1117431325	.1025875111	.0932374101
9	.1660505615	.1431842281	.1368500963	.1210685240
10	.2101108130	.1756679513	.1745774909	.1502773227
11	.2571877690	.2087130150	.2152458663	.1803442288
12	.3067904523	.2420421748	.2584106634	.2109195950
13	.3585281877	.2755165338	.3037036511	.2417867855
14	.4120921405	.3090810476	.3508240808	.2728239294
15	.4672377727	.3427254097	.3995278746	.3039711584
16	.5237699637	.3764590823	.4496170133	.3352056836
17	.5815311902	.4102971381	.5009301517	.3665245259
18	.6403925110	.4442533581	.5533348147	.3979335441
19	.7002468546	.4783376613	.6067211568	.4294410997
20	.7610040811	.5125558008	.6609970960	.4610548332
21	.8225873556	.5469100332	.7160845787	.4927803519
22	.8849304655	.5814000455	.7719167321	.5246209771
23	.9479758121	.6160238059	.8284356955	.5565779965
24	1.0116728809	.6507782208	.8855909601	.5886510980
25	1.0759770552	.6856595937	.9433380878	.6208388111
26	1.1408486808	.7206639226	1.0016377092	.6531388832
27	1.2062523162	.7557870845	1.0604547314	.6855485673
28	1.2721561238	.7910249459	1.1197577020	.7180648296
29	1.3385313708	.8263734284	1.1795182924	.7506844911
30	1.4053520158	.8618285479	1.2397108739	.7834043208
31	1.4725943636	.8973864397	1.3003121655	.8162210972
32	1.5402367756	.9330433734	1.3613009390	.8491316464
33	1.6082594247	.9687957629	1.4226577711	.8821328671
34	1.6766440873	1.0046401713	1.4843648320	.9152217464
35	1.7453739650	1.0405733135	1.5464057061	.9483953699
36	1.8144335314	1.0765920551	1.6087652376	.9816509273
37	1.8838083998	1.1126934097	1.6714293968	1.0149857163
38	1.9534852086	1.1488745350	1.7343851655	1.0483971432
39	2.0234515207	1.1851327274	1.7976204358	1.0818827228
40	2.0936957355	1.2214654158	1.8611239233	1.1154400766
41	2.1642070121	1.2578701554	1.9248850890	1.1490669305
42	2.2349752004	1.2943446208	1.9888940720	1.1827611111
43	2.3059907811	1.3308866001	2.0531416294	1.2165205425
44	2.3772448117	1.3674939877	2.1176190829	1.2503432417
45	2.4487288781	1.4041647792	2.1823182710	1.2842273149
46	2.5204350521	1.4408970646	2.2472315071	1.3181709528
47	2.5923558529	1.4776890233	2.3123515411	1.3521724267
48	2.6644842118	1.5145389187	2.3776715255	1.3862300845
49	2.7368134416	1.5514450936	2.4431849843	1.4203423462
50	2.8093372079	1.5884059650	2.5088857855	1.4545077004
51	2.8820495034	1.6254200203	2.5747681156	1.4887247007
52	2.9549446250	1.6624858130	2.6408264574	1.5229919622
53	3.0280171525	1.6996019591	2.7070555685	1.5573081579
54	3.1012619288	1.7367671335	2.7734504627	1.5916720161
55	3.1746740431	1.7739800667	2.8400063930	1.6260823170
56	3.2482488140	1.8112395415	2.9067188352	1.6605378904
57	3.3219817751	1.8485443905	2.9735834741	1.6950376125

DIRICHLET 1 (MEANS AND VARIANCES),   B= 4

N	P=1/ 4 MEAN	P=1/ 4 VARIANCE	P=1/ 5 MEAN	P=1/ 5 VARIANCE
4	.0937500000	.0849609375	.0384000000	.0369254400
5	.2343750000	.1794433594	.1152000000	.1019289600
6	.3808593750	.2358055115	.2150400000	.1687977984
7	.5126953125	.2498388290	.3225600000	.2185150464
8	.6613769531	.3008617759	.4335206400	.2584828947
9	.8267211914	.3739661537	.5496422400	.3017257280
10	.9968948364	.4356821915	.6721658880	.3519633870
11	1.1600303650	.4664442679	.7996809216	.4035478124
12	1.3302268982	.5157184529	.9303029514	.4522416714
13	1.5091645718	.5817035968	1.0633493545	.4992660862
14	1.6914245188	.6433461703	1.1990295695	.5476156865
15	1.8695487455	.6820002408	1.3373742632	.5977571596
16	2.0517170951	.7321238480	1.4779631738	.6481390424
17	2.2394021871	.7950538161	1.6203465875	.6978315129
18	2.4294736345	.8561518716	1.7643545927	.7473812208
19	2.6167784800	.8992122823	1.9099744854	.7975417701
20	2.8067816465	.9503411163	2.0571269817	.8482647111
21	3.0006552182	1.0117215406	2.2056409862	.8990542081
22	3.1963310500	1.0723536795	2.3553674215	.9497337635
23	3.3899972820	1.1182053869	2.5062341459	1.0005283978
24	3.5856286238	1.1701941345	2.6581990362	1.0516245116
25	3.7841641565	1.2306930967	2.8111976081	1.1029433731
26	3.9841171679	1.2909758314	2.9651487414	1.1543278052
27	4.1825273722	1.3387564296	3.1199864660	1.2057573657
28	4.3824515110	1.3914499559	3.2756677263	1.2573154034
29	4.5846575534	1.4514058016	3.4321553935	1.3090397601
30	4.7880141370	1.5114302190	3.5894062453	1.3608851381
31	4.9901387814	1.5606277753	3.7473759191	1.4128032800
32	5.1934763315	1.6139006530	3.9060275637	1.4647983989
33	5.3986677390	1.6735024113	4.0653316797	1.5168975179
34	5.6048163477	1.7333339294	4.2252603445	1.5691032639
35	5.8099520613	1.7836197799	4.3857847148	1.6213944015
36	6.0160879361	1.8373751378	4.5468773082	1.6737564281
37	6.2237685000	1.8967365181	4.7085141588	1.7261931827
38	6.4322612283	1.9564218369	4.8706740023	1.7787115862
39	6.6399021658	2.0075721301	5.0333364087	1.8313081200
40	6.8483863033	2.0617351352	5.1964813832	1.8839734073
41	7.0581832714	2.1209285266	5.3600902146	1.9367027740
42	7.2686804350	2.1805014356	5.5241459113	1.9894973930
43	7.4784483455	2.2323565807	5.6886327093	2.0423578364
44	7.6889397164	2.2868686995	5.8535354916	2.0952807878
45	7.9005647767	2.3459422731	6.0188397700	2.1482619508
46	8.1128015170	2.4054276674	6.1845319526	2.2012993292
47	8.3244047581	2.4578696947	6.3505993760	2.2543925996
48	8.5366376404	2.5126844345	6.5170300739	2.3075407074
49	8.7498624108	2.5716713961	6.6838125984	2.3607413265
50	8.9636274826	2.6310879548	6.8509360365	2.4139922162
51	9.1768355557	2.6840270683	7.0183900739	2.4672921034
52	9.3905981343	2.7391069328	7.1861649596	2.5206401298
53	9.6052380835	2.7980308343	7.3542514041	2.5740350482
54	9.8203597975	2.8573928109	7.5226405208	2.6274752661
55	10.0349867728	2.9107589873	7.6913238292	2.6809593861
56	10.2501069723	2.9660733037	7.8602932582	2.7344863863
57	10.4660104853	3.0249512447	8.0295411133	2.7880553298
58	10.6823470600	3.0842696821	8.1990600310	2.8416651270

DIRICHLET 1 (MEANS AND VARIANCES),   B= 4

	P=1/ 6		P=1/ 7	
N	MEAN	VARIANCE	MEAN	VARIANCE
4	.0185185185	.0181755830	.0099958351	.0098959184
5	.0617283951	.0579180003	.0356994110	.0344249630
6	.1260288066	.1101455465	.0775187209	.0715095688
7	.2040466392	.1624116083	.1325978121	.1150156324
8	.2898662551	.2088444951	.1972356028	.1592079910
9	.3805869913	.2507439627	.2684012857	.2012329756
10	.4753923913	.2919041821	.3441846333	.2409767134
11	.5740955075	.3345674774	.4235792203	.2796090588
12	.6763558302	.3786643057	.5060714717	.3182553406
13	.7816319399	.4231821918	.5913254001	.3574321424
14	.8894107892	.4675096467	.6790386862	.3971320213
15	.9993451038	.5117259519	.7689197871	.4371489167
16	1.1112276416	.5561673030	.8607089564	.4773255370
17	1.2249044844	.6010040254	.9541941461	.5176246533
18	1.3402228418	.6461733855	1.0492089138	.5580829834
19	1.4570299699	.6915423479	1.1456198245	.5987429897
20	1.5751910409	.7370471242	1.2433138941	.6396179097
21	1.6945972725	.7827036031	1.3421912112	.6806940512
22	1.8151601130	.8285447893	1.4421623137	.7219488275
23	1.9368017024	.8745752119	1.5431476738	.7633647512
24	2.0594499201	.9207730529	1.6450772032	.8049327895
25	2.1830384323	.9671146347	1.7478891238	.8466487214
26	2.3075079927	1.0135891047	1.8515285100	.8885086078
27	2.4328063518	1.0601953268	1.9559459851	.9305067662
28	2.5588868358	1.1069322725	2.0610968004	.9726362251
29	2.6857068821	1.1537943804	2.1669402436	1.0148900430
30	2.8132272945	1.2007735025	2.2734392164	1.0572621698
31	2.9414120612	1.2478625454	2.3805598569	1.0997475144
32	3.0702282416	1.2950567816	2.4882711717	1.1423415730
33	3.1996456733	1.3423527626	2.5965446898	1.1850400715
34	3.3296365743	1.3897469013	2.7053541640	1.2278388323
35	3.4601752046	1.4372350981	2.8146753260	1.2707338211
36	3.5912376502	1.4848132255	2.9244856842	1.3137212374
37	3.7228016812	1.5324776209	3.0347643522	1.3567975549
38	3.8548466178	1.5802251333	3.1454918958	1.3999595000
39	3.9873531833	1.6280528761	3.2566501957	1.4432040032
40	4.1203033594	1.6759580277	3.3682223240	1.4865281601
41	4.2536802652	1.7239378198	3.4801924364	1.5299292136
42	4.3874680599	1.7719896239	3.5925456760	1.5734045493
43	4.5216518613	1.8201110046	3.7052680900	1.6169516945
44	4.6562176689	1.8682996969	3.8183465544	1.6605683116
45	4.7911522913	1.9165535557	3.9317687069	1.7042521877
46	4.9264432804	1.9648705254	4.0455228862	1.7480012233
47	5.0620788722	2.0132486415	4.1595980771	1.7918134228
48	5.1980479369	2.0616860396	4.2739838619	1.8356868872
49	5.3343399319	2.1101809559	4.3886703752	1.8796198088
50	5.4709448601	2.1587317157	4.5036482642	1.9236104658
51	5.6078532304	2.2073367218	4.6189086520	1.9676572170
52	5.7450560227	2.2559944471	4.7344431040	2.0117584967
53	5.8825446553	2.3047034328	4.8502435981	2.0559128097
54	6.0203109559	2.3534622867	4.9663024968	2.1001187271
55	6.1583471348	2.4022696791	5.0826125216	2.1443748820
56	6.2966457605	2.4511243384	5.1991667300	2.1886799661
57	6.4351997366	2.5000250466	5.3159584940	2.2330327262
58	6.5740022813	2.5489706362	5.4329814806	2.2774319614

DIRICHLET 1 (MEANS AND VARIANCES),   B= 4

	P=1/ 8		P=1/ 9	
N	MEAN	VARIANCE	MEAN	VARIANCE
4	.0058593750	.0058250427	.0036579790	.0036445982
5	.0219726563	.0214898586	.0142254738	.0140231097
6	.0498962402	.0474066054	.0334185733	.0323017722
7	.0889205933	.0810137214	.0614680965	.0576897696
8	.1371660233	.1186519127	.0975996290	.0881910235
9	.1925235987	.1572607071	.1405635519	.1215469596
10	.2532031015	.1951463777	.1890404580	.1559276693
11	.3179140482	.2318783759	.2418757752	.1902165559
12	.3858168712	.2677618201	.2981681830	.2239614405
13	.4563849994	.3033080549	.3572632099	.2571567033
14	.5292710871	.3389154082	.4187021234	.2900043338
15	.6042164830	.3747753801	.4821618021	.3227381885
16	.6810054032	.4109181051	.5474046140	.3555335392
17	.7594487431	.4472995042	.6142442901	.3884852322
18	.8393812298	.4838687303	.6825261854	.4216242355
19	.9206613056	.5205974584	.7521173982	.4549458549
20	1.0031694561	.5574799748	.8229022654	.4884334886
21	1.0868047461	.5945212144	.8947800827	.5220721984
22	1.1714808056	.6317257193	.9676633798	.5558531298
23	1.2571223818	.6690926443	1.0414761630	.5897725147
24	1.3436628956	.7066160701	1.1161521094	.6238288702
25	1.4310428912	.7442876508	1.1916328807	.6580206277
26	1.5192090440	.7820990741	1.2678666964	.6923449881
27	1.6081134251	.8200432609	1.3448072144	.7267978502
28	1.6977128542	.8581144171	1.4224126846	.7613743036
29	1.7879682952	.8963075478	1.5006453109	.7960692143
30	1.8788443055	.9346179706	1.5794707515	.8308776382
31	1.9703085607	.9730410730	1.6588577069	.8657950029
32	2.0623314619	1.0115723011	1.7387775669	.9008171135
33	2.1548858160	1.0502072560	1.8192041015	.9359400783
34	2.2479465711	1.0889417861	1.9001131902	.9711602258
35	2.3414905915	1.1277720229	1.9814825867	1.0064740511
36	2.4354964617	1.1666943658	2.0632917171	1.0418781947
37	2.5299443124	1.2057054421	2.1455215050	1.0773694427
38	2.6248156680	1.2448020698	2.2281542220	1.1129447338
39	2.7200933111	1.2839812325	2.3111733575	1.1486011637
40	2.8157611650	1.3232400684	2.3945635040	1.1843359821
41	2.9118041896	1.3625758642	2.4783102575	1.2201465855
42	3.0082082900	1.4019860502	2.5624001270	1.2560305059
43	3.1049602355	1.4414681933	2.6468204560	1.2919853998
44	3.2020475867	1.4810199879	2.7315593510	1.3280090393
45	3.2994586312	1.5206392460	2.8166056185	1.3640993035
46	3.3971823247	1.5603238886	2.9019487084	1.4002541726
47	3.4952082392	1.6000719377	2.9875786628	1.4364717219
48	3.5935265152	1.6398815099	3.0734860707	1.4727501165
49	3.6921278194	1.6797508107	3.1596620265	1.5090876064
50	3.7910033062	1.7196781291	3.2460980926	1.5454825211
51	3.8901445823	1.7596618327	3.3327862659	1.5819332654
52	3.9895436752	1.7997003628	3.4197189466	1.6184383141
53	4.0891930042	1.8397922299	3.5068889107	1.6549962089
54	4.1890853537	1.8799360095	3.5942892837	1.6916055534
55	4.2892138492	1.9201303383	3.6819135183	1.7282650105
56	4.3895719349	1.9603739107	3.7697553718	1.7649732984
57	4.4901533535	2.0006654754	3.8578088875	1.8017291881
58	4.5909521275	2.0410038327	3.9460683763	1.8385315001

DIRICHLET 1 (MEANS AND VARIANCES),   B= 4

	P=1/10		P=1/11	
N	MEAN	VARIANCE	MEAN	VARIANCE
4	.0024000000	.0023942400	.0016392323	.0016365452
5	.0096000000	.0095078400	.0067059503	.0066609805
6	.0231600000	.0226236144	.0165277967	.0162546286
7	.0436800000	.0417720576	.0318116981	.0307997139
8	.0709934400	.0660037715	.0527044364	.0499501907
9	.1044540000	.0938760019	.0789425146	.0728709029
10	.1432002120	.1239186313	.1100162874	.0985218777
11	.1863488352	.1549434332	.1453097375	.1258978488
12	.2331089629	.1861671434	.1842010741	.1541795372
13	.2828304576	.2171897874	.2261237933	.1827937710
14	.3350080252	.2478951381	.2705953016	.2114040529
15	.3892613554	.2783328082	.3172227798	.2398613724
16	.4453067534	.3086195356	.3656955242	.2681419003
17	.5029297582	.3388746764	.4157710715	.2962897641
18	.5619631964	.3691888652	.4672600542	.3243740321
19	.6222716728	.3996167542	.5200125981	.3524619058
20	.6837416235	.4301830598	.5739074527	.3806057888
21	.7462753747	.4608931861	.6288440020	.4088400700
22	.8097876981	.4917430412	.6847367454	.4371834094
23	.8742037169	.5227257545	.7415116422	.4656432023
24	.9394574486	.5538351134	.7991037258	.4942200928
25	1.0054906176	.5850665818	.8574555127	.5229114835
26	1.0722515925	.6164170029	.9165158742	.5517137666
27	1.1396944247	.6478838707	.9762391669	.5806234462
28	1.2077779991	.6794646967	1.0365845112	.6096374986
29	1.2764653115	.7111566653	1.0975151668	.6387533207
30	1.3457228739	.7429565664	1.1589979852	.6679685278
31	1.4155202318	.7748609003	1.2210029339	.6972807608
32	1.4858295751	.8068660446	1.2835026879	.7266875676
33	1.5566254206	.8389684043	1.3464722835	.7561863677
34	1.6278843470	.8711645110	1.4098888284	.7857744761
35	1.6995847701	.9034510690	1.4737312584	.8154491572
36	1.7717067483	.9358249619	1.5379801334	.8452076837
37	1.8442318131	.9682832383	1.6026174639	.8750473843
38	1.9171428199	1.0008230930	1.6676265626	.9049656758
39	1.9904238174	1.0334418500	1.7329919154	.9349600788
40	2.0640599332	1.0661369512	1.7986990688	.9650282218
41	2.1380372725	1.0989059507	1.8647345306	.9951678383
42	2.2123428297	1.1317465113	1.9310856821	1.0253767610
43	2.2869644103	1.1646564016	1.9977407002	1.0556529153
44	2.3618905607	1.1976334935	2.0646884882	1.0859943136
45	2.4371105072	1.2306757575	2.1319186143	1.1163990511
46	2.5126140998	1.2637812585	2.1994212562	1.1468653018
47	2.5883917630	1.2969481500	2.2671871522	1.1773913159
48	2.6644344511	1.3301746692	2.3352075571	1.2079754170
49	2.7407336076	1.3634591318	2.4034742022	1.2386159988
50	2.8172811290	1.3967999273	2.4719792598	1.2693115223
51	2.8940693315	1.4301955150	2.5407153107	1.3000605130
52	2.9710909213	1.4636444201	2.6096753147	1.3308615571
53	3.0483389671	1.4971452300	2.6788525840	1.3617132989
54	3.1258068757	1.5306965913	2.7482407588	1.3926144376
55	3.2034883684	1.5642972061	2.8178337852	1.4235637243
56	3.2813774614	1.5979458297	2.8876258950	1.4545599596
57	3.3594684457	1.6316412671	2.9576115865	1.4856019910
58	3.4377558706	1.6653823709	3.0277856084	1.5166887104

DIRICHLET 1 (MEANS AND VARIANCES),   B= 4

	P=1/12		P=1/13	
N	MEAN	VARIANCE	MEAN	VARIANCE
4	.0011574074	.0011560678	.0008403067	.0008396006
5	.0048225309	.0047992741	.0035551438	.0035425047
6	.0120965149	.0119501892	.0090494569	.0089675642
7	.0236772870	.0231166730	.0179650461	.0176423032
8	.0398598921	.0382828025	.0306552927	.0297217242
9	.0606105335	.0570189468	.0472175953	.0450322941
10	.0856653655	.0786467080	.0675511580	.0631640317
11	.1146259559	.1024038218	.0914201950	.0835777403
12	.1470374622	.1275737439	.1185112425	.1057026749
13	.1824448426	.1535655898	.1484792020	.1290102861
14	.2204280129	.1799446337	.1809806463	.1530592964
15	.2606197258	.2064373489	.2156952633	.1775143944
16	.3027109137	.2328875535	.2523375165	.2021447423
17	.3464480325	.2592396317	.2906609991	.2268096968
18	.3916261231	.2854952529	.3304578643	.2514385021
19	.4380802874	.3116868413	.3715553365	.2760091049
20	.4856772998	.3378572914	.4138108369	.3005293846
21	.5343082809	.3640475075	.4571067811	.3250224204
22	.5838827885	.3902902777	.5013456982	.3495161626
23	.6343243226	.4166085529	.5464460089	.3740370729
24	.6855670489	.4430163147	.5923385769	.3986068971
25	.7375534754	.4695206064	.6389640078	.4232416326
26	.7902328231	.4961237718	.6862706003	.4479518375
27	.8435598701	.5228253766	.7342128221	.4727436068
28	.8974941061	.5496236052	.7827501856	.4976197484
29	.9519990831	.5765161345	.8318464144	.5225808770
30	1.0070418921	.6035005882	.8814688153	.5476262895
31	1.0625927231	.6305747085	.9315877895	.5727545869
32	1.1186244863	.6577363695	.9821764413	.5979640687
33	1.1751124824	.6849835239	1.0332102532	.6232529489
34	1.2320341161	.7123141421	1.0846668095	.6486194519
35	1.2893686481	.7397261697	1.1365255583	.6740618388
36	1.3470969808	.7672175111	1.1887676028	.6995783982
37	1.4052014755	.7947860360	1.2413755199	.7251674290
38	1.4636657951	.8224295992	1.2943332012	.7508272259
39	1.5224747693	.8501460652	1.3476257147	.7765560733
40	1.5816142785	.8779333315	1.4012391842	.8023522476
41	1.6410711524	.9057893473	1.4551606844	.8282140258
42	1.7008330828	.9337121257	1.5093781490	.8541396966
43	1.7608885450	.9616997502	1.5638802914	.8801275726
44	1.8212267297	.9897503769	1.6186565339	.9061760004
45	1.8818374817	1.0178622340	1.6736969463	.9322833691
46	1.9427112454	1.0460336196	1.7289921905	.9584481152
47	2.0038390162	1.0742628982	1.7845334721	.9846687263
48	2.0652122959	1.1025484983	1.8403124964	1.0109437415
49	2.1268230541	1.1308889093	1.8963214291	1.0372717519
50	2.1886636922	1.1592826789	1.9525528615	1.0636513990
51	2.2507270113	1.1877284107	2.0089997778	1.0900813733
52	2.3130061831	1.2162247623	2.0656555268	1.1165604126
53	2.3754947237	1.2447704430	2.1225137954	1.1430873002
54	2.4381864694	1.2733642116	2.1795685849	1.1696608629
55	2.5010755550	1.3020048745	2.2368141894	1.1962799698
56	2.5641563937	1.3306912834	2.2942451756	1.2229435304
57	2.6274236590	1.3594223335	2.3518563654	1.2496504929
58	2.6908722676	1.3881969613	2.4096428187	1.2763998431

DIRICHLET 1 (MEANS AND VARIANCES),   B= 5

	P=1/ 5		P=1/ 6	
N	MEAN	VARIANCE	MEAN	VARIANCE
5	.0384000000	.0369254400	.0154320988	.0151939491
6	.1152000000	.1019289600	.0540123457	.0510950122
7	.2150400000	.1687977984	.1140260631	.1010241200
8	.3225600000	.2185150464	.1890432099	.1533058747
9	.4270694400	.2446811334	.2718121285	.1979302953
10	.5341593600	.2720574581	.3580818473	.2336100953
11	.6489415680	.3129722493	.4469844430	.2655268755
12	.7716741120	.3635360249	.5392633065	.2992686176
13	.8990034493	.4125717815	.6355445086	.3370260056
14	1.0262376284	.4497347190	.7355242262	.3771513141
15	1.1544808784	.4840990904	.8383004543	.4168602428
16	1.2858941139	.5240842485	.9430858686	.4548737080
17	1.4212244053	.5705129637	1.0495684232	.4918844798
18	1.5593594702	.6179440560	1.1577976081	.5293373881
19	1.6978969545	.6586766025	1.2678508130	.5679741381
20	1.8371434887	.6968063447	1.3796264074	.6074325235
21	1.9781875044	.7377374428	1.4928824256	.6469107825
22	2.1215823784	.7829805411	1.6073916265	.6859828044
23	2.2668502228	.8295707569	1.7230416002	.7248427442
24	2.4125757805	.8719005655	1.8398128281	.7639398105
25	2.5588644439	.9121615551	1.9576933811	.8035003200
26	2.7063516293	.9540366717	2.0766238648	.8433899228
27	2.8554270276	.9989695591	2.1965083183	.8833331024
28	3.0058466070	1.0451649822	2.3172590665	.9231940982
29	3.1566890813	1.0883996002	2.4388247982	.9630535106
30	3.3079959775	1.1300008186	2.5611819282	1.0030697848
31	3.4601791154	1.1726357087	2.6843079426	1.0433103251
32	3.6135204381	1.2175563760	2.8081653878	1.0837117457
33	3.7678792789	1.2635823625	2.9327070219	1.1241693741
34	3.9226116154	1.3074151805	3.0578908084	1.1646401211
35	4.0777352165	1.3499343844	3.1836881157	1.2051618141
36	4.2335375704	1.3931685440	3.3100795521	1.2457938533
37	4.3902300094	1.4381847898	3.4370455884	1.2865540940
38	4.5477239950	1.4841562969	3.5645617344	1.3274105011
39	4.7055451442	1.5284256876	3.6926011468	1.3683209232
40	4.8637013076	1.5716154846	3.8211400452	1.4092720290
41	5.0224043434	1.6153287903	3.9501601027	1.4502818502
42	5.1818178302	1.6604779648	4.0796463723	1.4913728840
43	5.3418822629	1.7064550050	4.2095837741	1.5325479996
44	5.5022341053	1.7510640630	4.3399556514	1.5737906782
45	5.6628767047	1.7947680005	4.4707450669	1.6150832606
46	5.8239726543	1.8388721029	4.6019368367	1.6564218175
47	5.9856520585	1.8841634901	4.7335181743	1.6978146880
48	6.1478728682	1.9301777634	4.8654776670	1.7392700566
49	6.3103477849	1.9750626731	4.9978039149	1.7807867901
50	6.4730779877	2.0191755720	5.1304851241	1.8223562576
51	6.6361922294	2.0636046630	5.2635097342	1.8639707738
52	6.7997963595	2.1090360214	5.3968671916	1.9056291919
53	6.9638594740	2.1551040842	5.5305480805	1.9473351114
54	7.1281488874	2.2002198540	5.6645436235	1.9890911339
55	7.2926646979	2.2446674623	5.7988451453	2.0308954724
56	7.4575114178	2.2893715268	5.9334439761	2.0727435232
57	7.6227768050	2.3349360062	6.0683317478	2.1146317547
58	7.7884374215	2.3810660542	6.2035006843	2.1565597243
59	7.9543010154	2.4263793332	6.3389435977	2.1985287686

DIRICHLET 1 (MEANS AND VARIANCES),   B= 5

N	P=1/ 7 MEAN	VARIANCE	P=1/ 8 MEAN	VARIANCE
5	.0071398822	.0070889043	.0036621094	.0036486983
6	.0275395456	.0267811190	.0151062012	.0148780039
7	.0632389566	.0592397909	.0368499756	.0354920549
8	.1127809963	.1000614431	.0693941116	.0645785689
9	.1727934754	.1429358903	.1116576791	.0991902418
10	.2397465981	.1830710690	.1617412269	.1357922264
11	.3110953524	.2189412802	.2176507562	.1716342583
12	.3855396715	.2517388692	.2777652856	.2054159767
13	.4626793438	.2836802361	.3410023796	.2371740861
14	.5424841655	.3164567816	.4067567324	.2677083252
15	.6249015188	.3505771067	.4747320266	.2979469337
16	.7097183050	.3856134958	.5447708279	.3285271771
17	.7966152089	.4208625801	.6167410573	.3596725190
18	.8852825338	.4558855570	.6904900712	.3912940939
19	.9754966098	.4906667678	.7658469218	.4231795794
20	1.0671258628	.5254504548	.8426442692	.4551510344
21	1.1600913539	.5604822972	.9207378310	.4871355009
22	1.2543232315	.5958515087	1.0000136699	.5191542418
23	1.3497394569	.6314923150	1.0803842047	.5512708427
24	1.4462487638	.6672826942	1.1617789430	.5835404830
25	1.5437644618	.7031410447	1.2441358689	.6159843957
26	1.6422148654	.7390601005	1.3273965824	.6485915080
27	1.7415443437	.7750799778	1.4115052420	.6813352899
28	1.8417075151	.8112405149	1.4964096235	.7141913155
29	1.9426625583	.8475515954	1.5820624233	.7471465752
30	2.0443678664	.8839942918	1.6684217137	.7801992704
31	2.1467825135	.9205408582	1.7554504111	.8133530647
32	2.2498684097	.9571735448	1.8431152134	.8466109382
33	2.3535918016	.9938901577	1.9313855704	.8799719365
34	2.4579231499	1.0306973724	2.0202330211	.9134313292
35	2.5628358839	1.0676005953	2.1096309328	.9469827498
36	2.6683051107	1.1045983817	2.1995544773	.9806204346
37	2.7743070035	1.1416835653	2.2899806471	1.0143403620
38	2.8808188912	1.1788479995	2.3808881887	1.0481401241
39	2.9878196132	1.2160864470	2.4722574375	1.0820180824
40	3.0952897030	1.2533972655	2.5640701040	1.1159725198
41	3.2032112496	1.2907805102	2.6563090771	1.1500012285
42	3.3115675676	1.3282356155	2.7489582806	1.1841015807
43	3.4203428965	1.3657603730	2.8420025852	1.2182708570
44	3.5295222591	1.4033514618	2.9354277523	1.2525065594
45	3.6390914610	1.4410056427	3.0292203866	1.2868065463
46	3.7490371315	1.4787205723	3.1233678804	1.3211689807
47	3.8593467174	1.5164947985	3.2178583501	1.3555921867
48	3.9700084097	1.5543272084	3.3126805703	1.3900745203
49	4.0810110404	1.5922164836	3.4078239149	1.4246143153
50	4.1923439948	1.6301609324	3.5032783083	1.4592099027
51	4.3039971650	1.6681586870	3.5990341870	1.4938596636
52	4.4159609336	1.7062080083	3.6950824672	1.5285620731
53	4.5282261632	1.7443074489	3.7914145149	1.5633157130
54	4.6407841744	1.7824558063	3.8880221171	1.5981192557
55	4.7536267094	1.8206519620	3.9848974527	1.6329714369
56	4.8667458929	1.8588947538	4.0820330657	1.6678710332
57	4.9801341994	1.8971829560	4.1794218401	1.7028168545
58	5.0937844314	1.9355153445	4.2770569780	1.7378077469
59	5.2076897044	1.9738907728	4.3749319807	1.7728426001

DIRICHLET 1 (MEANS AND VARIANCES),   B= 5

N	P=1/ 9 MEAN	VARIANCE	P=1/10 MEAN	VARIANCE
5	.0020322105	.0020280807	.0012000000	.0011985600
6	.0088062457	.0087286957	.0054000000	.0053708400
7	.0224797610	.0219744213	.0142800000	.0140760816
8	.0441399474	.0421916124	.0289800000	.0281401596
9	.0738072477	.0683597379	.0499741200	.0474767073
10	.1107206284	.0985266165	.0771422400	.0712139948
11	.1537136891	.1305362919	.1099425360	.0980214948
12	.2015495244	.1626439287	.1476087030	.1264898118
13	.2531456115	.1938277963	.1893212641	.1554369193
14	.3076760023	.2237858854	.2343275915	.1840735661
15	.3645754882	.2527245096	.2820059410	.2120251032
16	.4234862140	.2810804501	.3318825048	.2392503851
17	.4841855337	.3092882711	.3836169137	.2659136611
18	.5465221423	.3376473497	.4369721293	.2922584085
19	.6103731349	.3662897005	.4917814350	.3185136760
20	.6756228952	.3952165628	.5479204642	.3448435408
21	.7421579422	.4243621702	.6052876210	.3713349897
22	.8098699592	.4536512633	.6637928449	.3980113459
23	.8786606180	.4830331112	.7233527690	.4248568000
24	.9484445814	.5124903405	.7838897387	.4518405499
25	1.0191496882	.5420305843	.8453324686	.4789339661
26	1.0907150017	.5716717616	.9076168501	.5061189439
27	1.1630880014	.6014294925	.9706861917	.5333889180
28	1.2362220158	.6313106809	1.0344907666	.5607455330
29	1.3100744667	.6613131975	1.0989868771	.5881939678
30	1.3846059758	.6914292452	1.1641357423	.6157390232
31	1.4597800734	.7216495550	1.2299024695	.6433829324
32	1.5355631660	.7519664181	1.2962552521	.6711248937
33	1.6119245031	.7823748406	1.3631648236	.6989617686
34	1.6888360216	.8128721481	1.4306041237	.7268892465
35	1.7662720713	.8434568478	1.4985480973	.7549029101
36	1.8442090859	.8741275205	1.5669735542	.7829989038
37	1.9226252705	.9048821854	1.6358590419	.8111741546
38	2.0015003540	.9357182058	1.7051847064	.8394262628
39	2.0808154126	.9666325559	1.7749321434	.8677532368
40	2.1605527506	.9976221901	1.8450842466	.8961532272
41	2.2406958121	1.0286843204	1.9156250663	.9246243496
42	2.3212291037	1.0598165213	1.9865396862	.9531646202
43	2.4021381142	1.0910166844	2.0578141209	.9817719800
44	2.4834092321	1.1222829006	2.1294352318	1.0104443632
45	2.5650296644	1.1536133443	2.2013906581	1.0391797673
46	2.6469873611	1.1850062061	2.2736687572	1.0679763002
47	2.7292709505	1.2164596810	2.3462585510	1.0968321963
48	2.8118696842	1.2479719937	2.4191496769	1.1257458066
49	2.8947733902	1.2795414336	2.4923323405	1.1547155765
50	2.9779724332	1.3111663799	2.5657972728	1.1837400219
51	3.0614576773	1.3428453078	2.6395356899	1.2128177121
52	3.1452204520	1.3745767794	2.7135392573	1.2419472605
53	3.2292525191	1.4063594270	2.7878000570	1.2711273243
54	3.3135460429	1.4381919380	2.8623105589	1.3003566063
55	3.3980935622	1.4700730452	2.9370635942	1.3296338591
56	3.4828879651	1.5020015239	3.0120523324	1.3589578860
57	3.5679224662	1.5339761931	3.0872702590	1.3883275408
58	3.6531905857	1.5659959172	3.1627111558	1.4177417249
59	3.7386861315	1.5980596069	3.2383690821	1.4471993834

DIRICHLET 1 (MEANS AND VARIANCES),   B= 5

	P=1/11		P=1/12	
N	MEAN	VARIANCE	MEAN	VARIANCE
5	.0007451056	.0007445504	.0004822531	.0004820205
6	.0034545805	.0034426463	.0022907022	.0022854548
7	.0093969515	.0093086488	.0063764575	.0063357983
8	.0195854726	.0192018819	.0135851364	.0134005805
9	.0346342034	.0334346754	.0245305093	.0239287634
10	.0547422831	.0517543096	.0395485905	.0379881624
11	.0797595163	.0734649741	.0587107498	.0552928977
12	.1092905691	.0976278073	.0818725115	.0752959701
13	.1428051201	.1232705805	.1087379719	.0973130736
14	.1797327652	.1495536836	.1389248413	.1206447125
15	.2195321013	.1758655446	.1720204636	.1446724863
16	.2617318815	.2018465580	.2076239581	.1689176435
17	.3059477988	.2273583293	.2453734027	.1930614086
18	.3518813750	.2524223925	.2849595586	.2169345503
19	.3993080652	.2771513317	.3261290630	.2404874153
20	.4480608006	.3016889513	.3686804855	.2637517915
21	.4980134990	.3261682445	.4124564276	.2868037507
22	.5490672199	.3506889017	.4573342233	.3097333495
23	.6011400346	.3753112846	.5032170160	.3326238094
24	.6541605372	.4000614407	.5500262348	.3555401783
25	.7080642757	.4249414774	.5976958674	.3785257948
26	.7627921594	.4499407083	.6461684926	.4016041056
27	.8182899600	.4750446866	.6953927813	.4247833518
28	.8745082442	.5002409204	.7453220734	.4480620764
29	.9314023311	.5255213503	.7959136452	.4714340643
30	.9889320943	.5508824125	.8471283520	.4948919872
31	1.0470615819	.5763237544	.8989304237	.5184295804
32	1.1057585154	.6018465478	.9512872824	.5420425379
33	1.1649937523	.6274520451	1.0041693235	.5657285001
34	1.2247407833	.6531406806	1.0575496502	.5894865377
35	1.2849753109	.6789117409	1.1114037796	.6133164764
36	1.3456749214	.7047635003	1.1657093439	.6372182924
37	1.4068188430	.7306935137	1.2204458116	.6611916996
38	1.4683877699	.7566990600	1.2755942433	.6852359510
39	1.5303637296	.7827774370	1.3311370879	.7093498205
40	1.5927299743	.8089261586	1.3870580189	.7335317006
41	1.6554708835	.8351430212	1.4433418041	.7577797507
42	1.7185718722	.8614260785	1.4999742021	.7820920444
43	1.7820193012	.8877735701	1.5569418760	.8064666832
44	1.8458003928	.9141838430	1.6142323205	.8309018663
45	1.9099031526	.9406552921	1.6718337957	.8553959181
46	1.9743162988	.9671863272	1.7297352672	.8799472837
47	2.0390292003	.9937753669	1.7879263494	.9045545089
48	2.1040318226	1.0204208484	1.8463972542	.9292162128
49	2.1693146806	1.0471212451	1.9051387419	.9539310657
50	2.2348687976	1.0738750825	1.9641420783	.9786977730
51	2.3006856685	1.1006809483	2.0233989946	1.0035150679
52	2.3667572263	1.1275374956	2.0829016524	1.0283817104
53	2.4330758121	1.1544434405	2.1426426120	1.0532964908
54	2.4996341461	1.1813975559	2.2026148038	1.0782582339
55	2.5664253025	1.2083986647	2.2628115033	1.1032658031
56	2.6334426853	1.2354456338	2.3232263079	1.1283181032
57	2.7006800060	1.2625373700	2.3838531154	1.1534140802
58	2.7681312641	1.2896728169	2.4446861053	1.1785527211
59	2.8357907286	1.3168509541	2.5057197206	1.2037330506

DIRICHLET 1 (MEANS AND VARIANCES).   B= 5

	P=1/13		P=1/14	
N	MEAN	VARIANCE	MEAN	VARIANCE
5	.0003231949	.0003230904	.0002231213	.0002230715
6	.0015662522	.0015637990	.0010996694	.0010984601
7	.0044444078	.0044246551	.0031715102	.0031614517
8	.0096446533	.0095516340	.0069907552	.0069418845
9	.0177244803	.0174103231	.0130417419	.0128716549
10	.0290603086	.0282174523	.0216934139	.0212235938
11	.0438368511	.0419286382	.0331794750	.0320851688
12	.0620650281	.0582731761	.0476000238	.0453643754
13	.0836164229	.0768199030	.0649376312	.0608206462
14	.1082643169	.0970557040	.0850814469	.0781109895
15	.1357239691	.1184601126	.1078541403	.0968411600
16	.1656874244	.1405643185	.1330378764	.1166131636
17	.1978504379	.1629888487	.1603968777	.1370630860
18	.2319309110	.1854595871	.1896952914	.1578862834
19	.2676795039	.2078056572	.2207100035	.1788496932
20	.3048838273	.2299447032	.2532387041	.1997930122
21	.3433679306	.2518614870	.2871039112	.2206216171
22	.3829887782	.2735849363	.3221538632	.2412944372
23	.4236311766	.2951673860	.3582612142	.2618097423
24	.4652022764	.3166681975	.3953203872	.2821911915
25	.5076264098	.3381425497	.4332442884	.3024757373
26	.5508406961	.3596351498	.4719609105	.3227042313
27	.5947915852	.3811779609	.5114101754	.3429149615
28	.6394323219	.4027907626	.5515412196	.3631398993
29	.6847212046	.4244833592	.5923101989	.3834031590
30	.7306204610	.4462584401	.6336786079	.4037210616
31	.7770955583	.4681143717	.6756120545	.4241031965
32	.8241147907	.4900474901	.7180794045	.4445539613
33	.8716490191	.5120537162	.7610522033	.4650741856
34	.9196714814	.5341294971	.8045042931	.4856625778
35	.9681576225	.5562721939	.8484115537	.5063168604
36	1.0170849219	.5784800843	.8927517167	.5270345512
37	1.0664327118	.6007521469	.9375042168	.5478134196
38	1.1161819924	.6230877673	.9826500593	.5686516817
39	1.1663152508	.6454864617	1.0281716934	.5895480138
40	1.2168162909	.6679476716	1.0740528864	.6105014604
41	1.2676700826	.6904706465	1.1202786019	.6315112981
42	1.3188626294	.7130544053	1.1668348820	.6525769013
43	1.3703808582	.7356977586	1.2137087380	.6736976341
44	1.4222125253	.7583993631	1.2608880515	.6948727801
45	1.4743461403	.7811577888	1.3083614852	.7161015101
46	1.5267709001	.8039715803	1.3561184074	.7373828804
47	1.5794766334	.8268393043	1.4041488245	.7587158518
48	1.6324537517	.8497595797	1.4524433250	.7800993189
49	1.6856932052	.8727310904	1.5009930303	.8015321429
50	1.7391864426	.8957525874	1.5497895525	.8230131797
51	1.7929253744	.9188228819	1.5988249569	.8445413025
52	1.8469023386	.9419408359	1.6480917290	.8661154159
53	1.9011100694	.9651053526	1.6975827444	.8877344639
54	1.9555416688	.9883153688	1.7472912415	.9093974315
55	2.0101905797	1.0115698499	1.7972107966	.9311033432
56	2.0650505629	1.0348677878	1.8473353006	.9528512591
57	2.1201156746	1.0582082001	1.8976589376	.9746402712
58	2.1753802474	1.0815901314	1.9481761656	.9964694991
59	2.2308388718	1.1050126540	1.9988816981	1.0183380871

DIRICHLET 1 (MEANS AND VARIANCES),   B= 6

N	P=1/ 6 MEAN	VARIANCE	P=1/ 7 MEAN	VARIANCE
6	.0154320988	.0151939491	.00611989890	.0060824459
7	.0540123457	.0510950122	.0244795961	.0238803455
8	.1140260631	.1010241200	.0577019051	.0543723952
9	.1890432099	.1533058747	.1049125547	.0939059105
10	.2718121285	.1979302953	.1630962010	.1364958302
11	.3562064186	.2293234060	.2284524404	.1762619229
12	.4412539665	.2534254753	.2978472753	.2102157360
13	.5287574329	.2789714878	.3695875626	.2390178739
14	.6205073138	.3108019172	.4432914806	.2655294524
15	.7169614914	.3484252168	.5192705520	.2927301037
16	.8171599876	.3877207221	.5979017902	.3222733971
17	.9192088504	.4234167366	.6792352533	.3540379277
18	1.0222059602	.4550009689	.7629487158	.3867363216
19	1.1263487512	.4853747353	.8485531567	.4190174074
20	1.2322705887	.5176452794	.9356358458	.4502701516
21	1.3403879147	.5529985670	1.0239863049	.4807538455
22	1.4505601371	.5902311931	1.1135762721	.5111962920
23	1.5620547630	.6264170298	1.2044492761	.5422242505
24	1.6743827855	.6604910155	1.2966104120	.5740036636
25	1.7875145992	.6934583820	1.3899820322	.6062612257
26	1.9017022567	.7270479654	1.4844305564	.6385732381
27	2.0171861655	.7623296345	1.5798244042	.6706690507
28	2.1339652104	.7989955268	1.6760776421	.7025599543
29	2.2516937407	.8353620998	1.7731559802	.7344573335
30	2.3700956734	.8705554858	1.8710514326	.7665825167
31	2.4891169874	.9049471798	1.9697502357	.7990174064
32	2.6088740479	.9395517852	2.0692161940	.8316831009
33	2.7295084988	.9751706171	2.1693954729	.8644286089
34	2.8510428573	1.0117799028	2.2702336091	.8971455678
35	2.9732884003	1.0483345648	2.3716902348	.9298280494
36	3.0960769984	1.0841802941	2.4737422732	.9625505170
37	3.2193620231	1.1194544591	2.5763763076	.9953972261
38	3.3432041848	1.1547960691	2.6795778253	1.0284013637
39	3.4676920995	1.1907961700	2.7833251310	1.0615323862
40	3.5928488736	1.2275215492	2.8875906854	1.0947292133
41	3.7185599271	1.2642769121	2.9923470396	1.1279468727
42	3.8447156993	1.3005723413	3.0975721763	1.1611821254
43	3.9712798736	1.3364514264	3.2032507167	1.1944651604
44	4.0982873558	1.3723428953	3.3093710958	1.2278302928
45	4.2257968891	1.4086906234	3.4159215098	1.2612901342
46	4.3538279050	1.4455862237	3.5228876154	1.2948301602
47	4.4823051641	1.4825400040	3.6302530907	1.3284231526
48	4.6111532027	1.5191777637	3.7380019749	1.3620494906
49	4.7403441514	1.5555035969	3.8461207402	1.3957080897
50	4.8698096907	1.5918206865	3.9545986961	1.4294123202
51	4.9998607325	1.6284706953	4.0634267848	1.4631767586
52	5.1302428961	1.6655469343	4.1725959206	1.4970057890
53	5.2609937665	1.7026883732	4.2820960883	1.5308916074
54	5.3920589900	1.7396024407	4.3919166359	1.5648211684
55	5.5234170003	1.7762763942	4.5020472760	1.5987854822
56	5.6550821220	1.8129340464	4.6124789318	1.6327843079
57	5.7870840397	1.8498448102	4.7232038498	1.6668238425
58	5.9194352307	1.8870958135	4.8342150384	1.7009103633
59	6.0520974803	1.9244109307	4.9455055442	1.7350450504
60	6.1850301972	1.9615562750	5.0570680907	1.7692234009

DIRICHLET 1 (MEANS AND VARIANCES),   B= 6

N	MEAN (P=1/8)	VARIANCE	MEAN (P=1/9)	VARIANCE
6	.0027465820	.0027390383	.0013548070	.0013529715
7	.0120162964	.0118719050	.0063224328	.0062824596
8	.0306415558	.0297026508	.0170939849	.0168017806
9	.0597059727	.0561411695	.0351246266	.0338908872
10	.0986697525	.0889340324	.0609130407	.0572026421
11	.1458941493	.1246090465	.0940964173	.0852422815
12	.1993122592	.1598047072	.1337175982	.1158902023
13	.2570311393	.1923819934	.1785492333	.1470521832
14	.3176915894	.2218223954	.2273822762	.1771880081
15	.3805429568	.2489016972	.2792194648	.2055577783
16	.4453097416	.2749837449	.3333564669	.2321716299
17	.5119700789	.3013172177	.3893685100	.2575387045
18	.5805529056	.3285984809	.4470392913	.2823500374
19	.6510171990	.3568918914	.5062716645	.3072110712
20	.7232243406	.3858327028	.5670106089	.3324897574
21	.7969749637	.4149409336	.6291944956	.3582904892
22	.8720671300	.4438782097	.6927365585	.3845212087
23	.9483407257	.4725551620	.7575288063	.4110011076
24	1.0256923282	.5010900330	.8234568017	.4375595511
25	1.1040637132	.5296845987	.8904149418	.4640956104
26	1.1834179902	.5585005316	.9583159331	.4905909996
27	1.2637181452	.5875936951	1.0270927553	.5170878193
28	1.3449165934	.6169193870	1.0966948334	.5436512414
29	1.4269563400	.6463842592	1.1670816589	.5703360469
30	1.5097787805	.6759057899	1.2382168945	.5971682831
31	1.5933317828	.7054482316	1.3100647457	.6241439868
32	1.6775737506	.7350245525	1.3825889304	.6512398644
33	1.7624727866	.7646735737	1.4557535099	.6784279305
34	1.8480027873	.7944310038	1.5295244315	.7056871597
35	1.9341392410	.8243106431	1.6038707917	.7330085823
36	2.0208567785	.8543022531	1.6787653099	.7603939409
37	2.1081290203	.8843823587	1.7541840032	.7878505119
38	2.1959299609	.9145287881	1.8301053851	.8153853811
39	2.2842356087	.9447304579	1.9065095987	.8430016477
40	2.3730248808	.9749887207	1.9833777901	.8706974821
41	2.4622794481	1.0053119400	2.0606918361	.8984675295
42	2.5519828625	1.0357079817	2.1384343664	.9263053589
43	2.6421195749	1.0661790951	2.2165889272	.9542056317
44	2.7326743324	1.0967211945	2.2951401323	.9821651860
45	2.8236321080	1.1273267164	2.3740737076	1.0101829173
46	2.9149784020	1.1579885849	2.4533764140	1.0382588710
47	3.0066996101	1.1887028966	2.5330358905	1.0663931594
48	3.0987832129	1.2194692495	2.6130404829	1.0945852054
49	3.1912177064	1.2502891638	2.6933791107	1.1228335311
50	3.2839923582	1.2811639329	2.7740411936	1.1511360152
51	3.3770969398	1.3120931949	2.8550166319	1.1794903729
52	3.4705215624	1.3430747961	2.9362958122	1.2078945875
53	3.5642566527	1.3741056810	3.0178696149	1.2363471267
54	3.6582930233	1.4051830690	3.0997294030	1.2648469113
55	3.7526219576	1.4363052095	3.1818669913	1.2933931228
56	3.8472352433	1.4674714178	3.2642746015	1.3219849747
57	3.9421251364	1.4986815533	3.3469448167	1.3506215561
58	4.0372842798	1.5299353594	3.4298705436	1.3793017916
59	4.1327056208	1.5612320560	3.5130449874	1.4080245010
60	4.2283823600	1.5925703392	3.5964616373	1.4367885040

DIRICHLET 1 (MEANS AND VARIANCES),   B= 6

N	P=1/10 MEAN	VARIANCE	P=1/11 MEAN	VARIANCE
6	.0007200000	.0007194816	.0004064212	.0004062561
7	.0035280000	.0035155532	.0020690535	.0020647726
8	.0099792000	.0098796156	.0060660888	.0060292913
9	.0213796800	.0209225893	.0134402982	.0132596565
10	.0385363440	.0370512942	.0250006101	.0243755796
11	.0616930776	.0578870418	.0412209561	.0395217888
12	.0906003410	.0824068880	.0622260905	.0583587737
13	.1246599114	.1092365745	.0878394506	.0801631384
14	.1630933720	.1369888425	.1176660286	.1039976927
15	.2050944522	.1645430191	.1511862728	.1288997145
16	.2499403601	.1912008322	.1878425578	.1540402966
17	.2970531046	.2167060437	.2271064848	.1788247446
18	.3460149606	.2411578558	.2685219771	.2029254301
19	.3965504928	.2648688848	.3117247514	.2262566734
20	.4484905120	.2882185727	.3564425911	.2489119532
21	.5017319622	.3115391714	.4024827477	.2710865079
22	.5562036947	.3350520675	.4497129647	.2930050234
23	.6118431506	.3588540462	.4980415640	.3148673285
24	.6685845938	.3829404465	.5474003182	.3368174293
25	.7263565590	.4072467191	.5977319996	.3589347224
26	.7850848569	.4316909908	.6489829264	.3812419187
27	.8446975561	.4562056581	.7010997477	.4037223714
28	.9051293262	.4807530244	.7540291661	.4263398099
29	.9663238066	.5053261355	.8077192132	.4490552322
30	1.0282338154	.5299398040	.8621209347	.4718380864
31	1.0908199638	.5546179696	.9171897298	.4946711389
32	1.1540485352	.5793825317	.9728860006	.5175501096
33	1.2178894156	.6042466051	1.0291750906	.5404800280
34	1.2823145733	.6292128029	1.0860266996	.5634703938
35	1.3472972523	.6542753974	1.1434140330	.5865308046
36	1.4128117762	.6794243867	1.2013129338	.6096680279
37	1.4788337219	.7046495491	1.2597011616	.6328847935
38	1.5453402024	.7299432028	1.3185578989	.6561800466
39	1.6123100690	.7553012213	1.3778634804	.6795501082
40	1.6797239430	.7807225566	1.4375992936	.7029901372
41	1.7475640825	.8062079220	1.4977477752	.7264954089
42	1.8158141462	.8317583414	1.5582924386	.7500621334
43	1.8844589317	.8573740900	1.6192178826	.7736877499
44	1.9534841526	.8830542483	1.6805097605	.7973707928
45	2.0228762876	.9087968221	1.7421547101	.8211105111
46	2.0926224994	.9345992059	1.8041402551	.8449064254
47	2.1627106068	.9604587241	1.8664546988	.8687579674
48	2.2331290782	.9863730440	1.9290870246	.8926642739
49	2.3038670248	1.0123403602	1.9920268142	.9166241429
50	2.3749141783	1.0383593603	2.0552641864	.9406361112
51	2.4462608513	1.0644290507	2.1187897552	.9646985933
52	2.5178978857	1.0905485460	2.1825946003	.9888100218
53	2.5898165983	1.1167169027	2.2466702446	1.0129689498
54	2.6620087315	1.1429330421	2.3110086340	1.0371740981
55	2.7344664150	1.1691957584	2.3756021149	1.0614243505
56	2.8071821369	1.1955037840	2.4404434106	1.0857187156
57	2.8801487239	1.2218558707	2.5055255949	1.1100562747
58	2.9533593259	1.2482508530	2.5708420665	1.1344361363
59	3.0268074014	1.2746876770	2.6363865246	1.1588574061
60	3.1004867021	1.3011653945	2.7021529469	1.1833191774

DIRICHLET 1 (MEANS AND VARIANCES),   B= 6

N	P=1/12 MEAN	VARIANCE	P=1/13 MEAN	VARIANCE
6	.0002411265	.0002410684	.0001491669	.0001491446
7	.0012659144	.0012643118	.0008032062	.0008025611
8	.0038211859	.0038065845	.0024837608	.0024775917
9	.0087031612	.0086274162	.0057887881	.0057552781
10	.0166168352	.0163407160	.0112975691	.0111699340
11	.0280817971	.0272932098	.0194953769	.0191153072
12	.0433898379	.0415088387	.0307271193	.0297836060
13	.0626063883	.0587013785	.0451795919	.0431441788
14	.0856029075	.0783433035	.0628871368	.0589604826
15	.1121067603	.0997687246	.0837536966	.0768372193
16	.1417567169	.1222844255	.1075841705	.0962851458
17	.1741549041	.1452660210	.1341188592	.1167898379
18	.2089091896	.1682240994	.1630661544	.1378724463
19	.2456630668	.1908347593	.1941301878	.1591342947
20	.2841127271	.2129373832	.2270316729	.1802817369
21	.3240129199	.2345081055	.2615214880	.2011318555
22	.3651743142	.2556197808	.2973875444	.2216026131
23	.4074554561	.2763988164	.3344561411	.2416926717
24	.4507522175	.2969869278	.3725893172	.2614563831
25	.4949870491	.3175127273	.4116797479	.2809787493
26	.5400996092	.3380749131	.4516445539	.3003538667
27	.5860395812	.3587362760	.4924190962	.3196688798
28	.6327618716	.3795260746	.5339514823	.3389940848
29	.6802239270	.4004475988	.5761981779	.3583787275
30	.7283846678	.4214878078	.6191208388	.3778513264
31	.7772044546	.4426265443	.6626842754	.3974230233
32	.8266455608	.4638437188	.7068553417	.4170924586
33	.8766727478	.4851237743	.7516024875	.4368509051
34	.9272536926	.5064574980	.7968957174	.4566867496
35	.9783591563	.5278417508	.8427067422	.4765888117
36	1.0299628937	.5492779099	.8890091661	.4965483439
37	1.0820413636	.5707698162	.9357786128	.5165598284
38	1.1345733319	.5923218487	.9829927497	.5366208514
39	1.1875394535	.6139375120	1.0306312082	.5567314013
40	1.2409218967	.6356186811	1.0786754210	.5768929240
41	1.2947040476	.6573664532	1.1271084127	.5971073978
42	1.3488703041	.6791764421	1.1759145745	.6173766009
43	1.4034059484	.7010493043	1.2250794537	.6377016492
44	1.4582970773	.7229813053	1.2745895720	.6580828072
45	1.5135305664	.7449697866	1.3244322831	.6785195193
46	1.5690940471	.7670124614	1.3745956683	.6990105875
47	1.6249758816	.7891075314	1.4250684634	.7195544133
48	1.6811651292	.8112536550	1.4758400090	.7401492395
49	1.7376515017	.8334498268	1.5269002163	.7607933439
50	1.7944253122	.8556952230	1.5782395384	.7814851664
51	1.8514774207	.8779890626	1.6298489445	.8022233638
52	1.9087991814	.9003305111	1.6817198921	.8230068082
53	1.9663823958	.9227186348	1.7338442975	.8438345460
54	2.0242192730	.9451524023	1.7862145039	.8647057396
55	2.0823023971	.9676307159	1.8388232492	.8856196092
56	2.1406247011	.9901524566	1.8916636355	.9065753858
57	2.1991794457	1.0127165270	1.9447290992	.9275722826
58	2.2579602006	1.0353218837	1.9980133857	.9486094833
59	2.3169608282	1.0579675525	2.0515105262	.9696861460
60	2.3761754669	1.0806526296	2.1052148181	.9908014153

DIRICHLET 1 (MEANS AND VARIANCES),   B= 6

| | P=1/14 | | P=1/15 | |
N	MEAN	VARIANCE	MEAN	VARIANCE
6	.0000956234	.0000956143	.0000632099	.0000632059
7	.0005259288	.0005256522	.0003539753	.0003538500
8	.0016597494	.0016569946	.0011366540	.0011353621
9	.0039444662	.0039289074	.0027468484	.0027393032
10	.0078432986	.0077817813	.0055505164	.0055197082
11	.0137788943	.0135890364	.0099031069	.0098050354
12	.0220921527	.0216043536	.0161160170	.0158564063
13	.0330187949	.0319310058	.0244335931	.0238376925
14	.0466823359	.0445153947	.0350211041	.0338002878
15	.0631001804	.0591627676	.0479624518	.0456829286
16	.0821988647	.0755698256	.0632655578	.0593247653
17	.1038345245	.0933670075	.0808730864	.0744882769
18	.1278151657	.1121628567	.1006762466	.0908877369
19	.1539220399	.1315838984	.1225296977	.1082188550
20	.1819282415	.1513053989	.1462659906	.1261858033
21	.2116134292	.1710706987	.1717084148	.1445228877
22	.2427742748	.1906989826	.1986815564	.1630093550
23	.2752307862	.2100830321	.2270192537	.1814770156
24	.3088290398	.2291795389	.2565699550	.1998113236
25	.3434410794	.2479949456	.2871997108	.2179472106
26	.3789628153	.2665696358	.3187931862	.2358612981
27	.4153107249	.2849627957	.3512531502	.2535621512
28	.4524180436	.3032395736	.3844989111	.2710800596
29	.4902309778	.3214614394	.4184641297	.2884575204
30	.5287053060	.3396799928	.4530943765	.3057412197
31	.5678035711	.3579339624	.4883447139	.3229759416
32	.6074929395	.3762488048	.5241774974	.3402005075
33	.6477436974	.3946381478	.5605605130	.3574455994
34	.6885282963	.4131063091	.5974654940	.3747331491
35	.7298208229	.4316512072	.6348670162	.3920768962
36	.7715967652	.4502671423	.6727417276	.4094836936
37	.8138329575	.4689471003	.7110678541	.4269551807
38	.8565076100	.4876844143	.7498249142	.4444895123
39	.8996003551	.5064737561	.7889935757	.4620829178
40	.9430922709	.5253115399	.8285505996	.4797309560
41	.9869658615	.5441958782	.8684938260	.4974294081
42	1.0312049933	.5631262522	.9087921708	.5151748133
43	1.0757947949	.5821030478	.9494356139	.5329646979
44	1.1207215341	.6011270811	.9904101690	.5507975697
45	1.1659724870	.6201991977	1.0317028342	.5686727578
46	1.2115358105	.6393199884	1.0733015258	.5865901740
47	1.2574004278	.6584896351	1.1151950004	.6045500565
48	1.3035559310	.6777078703	1.1573727731	.6225527406
49	1.3499925034	.6969740227	1.1998250361	.6405984829
50	1.3967008584	.7162871158	1.2425425840	.6586873475
51	1.4436721941	.7356459847	1.2855167463	.6768191544
52	1.4908981576	.7550493899	1.3287393309	.6949934774
53	1.5383708180	.7744961072	1.3722025771	.7132096784
54	1.5860826421	.7939849882	1.4158991172	.7314669619
55	1.6340264736	.8135149913	1.4598219463	.7497644355
56	1.6821955117	.8330851858	1.5039643968	.7681011674
57	1.7305832902	.8526947388	1.5483201179	.7864762323
58	1.7791836565	.8723428915	1.5928830574	.8048887437
59	1.8279907510	.8920289335	1.6376474449	.8233378714
60	1.8769989867	.9117521798	1.6826077762	.8418228472

DIRICHLET 1 (MEANS AND VARIANCES),   B= 7

	P=1/ 7		P=1/ 8	
N	MEAN	VARIANCE	MEAN	VARIANCE
7	.0061198990	.0060824459	.0024032593	.0023974836
8	.0244795961	.0238803455	.0108146667	.0106977097
9	.0577019051	.0543723952	.0281631947	.0273700291
10	.1049125547	.0939059105	.0557631254	.0526535993
11	.1630962010	.1364958302	.0933064241	.0846003353
12	.2284524404	.1762619229	.1393206837	.1199104308
13	.2973065453	.2089153634	.1917180140	.1549622171
14	.3675791367	.2344731409	.2484024194	.1870083769
15	.4389398911	.2563137927	.3077657552	.2149817420
16	.5120657225	.2785462172	.3689013748	.2394914774
17	.5877867925	.3039216413	.4315303658	.2621679632
18	.6665327269	.3330250018	.4957608707	.2848084862
19	.7481365254	.3645136943	.5618185180	.3086990258
20	.8318901291	.3958859441	.6298383047	.3342739796
21	.9170473525	.4253652601	.6997700526	.3611713099
22	1.0031868198	.4527753813	.7714006052	.3885988626
23	1.0902727402	.4792015272	.8444467456	.4157919743
24	1.1785046685	.5060893263	.9186552935	.4423391540
25	1.2681040919	.5344157099	.9938644151	.4682692892
26	1.3591420673	.5642429213	1.0700118260	.4939274900
27	1.4514495334	.5946822768	1.1471008826	.5197420548
28	1.5447516678	.6247098271	1.2251500334	.5460071848
29	1.6388288162	.6538389455	1.3041524643	.5727771642
30	1.7335978899	.6822777714	1.3840614859	.5998973895
31	1.8290995875	.7106622848	1.4648010279	.6271262712
32	1.9254260819	.7396244193	1.5462891228	.6542693785
33	2.0226334731	.7694201674	1.6284597661	.6812606572
34	2.1206714577	.7997338923	1.7112731327	.7081644287
35	2.2194214263	.8300217104	1.7947119734	.7351138145
36	2.3187678323	.8599100625	1.8787691405	.7622286277
37	2.4186489868	.8893858461	1.9634343871	.7895575015
38	2.5190669266	.9187394588	2.0486869942	.8170678021
39	2.6200616234	.9483521796	2.1344963254	.8446781153
40	2.7216665851	.9784505116	2.2208280434	.8723083210
41	2.8238635065	1.0089211122	2.3076515162	.8999188388
42	2.9265932485	1.0394621738	2.3949443682	.9275215333
43	3.0297903454	1.0698178305	2.4826924553	.9551622783
44	3.1334130964	1.0999270914	2.5708862385	.9828896020
45	3.2374546400	1.1299311536	2.6595161547	1.0107281968
46	3.3419333219	1.1600660033	2.7485695222	1.0386699158
47	3.4468687715	1.1905039461	2.8380301863	1.0666834866
48	3.5522528755	1.2212069595	2.9278804645	1.0947343961
49	3.6580524794	1.2519968868	3.0181038734	1.1228027721
50	3.7642276374	1.2826988170	3.1086870303	1.1508903242
51	3.8707499680	1.3132494590	3.1996198836	1.1790147137
52	3.9776114689	1.3437216434	3.2908944667	1.2071965970
53	4.0848209143	1.3742679187	3.3825031061	1.2354474763
54	4.1923900783	1.4050161012	3.4744371099	1.2637645967
55	4.3003146094	1.4359573827	3.5666865119	1.2921343240
56	4.4085741190	1.4669801476	3.6592407812	1.3205407299
57	4.5171426316	1.4979619451	3.7520899207	1.3489739251
58	4.6260002847	1.5288479777	3.8452252831	1.3774337494
59	4.7351398910	1.5596777219	3.9386397100	1.4059276433
60	4.8445657145	1.5905542271	4.0323270373	1.4344648320
61	4.9542850404	1.6215729961	4.1262813371	1.4630505546

DIRICHLET 1 (MEANS AND VARIANCES),   B= 7

N	P=1/ 9 MEAN	VARIANCE	P=1/10 MEAN	VARIANCE
7	.0010537388	.0010526284	.0005040000	.0005037460
8	.0051516119	.0051250728	.0026208000	.0026139314
9	.0144401243	.0142316071	.0077716800	.0077112810
10	.0305280705	.0295961074	.0172972800	.0169980841
11	.0541608251	.0512274301	.0321646248	.0311300617
12	.0852218412	.0779590790	.0528345418	.0500430530
13	.1229201522	.1078107884	.0792653622	.0729823645
14	.1660681286	.1385490486	.1110074926	.0986984508
15	.2133786995	.1682782654	.1473468300	.1257447145
16	.2637128392	.1958591147	.1874601680	.1527901610
17	.3162275663	.2210307652	.2305515267	.1788608200
18	.3704128343	.2442531638	.2759481849	.2034563215
19	.4260404917	.2663845189	.3231482287	.2265349953
20	.4830652905	.2883299103	.3718236641	.2483994633
21	.5415182679	.3107683520	.4217913862	.2695333619
22	.6014230459	.3340182333	.4729675964	.2904396035
23	.6627505670	.3580483113	.5253202834	.3115179342
24	.7254121816	.3825951970	.5788304988	.3330010341
25	.7892795180	.4073226818	.6334679602	.3549495605
26	.8542149653	.4319598841	.6891815021	.3772918010
27	.9200984882	.4563779773	.7459012165	.3998858161
28	.9868423294	.4805960108	.8035472977	.4225817117
29	1.0543919197	.5047327094	.8620405751	.4452674224
30	1.1227166178	.5289354386	.9213109644	.4678901770
31	1.1917964138	.5533176851	.9813019027	.4904545477
32	1.2616103877	.5779254807	1.0419706090	.5130042457
33	1.3321304074	.6027375905	1.1032852742	.5355974838
34	1.4033206624	.6276907420	1.1652208439	.5582849146
35	1.4751413461	.6527141239	1.2277549709	.5810959764
36	1.5475537723	.6777578524	1.2908651957	.6040354325
37	1.6205244663	.7028061325	1.3545277401	.6270883831
38	1.6940268625	.7278739085	1.4187177105	.6502299341
39	1.7680405445	.7529923925	1.4834101531	.6734353086
40	1.8425489327	.7781918339	1.5485813187	.6966871599
41	1.9175366855	.8034889437	1.6142096290	.7199785588
42	1.9929878582	.8288827896	1.6802760819	.7433118717
43	2.0688852917	.8543587511	1.7467640964	.7666949730
44	2.1452111040	.8798970349	1.8136589711	.7901367025
45	2.2219477563	.9054812669	1.8809472146	.8136432188
46	2.2990790819	.9311036887	1.9486159696	.8372161952
47	2.3765908417	.9567656551	2.0166526684	.8608529634
48	2.4544706739	.9824743185	2.0850449431	.8845480617
49	2.5327075883	1.0082377300	2.1537807282	.9082953201
50	2.6112913023	1.0340607196	2.2228484520	.9320896632
51	2.6902117099	1.0599430687	2.2922372104	.9559281059
52	2.7694586490	1.0858802084	2.3619368572	.9798098144
53	2.8490219735	1.1118655896	2.4319379897	1.0037354464
54	2.9288918084	1.1378933669	2.5022318497	1.0277061676
55	3.0090588259	1.1639602133	2.5728101849	1.0517227550
56	3.0895144068	1.1900656964	2.6436651185	1.0757850658
57	3.1702506378	1.2162113742	2.7147890571	1.0998919608
58	3.2512601716	1.2423992657	2.7861746515	1.1240415986
59	3.3325360317	1.2686304738	2.8578148005	1.1482319085
60	3.4140714469	1.2949045074	2.9297026787	1.1724610398
61	3.4958597696	1.3212194387	3.0018317648	1.1967276387

DIRICHLET 1 (MEANS AND VARIANCES),   B= 7

	P=1/11		P=1/12	
N	MEAN	VARIANCE	MEAN	VARIANCE
7	.0002586317	.0002585648	.0001406572	.0001406374
8	.0014107183	.0014087282	.0007970572	.0007964219
9	.0043732268	.0043541017	.0025611323	.0025545729
10	.0101431813	.0100402972	.0061439825	.0061062339
11	.0195979087	.0192138307	.0122531466	.0121030070
12	.0333580298	.0322452716	.0214865261	.0210248553
13	.0517270963	.0490514038	.0342629683	.0330890173
14	.0746969244	.0691208809	.0507929167	.0482140573
15	.1019994320	.0916265264	.0710834789	.0660403432
16	.1331849566	.1155907966	.0949685938	.0859974497
17	.1677087348	.1400616926	.1221541965	.1074007919
18	.2050104035	.1642587950	.1522689559	.1295558599
19	.2445759408	.1876624236	.1849127194	.1518497046
20	.2859768024	.2100374272	.2196969454	.1738150679
21	.3288860878	.2313999327	.2562738313	.1951605773
22	.3730754883	.2519461444	.2943531829	.2157682733
23	.4183990286	.2719659280	.3337080112	.2356655881
24	.4647702314	.2917616634	.3741711670	.2549819675
25	.5121385917	.3115868673	.4156259891	.2739007281
26	.5604696470	.3316115919	.4579939909	.2926151048
27	.6097309780	.3519144742	.5012221982	.3112946153
28	.6598846418	.3724958901	.5452720495	.3300646438
29	.7108851411	.3933036916	.5901109694	.3489991961
30	.7626812299	.4142625428	.6357069708	.3681245330
31	.8152196381	.4352994539	.6820260530	.3874300881
32	.8684490425	.4563609229	.7290317766	.4068827176
33	.9223231324	.4774202118	.7766862373	.4264407705
34	.9768022252	.4984759253	.8249516756	.4460654355
35	1.0318534182	.5195447516	.8737921059	.4657280097
36	1.0874496315	.5406517988	.9231745615	.4854128756
37	1.1435680616	.5618215813	.9730697655	.5051168535
38	1.2001885552	.5830717050	1.0234522245	.5248461184
39	1.2572922810	.6044100727	1.0742998661	.5446120196
40	1.3148608957	.6258353336	1.1255934015	.5644269901
41	1.3728762184	.6473395654	1.1773156016	.5843013787
42	1.4313203054	.6689118908	1.2294506355	.6042416175
43	1.4901757481	.6905418440	1.2819835681	.6242497423
44	1.5494260220	.7122216886	1.3349000510	.6443239919
45	1.6090557517	.7339473729	1.3881861928	.6644600605
46	1.6690508278	.7557182457	1.4418285628	.6846525516
47	1.7293983654	.7775359464	1.4958142707	.7048962614
48	1.7900865470	.7994029895	1.5501310644	.7251870610
49	1.8511044084	.8213215046	1.6047674067	.7455222912
50	1.9124416276	.8432924288	1.6597125074	.7659007204
51	1.9740883599	.8653152503	1.7149563054	.7863221930
52	2.0360351385	.8873882258	1.7704894103	.8067871339
53	2.0982728388	.9095088878	1.8263030198	.8272960608
54	2.1607926865	.9316746270	1.8823888303	.8478492132
55	2.2235862873	.9538831692	1.9387389562	.8684463509
56	2.2866456550	.9761328439	1.9953458672	.8890867251
57	2.3499632231	.9984226227	2.0522023458	.9097691801
58	2.4135318338	1.0207519775	2.1093014650	.9304923290
59	2.4773447087	1.0431206427	2.1666365787	.9512547429
60	2.5413954072	1.0655283733	2.2242013197	.9720551046
61	2.6056777825	1.0879747668	2.2819895984	.9928923018

DIRICHLET 1 (MEANS AND VARIANCES),   B= 7

	P=1/13		P=1/14	
N	MEAN	VARIANCE	MEAN	VARIANCE
7	.0000803206	.0000803142	.0000478117	.0000478094
8	.0004695667	.0004693462	.0002868703	.0002867880
9	.0015541328	.0015517174	.0009733098	.0009723625
10	.0038343318	.0038196297	.0024588880	.0024528419
11	.0078529981	.0077913285	.0051510933	.0051245596
12	.0141219713	.0139225412	.0094649938	.0093754077
13	.0230630319	.0225311284	.0157785073	.0155295460
14	.0349705650	.0337479705	.0243980995	.0238029548
15	.0499965731	.0475002422	.0355377251	.0342760211
16	.0681548557	.0635269259	.0493108748	.0468858795
17	.0893394460	.0814209224	.0657338399	.0614379330
18	.1133519322	.1006881074	.0847374533	.0776331208
19	.1399325914	.1208118507	.1061843136	.0951054685
20	.1687910232	.1413121105	.1298886328	.1134635421
21	.1996330278	.1617908162	.1556362421	.1323296885
22	.2321816549	.1819589871	.1832028199	.1513721843
23	.2661915055	.2016448856	.2123690231	.1703272422
24	.3014563802	.2207856904	.2429318023	.1890098076
25	.3378111207	.2394072324	.2747117303	.2073138337
26	.3751289766	.2575971743	.3075566128	.2252040192
27	.4133160159	.2754768036	.3413419619	.2427017192
28	.4523040604	.2931756357	.3759690904	.2598679269
29	.4920434020	.3108116485	.4113616402	.2767859715
30	.5324962332	.3284784925	.4474613120	.2935460309
31	.5736313659	.3462396912	.4842234445	.3102328584
32	.6154204746	.3641288285	.5216129319	.3269174027
33	.6578358255	.3821540921	.5596007979	.3436523609
34	.7008492641	.4003053101	.5981615775	.3604712026
35	.7444321267	.4185617191	.6372715255	.3773898822
36	.7885557197	.4368990544	.6769075684	.3944103072
37	.8331920434	.4552950312	.7170468532	.4115246388
38	.8783145095	.4737327934	.7576667219	.4287196306
39	.9238984905	.4922023462	.7987449395	.4459804078
40	.9699216241	.5107003186	.8402600316	.4632933278
41	1.0163638670	.5292285826	.8821916174	.4806477786
42	1.0632073425	.5477923033	.9245206669	.4980369630
43	1.1104360513	.5663979313	.9672296464	.5154578444
44	1.1580355198	.5850515192	1.0103025478	.5329105039
45	1.2059924530	.6037575764	1.0537248167	.5503971764
46	1.2542944373	.6225185236	1.0974832089	.5679212066
47	1.3029297189	.6413346821	1.1415656074	.5854861118
48	1.3518870639	.6602046534	1.1859608296	.6030948714
49	1.4011556883	.6791259119	1.2306584495	.6207494934
50	1.4507252414	.6980954434	1.2756486495	.6384508529
51	1.5005858189	.7171102972	1.3209221085	.6561987545
52	1.5507279863	.7361679734	1.3664699278	.6739921493
53	1.6011427980	.7552666167	1.4122835869	.6918294292
54	1.6518218015	.7744050338	1.4583549242	.7097087320
55	1.7027570255	.7935825797	1.5046761306	.7276282038
56	1.7539409518	.8127989704	1.5512397502	.7455861893
57	1.8053664773	.8320540788	1.5980386801	.7635813368
58	1.8570268708	.8513477589	1.6450661668	.7816126256
59	1.9089157305	.8706797241	1.6923157957	.7996793318
60	1.9610269465	.8900494905	1.7397814749	.8177809538
61	2.0133546706	.9094563777	1.7874574144	.8359171231

DIRICHLET 1 (MEANS AND VARIANCES), B= 7

	P=1/15		P=1/16	
N	MEAN	VARIANCE	MEAN	VARIANCE
7	.0000294979	.0000294971	.0000187755	.0000187751
8	.0001809207	.0001808880	.0001173466	.0001173329
9	.0006269296	.0006265366	.0004140870	.0004139155
10	.0016162250	.0016136129	.0010863732	.0010851930
11	.0034522277	.0034403098	.0023599473	.0023543780
12	.0064626380	.0064208723	.0044902120	.0044700500
13	.0109675213	.0108472347	.0077402582	.0076803466
14	.0172514548	.0169538888	.0123596319	.0122068903
15	.0255428654	.0248909096	.0185665644	.0182220479
16	.0360016735	.0347082199	.0265351217	.0258321507
17	.0487149461	.0463522870	.0363877096	.0350682543
18	.0636993975	.0596746507	.0481926464	.0458849443
19	.0809091336	.0744501461	.0619660699	.0581666781
20	.1002469018	.0904016153	.0776772031	.0717403080
21	.1215771693	.1072274127	.0952559307	.0863918367
22	.1447395618	.1246281234	.1146016679	.1018851867
23	.1695614746	.1423295270	.1355926163	.1179808101
24	.1958690015	.1600997784	.1580946566	.1344522813
25	.2234956536	.1777598324	.1819693146	.1510995166
26	.2522886501	.1951871351	.2070804212	.1677578458
27	.2821128212	.2123133998	.2332992693	.1843027370
28	.3128523635	.2291178165	.2605082242	.2006504683
29	.3444108206	.2456172987	.2886028741	.2167554004
30	.3767097292	.2618553722	.3174928960	.2326047286
31	.4096863839	.2778911252	.3471018761	.2482116699
32	.4432911404	.2937893303	.3773663451	.2636080047
33	.4774846109	.3096124756	.4082342923	.2788367673
34	.5122350241	.3254150776	.4396633974	.2939457025
35	.5475159340	.3412403149	.4716191847	.3089818994
36	.5833043762	.3571187638	.5040732577	.3239878184
37	.6195795023	.3730688408	.5370017252	.3389987435
38	.6563216635	.3890984628	.5703838853	.3540415559
39	.6935118809	.4052074216	.6042011942	.3691346195
40	.7311316157	.4213900134	.6384365142	.3842885120
41	.7691627508	.4376375491	.6730736128	.3995073173
42	.8075876962	.4539404833	.7080968716	.4147902070
43	.8463895500	.4702900099	.7434911547	.4301330774
44	.8855522582	.4866790773	.7792417883	.4455300589
45	.9250607398	.5031028598	.8153346080	.4609747757
46	.9649009586	.5195587775	.8517560342	.4764612894
47	1.0050599372	.5360461914	.8884931511	.4919847098
48	1.0455257187	.5525659048	.9255337680	.5075414980
49	1.0862872899	.5691195956	.9628664550	.5231295125
50	1.1273344792	.5857092756	1.0004805477	.5387478640
51	1.1686578447	.6023368455	1.0383661251	.5543966508
52	1.2102485647	.6190037822	1.0765139649	.5700766375
53	1.2520983393	.6357109679	1.1149154827	.5857889344
54	1.2941993096	.6524586458	1.1535626644	.6015347151
55	1.3365439947	.6692464775	1.1924479966	.6173149984
56	1.3791252473	.6860736658	1.2315644009	.6331305051
57	1.4219362249	.7029391096	1.2709051765	.6489815873
58	1.4649703712	.7198415604	1.3104639517	.6648682210
59	1.5082214067	.7367797581	1.3502346460	.6807900473
60	1.5516833217	.7537525323	1.3902114413	.6967464439
61	1.5953503713	.7707588641	1.4303887615	.7127366103

DIRICHLET 1 (MEANS AND VARIANCES),   B= 8

N	P=1/8 MEAN	VARIANCE	P=1/9 MEAN	VARIANCE
8	.0024032593	.0023974836	.0009366567	.0009357794
9	.0108146667	.0106977097	.0046832835	.0046613504
10	.0281631947	.0273700291	.0133560308	.0131776473
11	.0557631254	.0526535993	.0286200661	.0278009579
12	.0933064241	.0846003353	.0513229630	.0486889164
13	.1393206837	.1199104308	.0814552992	.0748203334
14	.1917180140	.1549622171	.1183129463	.1043149930
15	.2482475596	.1866207088	.1607355335	.1348996218
16	.3070882438	.2133657784	.2073824383	.1644631749
17	.3672075963	.2356569473	.2570030426	.1915634310
18	.4283470031	.2553551767	.3086355296	.2156938528
19	.4907548729	.2747289884	.3616909646	.2372309603
20	.5548547157	.2955839106	.4159261062	.2571356567
21	.6209783032	.3187762120	.4713429686	.2765581311
22	.6892136335	.3441251709	.5280624523	.2964831137
23	.7593608958	.3706103689	.5862097592	.3174953157
24	.8310424568	.3969701013	.6458371639	.3397032458
25	.9038725119	.4223113823	.7068959233	.3628198347
26	.9775864653	.4464126190	.7692526812	.3863503378
27	1.0520859834	.4696634916	.8327325176	.4098060793
28	1.1274082391	.4927718427	.8971665755	.4328670989
29	1.2036530906	.5164131728	.9624268960	.4554510563
30	1.2809031274	.5409541147	1.0284402764	.4776873829
31	1.3591604826	.5663104236	1.0951817640	.4998256008
32	1.4383411466	.5920651899	1.1626546975	.5221213025
33	1.5183148961	.6177320241	1.2308671779	.5447431956
34	1.5989581879	.6429939679	1.2998139899	.5677302504
35	1.6801923813	.6678132367	1.3694690248	.5910047096
36	1.7619950404	.6923970721	1.4397882563	.6144249289
37	1.8443861111	.7170660926	1.5107194985	.6378505835
38	1.9273985620	.7420912695	1.5822136868	.6611940716
39	2.0110447550	.7675554325	1.6542331315	.6844421632
40	2.0953018523	.7933405029	1.7267542943	.7076455769
41	2.1801191575	.8192212788	1.7997651751	.7308863077
42	2.2654378963	.8449927801	1.8732594146	.7542392251
43	2.3512111662	.8705634524	1.9472300402	.7777437717
44	2.4374153027	.8959806346	2.0216652964	.8013948005
45	2.5240494769	.9213901561	2.0965476732	.8251524797
46	2.6111250454	.9469543226	2.1718557878	.8489638832
47	2.6986487974	.9727593913	2.2475677606	.8727857719
48	2.7866125504	.9987808797	2.3236644327	.8965994720
49	2.8749930534	1.0249135059	2.4001311612	.9204133658
50	2.9637596610	1.0510358028	2.4769577383	.9442541474
51	3.0528844366	1.0770713932	2.5541368154	.9681522855
52	3.1423496968	1.1030200980	2.6316617404	.9921285344
53	3.2321500785	1.1289499500	2.7095247730	1.0161867328
54	3.3222885959	1.1549565768	2.7877163051	1.0403147045
55	3.4127679868	1.1811051085	2.8662251936	1.0644915268
56	3.5035841652	1.2073989475	2.9450398480	1.0886971639
57	3.5947247454	1.2337868092	3.0241494772	1.1129201567
58	3.6861721692	1.2601960771	3.1035499444	1.1371605333
59	3.7779090205	1.2865714342	3.1832189405	1.1614275162
60	3.8699227488	1.3129003491	3.2631655171	1.1857338113
61	3.9622077430	1.3392157024	3.3433792790	1.2100893884
62	4.0547638824	1.3655753482	3.4238546225	1.2344974060

DIRICHLET 1 (MEANS AND VARIANCES),   B= 8

N	P=1/10 MEAN	VARIANCE	P=1/11 MEAN	VARIANCE
8	.0004032000	.0004030374	.0001880958	.0001880604
9	.0021772800	.0021725395	.0010772758	.0010761153
10	.0066528000	.0066085403	.0034743311	.0034622601
11	.0151683840	.0149383041	.0083244040	.0082551083
12	.0287653766	.0279379297	.0165233165	.0162502965
13	.0480207087	.0457147203	.0287667443	.0279392188
14	.0730164315	.0676850322	.0454651821	.0433980993
15	.1034038317	.0927114793	.0667233267	.0622713244
16	.1385228141	.1193505900	.0923691897	.0838406798
17	.1775473784	.1461539846	.1220160855	.1071593142
18	.2196302902	.1719485874	.1551418283	.1312192808
19	.2640214280	.1960233398	.1911707991	.1551155002
20	.3101419795	.2181832346	.2295464477	.1781717775
21	.3576093717	.2386777678	.2697851848	.2000069658
22	.4062199995	.2580455831	.3115072805	.2205366740
23	.4559042128	.2769291193	.3544451617	.2399214136
24	.5066700514	.2959073027	.3984332707	.2584814634
25	.5585503262	.3153799376	.4433858063	.2766012818
26	.6115633452	.3355197658	.4892692073	.2946435731
27	.6656920759	.3562899079	.5360754535	.3128871737
28	.7208811042	.3775089324	.5838005925	.3314955605
29	.7770466718	.3989369644	.6324308226	.3505155086
30	.8340931327	.4203560127	.6819364454	.3698995270
31	.8919294128	.4416248171	.7322724052	.3895421517
32	.9504808490	.4626994953	.7833831881	.4093193562
33	1.0096942856	.4836222520	.8352095970	.4291219711
34	1.0695367009	.5044885058	.8876952330	.4488772798
35	1.1299893531	.5254064673	.9407911984	.4685568307
36	1.1910401881	.5464623058	.9944583579	.4881720014
37	1.2526770543	.5676996610	1.0486672337	.5077612544
38	1.3148834051	.5891162661	1.1033961479	.5273740240
39	1.3776370360	.6106749014	1.1586284717	.5470558307
40	1.4409113913	.6323223005	1.2143498336	.5668378847
41	1.5046783453	.6540086552	1.2705459278	.5867326392
42	1.5689112034	.6757017403	1.3272012584	.6067349955
43	1.6335869151	.6973925462	1.3842988465	.6268275642
44	1.6986869785	.7190925225	1.4418206877	.6469877537
45	1.7641970407	.7408250678	1.4997486243	.6671945060
46	1.8301056037	.7626151206	1.5580652779	.6874330723
47	1.8964024268	.7844804954	1.6167547699	.7076970694
48	1.9630771744	.8064273164	1.6758030796	.7279879103
49	2.0301186590	.8284501384	1.7351980211	.7483123460
50	2.0975147607	.8505357510	1.7949289262	.7686791785
51	2.1652528769	.8726687011	1.8549861741	.7890961826
52	2.2333206214	.8948364158	1.9153607167	.8095679947
53	2.3017064760	.9170323580	1.9760437137	.8300953100
54	2.3704001824	.9392565884	2.0370263372	.8506753132
55	2.4393927923	.9615140687	2.0982997471	.8713029605
56	2.5086764232	.9838117048	2.1598551938	.8919725860
57	2.5782438558	1.0061553484	2.2216841822	.9126793333
58	2.6480881285	1.0285477466	2.2837786321	.9334200647
59	2.7182022566	1.0509879341	2.3461309834	.9541936146
60	2.7885791341	1.0734719887	2.4087342265	.9750004566
61	2.8592116048	1.0959946380	2.4715818599	.9958420004
62	2.9300926368	1.1185510203	2.5346677983	1.0167197857

DIRICHLET 1 (MEANS AND VARIANCES),   B= 8
P=1/12   P=1/13

N	MEAN	VARIANCE	MEAN	VARIANCE
8	.0000937714	.0000937626	.0000494281	.0000494256
9	.0005626286	.0005623120	.0003079749	.0003078801
10	.0018949644	.0018913735	.0010748413	.0010736860
11	.0047276431	.0047052925	.0027729893	.0027652998
12	.0097446454	.0096496873	.0058991485	.0058643486
13	.0175725658	.0172637708	.0109593472	.0108392399
14	.0286995008	.0278758395	.0184077685	.0180689226
15	.0434280911	.0415420920	.0286001522	.0277821835
16	.0618622166	.0580361669	.0417662797	.0400221032
17	.0839211992	.0768867815	.0580013093	.0546396202
18	.1093734751	.0974530855	.0772730395	.0713150875
19	.1378815219	.1190215422	.0994409984	.0896024238
20	.1690503542	.1409053108	.1242829393	.1089872886
21	.2024729501	.1625282978	.1515245029	.1289491571
22	.2377675608	.1834810851	.1808683137	.1490175352
23	.2746038666	.2035432033	.2120195471	.1688146836
24	.3127170510	.2226734724	.2447059586	.1880804769
25	.3519107262	.2409756374	.2786913753	.2066785606
26	.3920509896	.2586494665	.3137826054	.2245860291
27	.4330546171	.2759379295	.3498304869	.2418709404
28	.4748745222	.2930795943	.3867263332	.2586629712
29	.5174852467	.3102726910	.4243952928	.2751224795
30	.5608705551	.3276540956	.4627881647	.2914124408
31	.6050143571	.3452933816	.5018730300	.3076764359
32	.6498953523	.3631995565	.5416277530	.3240243920
33	.6954850921	.3813364534	.5820340327	.3405263518
34	.7417486883	.3996421169	.6230733050	.3572133556
35	.7886471681	.4180478289	.6647244791	.3740837030
36	.8361404809	.4364934425	.7069632401	.3911124426
37	.8841903321	.4549371068	.7497625116	.4082619336
38	.9327622913	.4733589290	.7930936167	.4254916302
39	.9818269211	.4917593569	.8369277059	.4427657716
40	1.0313599348	.5101538850	.8812371044	.4600582807
41	1.0813415853	.5285660189	.9259963467	.4773547705
42	1.1317555866	.5470203221	.9711827838	.4946520426
43	1.1825878862	.5655369287	1.0167767542	.5119557759
44	1.2338255550	.5841282883	1.0627613895	.5292772401
45	1.2854559688	.6027982833	1.1091221681	.5466298311
46	1.3374663542	.6215433433	1.1558463461	.5640260711
47	1.3898436777	.6403548547	1.2029223853	.5814754847
48	1.4425747946	.6592220600	1.2503394637	.5989835158
49	1.4956467436	.6781347206	1.2980871241	.6165514298
50	1.5490470710	.6970850332	1.3461550737	.6341769875
51	1.6027640971	.7160685652	1.3945331227	.6518555873
52	1.6567870690	.7350842351	1.4432112255	.6695815625
53	1.7111061878	.7541335629	1.4921795837	.6873493612
54	1.7657125258	.7732195244	1.5414287667	.7051544230
55	1.8205978741	.7923453501	1.5909498170	.7229936625
56	1.8757545617	.8115135480	1.6407343181	.7408655627
57	1.9311752903	.8307253167	1.6907744166	.7587699526
58	1.9868530090	.8499803905	1.7410628003	.7767075833
59	2.0427808438	.8692772558	1.7915926434	.7946796298
60	2.0989520778	.8886136062	1.8423575335	.8126872305
61	2.1553601703	.9079868799	1.8933513941	.8307311423
62	2.2119987962	.9273947404	1.9445684159	.8488115516

DIRICHLET 1 (MEANS AND VARIANCES),   8= 8

	P=1/14		P=1/15	
N	MEAN	VARIANCE	MEAN	VARIANCE
8	.0000273210	.0000273202	.0000157322	.0000157320
9	.0001756349	.0001756040	.0001038328	.0001038220
10	.0006314491	.0006310504	.0003828177	.0003826712
11	.0016756999	.0016728919	.0010406350	.0010395520
12	.0036616390	.0036482314	.0023268177	.0023214037
13	.0069778418	.0069291515	.0045325885	.0045120442
14	.0120067935	.0118626304	.0079645526	.0079011186
15	.0190875047	.0187231719	.0129175547	.0127506915
16	.0284873179	.0276758657	.0196509376	.0192648032
17	.0403849958	.0387548378	.0283707494	.0275661226
18	.0548649419	.0518592066	.0392189219	.0376823869
19	.0719210363	.0667659614	.0522693137	.0495437791
20	.0914678747	.0831568246	.0675297750	.0629909010
21	.1133569485	.1006544076	.0849489862	.0777916774
22	.1373953471	.1188620581	.1044266482	.0936645294
23	.1633648018	.1374018425	.1258255979	.1103046052
24	.1910392727	.1559460236	.1489845338	.1274097942
25	.2201997595	.1742389263	.1737302309	.1447036521
26	.2506455322	.1921078654	.1998883736	.1619531074
27	.2822014825	.2094634967	.2272924114	.1789797413
28	.3147217387	.2262912888	.2557901199	.1956643765
29	.3480900261	.2426366557	.2852478054	.2119455240
30	.3822174802	.2585866232	.3155523028	.2278128435
31	.4170387233	.2742507979	.3466110762	.2432971262
32	.4525070114	.2897439594	.3783508343	.2584584217
33	.4885891684	.3051719548	.4107151121	.2733738328
34	.5252608776	.3206218486	.4436612657	.2881262552
35	.5625027225	.3361565761	.4771572807	.3027950017
36	.6002971899	.3518137577	.5111787157	.3174488694
37	.6386266845	.3676078919	.5457060201	.3321418471
38	.6774724758	.3835348894	.5807223662	.3469113454
39	.7168144105	.3995778419	.6162120564	.3617785840
40	.7566311756	.4157129943	.6521594917	.3767506220
41	.7969008895	.4319150957	.6885486357	.3918234392
42	.8376018209	.4481615633	.7253628753	.4069854917
43	.8787130729	.4644351820	.7625851627	.4222212327
44	.9202151265	.4807253230	.8001983261	.4375142043
45	.9620901847	.4970278726	.8381854483	.4528494432
46	1.0043223062	.5133441954	.8765302342	.4682150784
47	1.0468973545	.5296795214	.9152173102	.4836031226
48	1.0898028063	.5460411333	.9542324274	.4990095556
49	1.1330274775	.5624366756	.9935625572	.5144338605
50	1.1765612199	.5788728104	1.0331958877	.5298782034
51	1.2203946344	.5953543432	1.0731217416	.5453464509
52	1.2645188338	.6118838418	1.1133304384	.5608431896
53	1.3089252711	.6284616904	1.1538131302	.5763728804
54	1.3536056361	.6450864704	1.1945616329	.5919392256
55	1.3985518151	.6617555317	1.2355682723	.6075447858
56	1.4437558950	.6784656214	1.2768257578	.6231908400
57	1.4892101983	.6952134584	1.3183270897	.6388774514
58	1.5349073281	.7119961755	1.3600654997	.6546036825
59	1.5808402113	.7288115902	1.4020344216	.6703678970
60	1.6270021285	.7456583014	1.4442274847	.6861680870
61	1.6733867263	.7625356359	1.4866385230	.7020021740
62	1.7199880114	.7794434895	1.5292615906	.7178682500

DIRICHLET 1 (MEANS AND VARIANCES),   B= 8

N	P=1/16 MEAN	VARIANCE	P=1/17 MEAN	VARIANCE
8	.0000093877	.0000093876	.0000057800	.0000057800
9	.0000633672	.0000633632	.0000397801	.0000397785
10	.0002387271	.0002386701	.0001527005	.0001526771
11	.0006625502	.0006621112	.0004315248	.0004313386
12	.0015112487	.0015089649	.0010015907	.0010005875
13	.0030007304	.0029917260	.0020224377	.0020183474
14	.0053704943	.0053416521	.0036786664	.0036651338
15	.0088650584	.0087864691	.0061677594	.0061297182
16	.0137157656	.0135276522	.0096868982	.0095930656
17	.0201252063	.0197202828	.0144204968	.0142125853
18	.0282555827	.0274578097	.0205296857	.0201084606
19	.0382215432	.0367632309	.0281444681	.0273534194
20	.0500874032	.0475873364	.0373588308	.0359668300
21	.0638682645	.0598137990	.0482287491	.0459134875
22	.0795342962	.0732701331	.0607727858	.0571069523
23	.0970173291	.0877429621	.0749748383	.0694168450
24	.1162188977	.1029956876	.0907885104	.0826791552
25	.1370189111	.1187865749	.1081425671	.0967083962
26	.1592842400	.1348854413	.1269469520	.1113103713
27	.1828766371	.1510875027	.1470989015	.1262943818
28	.2076595639	.1672234247	.1684887638	.1414838986
29	.2335036542	.1831651544	.1910052180	.1567249823
30	.2602906932	.1988276045	.2145396792	.1718920395
31	.2879161265	.2141666660	.2389897701	.1868908076
32	.3162902176	.2291743118	.2642618165	.2016587198
33	.3453380524	.2438717047	.2902724020	.2161630131
34	.3749986375	.2583012580	.3169490661	.2303970776
35	.4052233542	.2725185253	.3442302763	.2443756141
36	.4359740289	.2865846585	.3720648250	.2581291737
37	.4672208523	.3005599794	.4004108104	.2716986032
38	.4989403380	.3144990131	.4292343576	.2851298388
39	.5311134704	.3284471264	.4585082217	.2984693843
40	.5637241349	.3424387435	.4882103903	.3117606905
41	.5967578820	.3564969670	.5183227805	.3250415437
42	.6302010355	.3706343402	.5488300940	.3383424674
43	.6640401206	.3848544284	.5797188703	.3516860577
44	.6982615673	.3991538867	.6109767539	.3650871132
45	.7328516313	.4135247056	.6425919681	.3785533797
46	.7677964678	.4279563697	.6745529804	.3920867171
47	.8030822993	.4424377343	.7068483249	.4056844961
48	.8386956228	.4569584916	.7394665515	.4193410542
49	.8746234144	.4715101717	.7723962624	.4330490689
50	.9108533026	.4860866809	.8056262039	.4468007456
51	.9473736910	.5006844295	.8391453834	.4605887536
52	.9841738240	.5153021334	.8729431874	.4744068807
53	1.0212437984	.5299403900	.9070094843	.4882504130
54	1.0585745277	.5446011287	.9413347005	.5021162641
55	1.0961576717	.5592870315	.9759098658	.5160029025
56	1.1339855448	.5740009965	1.0107266263	.5299101307
57	1.1720510153	.5887456985	1.0457772303	.5438387734
58	1.2103474069	.6035232757	1.0810544913	.5577903272
59	1.2488684109	.6183351518	1.1165517354	.5717666185
60	1.2876080134	.6331819834	1.1522627407	.5857695019
61	1.3265604407	.6480637105	1.1881816735	.5998006234
62	1.3657201224	.6629796820	1.2243030284	.6138612573

DIRICHLET 1 (MEANS AND VARIANCES),   B= 9

N	P=1/ 9		P=1/10	
	MEAN	VARIANCE	MEAN	VARIANCE
9	.0009366567	.0009357794	.0003628800	.0003627483
10	.0046832835	.0046613504	.0019958400	.0019918566
11	.0133560308	.0131776473	.0061871040	.0061488237
12	.0286200661	.0278009579	.0142702560	.0140666158
13	.0513229630	.0486889164	.0273158646	.0265697081
14	.0814552992	.0748203334	.0459502243	.0438388012
15	.1183129463	.1043149930	.0703098108	.0653663413
16	.1607355335	.1348996218	.1000944296	.0900755348
17	.2073383321	.1643491481	.1346726200	.1165359054
18	.2567808781	.1910110821	.1732140523	.1432359536
19	.3080060636	.2141936096	.2148323584	.1688694867
20	.3603673032	.2342157367	.2587158860	.1925660170
21	.4136220513	.2521348165	.3042222723	.2140016621
22	.4678274017	.2693131466	.3509225421	.2333703032
23	.5231948642	.2869974370	.3985954711	.2512432398
24	.5799531513	.3060217160	.4471849910	.2683726934
25	.6382473318	.3266704733	.4967385589	.2854955501
26	.6980830381	.3486859139	.5473432579	.3031780728
27	.7593297565	.3714411242	.5990726931	.3217257284
28	.8217723362	.3942137637	.6519532148	.3411677597
29	.8851813098	.4164444219	.7059525220	.3613098564
30	.9493725509	.4378885113	.7609879290	.3818320033
31	1.0142390605	.4586328312	.8169472230	.4023989782
32	1.0797520810	.4790029680	.8737134051	.4227523227
33	1.1459388320	.4994146099	.9311857351	.4427633060
34	1.2128482650	.5202220386	.9892923666	.4624409190
35	1.2805156347	.5416017309	1.0479931380	.4819016141
36	1.3489398264	.5635162102	1.1072738981	.5013157789
37	1.4180798975	.5857638020	1.1671356495	.5208493325
38	1.4878678781	.6080805780	1.2275825484	.5406173334
39	1.5582292795	.6302460734	1.2886123205	.5606605925
40	1.6291022077	.6521540112	1.3502112495	.5809479360
41	1.7004488400	.6738308225	1.4123541204	.6013988423
42	1.7722569939	.6954056545	1.4750080037	.6219161768
43	1.8445328710	.7170488039	1.5381379091	.6424177256
44	1.9172880433	.7388997747	1.6017122333	.6628576187
45	1.9905269899	.7610169059	1.6657064109	.6832331552
46	2.0642401760	.7833668560	1.7301040121	.7035774922
47	2.1384042480	.8058514591	1.7948954346	.7239427475
48	2.2129877570	.8283547550	1.8600750381	.7443802421
49	2.2879591631	.8507890276	1.9256378825	.7649244102
50	2.3632938075	.8731235289	1.9915771341	.7855847047
51	2.4389775584	.8953885412	2.0578827888	.8063466389
52	2.5150062357	.9176562033	2.1245418373	.8271801048
53	2.5913811777	.9400055222	2.1915395297	.8480511898
54	2.6681034557	.9624889152	2.2588611273	.8689332519
55	2.7451693993	.9851144849	2.3264934891	.8898139022
56	2.8225689760	1.0078487275	2.3944260080	.9106963365
57	2.9002871051	1.0306351505	2.4626507070	.9315954753
58	2.9783068771	1.0534191058	2.5311616046	.9525309629
59	3.0566131703	1.0761687177	2.5999536755	.9735197657
60	3.1351952745	1.0988847865	2.6690218071	.9945708138
61	3.2140476176	1.1215969680	2.7383600792	1.0156830943
62	3.2931683024	1.1443476657	2.8079615383	1.0368472804
63	3.3725564296	1.1671724098	2.8778184491	1.0580498318

DIRICHLET 1 (MEANS AND VARIANCES),   B= 9

	P=1/11		P=1/12	
N	MEAN	VARIANCE	MEAN	VARIANCE
9	.0001538965	.0001538729	.0000703286	.0000703236
10	.0009093887	.0009085617	.0004395536	.0004393604
11	.0030079779	.0029989300	.0015311117	.0015287674
12	.0073577810	.0073036441	.0039285102	.0039130771
13	.0148568555	.0146361294	.0082906116	.0082218774
14	.0262376736	.0255492581	.0152520773	.0150194514
15	.0419692715	.0402078517	.0253376449	.0246956487
16	.0622225894	.0583509387	.0389054795	.0373918432
17	.0868865344	.0793372645	.0561227008	.0529729432
18	.1156183988	.1022552828	.0769689130	.0710456388
19	.1479148580	.1260749004	.1012608664	.0910160274
20	.1831917531	.1498126460	.1286911122	.1121749498
21	.2208611428	.1726773534	.1588738848	.1337956998
22	.2603949842	.1941672463	.1913918680	.1552270534
23	.3013678250	.2141027497	.2258382105	.1759661700
24	.3434756045	.2325967218	.2618494433	.1957009731
25	.3865324735	.2499779425	.2991267782	.2143186467
26	.4304510281	.2666905970	.3374452835	.2318836930
27	.4752129931	.2831924920	.3766522269	.2485938730
28	.5208374075	.2998704578	.4166571397	.2647246310
29	.5673522384	.3169851141	.4574167900	.2805725396
30	.6147734999	.3346500851	.4989182953	.2964065695
31	.6630937492	.3528436477	.5411631783	.3124332503
32	.7122796842	.3714447778	.5841544216	.3287786235
33	.7622768771	.3902818369	.6278876796	.3454867626
34	.8130187243	.4091814509	.6723469030	.3625319987
35	.8644365438	.4280072898	.7175038840	.3798402034
36	.9164682647	.4466825092	.7633207099	.3973137560
37	.9690640769	.4651942841	.8097538788	.4148551511
38	1.0221884536	.4835830179	.8567588545	.4323853686
39	1.0758188977	.5019216605	.9042940648	.4498547907
40	1.1299424144	.5202917449	.9523236930	.4672462134
41	1.1845510065	.5387622626	1.0008189963	.4845710287
42	1.2396374266	.5573757204	1.0497582221	.5018607039
43	1.2951920902	.5761432480	1.0991254492	.5191561483
44	1.3512015833	.5950481622	1.1489088076	.5364974504
45	1.4076487318	.6140555081	1.1990985522	.5539159117
46	1.4645138385	.6331241424	1.2496853852	.5714294812
47	1.5217765026	.6522179519	1.3006592855	.5890418118
48	1.5794174275	.6713136402	1.3520089454	.6067444030
49	1.6374197532	.6904038185	1.4037217741	.6245207789
50	1.6957696688	.7094955201	1.4557843222	.6423514480
51	1.7544562910	.7286053650	1.5081829309	.6602184740
52	1.8134709824	.7477532138	1.5609044084	.6781087914
53	1.8728063883	.7669561964	1.6139365762	.6960158251
54	1.9324554875	.7862245635	1.6672685880	.7139394098
55	1.9924108864	.8055600897	1.7208909975	.7318843598
56	2.0526644798	.8249569730	1.7747956075	.7498582567
57	2.1132074827	.8444045484	1.8289751731	.7678690796
58	2.1740307416	.8638907843	1.8834230495	.7859232168
59	2.2351251811	.8834055116	1.9381328667	.8040242165
60	2.2964822337	.9029425845	1.9930982900	.8221724039
61	2.3580941375	.9225005938	2.0483128974	.8403652856
62	2.4199540431	.9420821975	2.1037701698	.8585985018
63	2.4820559327	.9616924876	2.1594635692	.8768670147

DIRICHLET 1 (MEANS AND VARIANCES),   B= 9
P=1/13   P=1/14

N	MEAN	VARIANCE	MEAN	VARIANCE
9	.0000342194	.0000342183	.0000175635	.0000175632
10	.0002237425	.0002236924	.0001191808	.0001191666
11	.0008129647	.0008123038	.0004484962	.0004482950
12	.0021699048	.0021651963	.0012374129	.0012358817
13	.0047516889	.0047291104	.0027958439	.0027880272
14	.0090492985	.0089674087	.0054842711	.0054541939
15	.0155282393	.0152871131	.0096773212	.0095836707
16	.0245780251	.0239739458	.0157267249	.0154793950
17	.0364775010	.0351468929	.0239299392	.0233572972
18	.0513778577	.0487383958	.0345079653	.0333172242
19	.0693018356	.0645013665	.0475934779	.0453289748
20	.0901559619	.0820402408	.0632286961	.0592344912
21	.1137520970	.1008601581	.0813714584	.0747650258
22	.1398345037	.1204260946	.1019075030	.0915702317
23	.1681088376	.1402227436	.1246668152	.1092547588
24	.1982698376	.1598064477	.1494419430	.1274172276
25	.2300251057	.1788425300	.1760063688	.1456865319
26	.2631131798	.1971244354	.2041313165	.1637512510
27	.2973150281	.2145744165	.2335997739	.1813793479
28	.3324589962	.2312283496	.2642169609	.1984269830
29	.3684199863	.2472091876	.2958169430	.2148368680
30	.4051141757	.2626944131	.3282655160	.2306278601
31	.4424908428	.2778827640	.3614598282	.2458783294
32	.4805228936	.2929646922	.3953254458	.2607061742
33	.5191975038	.3080997500	.4298116857	.2752482862
34	.5585079722	.3234026266	.4648860509	.2896418587
35	.5984474883	.3389380994	.5005285263	.3040093143
36	.6390051149	.3547238977	.5367263472	.3184479057
37	.6801639235	.3707395528	.5734696714	.3330243208
38	.7219009488	.3869388022	.6107483907	.3477739790
39	.7641884557	.4032630441	.6485501396	.3627042012
40	.8069959516	.4196536461	.6868594093	.3778001174
41	.8502924087	.4360614977	.7256575661	.3930320430
42	.8940482637	.4524529196	.7649235176	.4083631064
43	.9382369037	.4688117738	.8046347490	.4237561070
44	.9828355041	.4851382387	.8447684785	.4391788729
45	1.0278252162	.5014451471	.8853027241	.4546077254
46	1.0731908174	.5177529936	.9262171405	.4700289828
47	1.1189199920	.5340847119	.9674935514	.4854387119
48	1.1650024409	.5504611404	1.0091161619	.5008411307
49	1.2114289964	.5668977957	1.0510714887	.5162461726
50	1.2581908829	.5834032341	1.0933480743	.5316667336
51	1.3052792040	.5999789547	1.1359360698	.5471160655
52	1.3526846849	.6166205489	1.1788267715	.5626056638
53	1.4003976442	.6333196426	1.2220121818	.5781438578
54	1.4484081421	.6500661329	1.2654846490	.5937351660
55	1.4967062286	.6668502664	1.3092366147	.6093803567
56	1.5452822190	.6836642242	1.3532604762	.6250770609
57	1.5941269347	.7005030315	1.3975485539	.6408207324
58	1.6432318684	.7173647634	1.4420931391	.6566057402
59	1.6925892548	.7342501536	1.4868865925	.6724264002
60	1.7421920514	.7511617936	1.5319214623	.6882777997
61	1.7920338487	.7681031529	1.5771905959	.7041563310
62	1.8421087373	.7850776348	1.6226872257	.7200599096
63	1.8924111634	.8020878348	1.6684050172	.7359879070

DIRICHLET 1 (MEANS AND VARIANCES),   B= 9
P=1/15   P=1/16

N	MEAN	VARIANCE	MEAN	VARIANCE
9	.0000094393	.0000094393	.0000052806	.0000052806
10	.0000660754	.0000660710	.0000379543	.0000379529
11	.0002561208	.0002560552	.0001508890	.0001508662
12	.0007268293	.0007263010	.0004386984	.0004385059
13	.0016868224	.0016839770	.0010420080	.0010409222
14	.0033942821	.0033827609	.0021437893	.0021391935
15	.0061364444	.0060987885	.0039588108	.0039431387
16	.0102051106	.0101009663	.0067185751	.0066734358
17	.0158725993	.0156206599	.0106544923	.0105409741
18	.0233714160	.0228252099	.0159817433	.0157263325
19	.0328795360	.0317986650	.0228856381	.0223169486
20	.0445119824	.0425318305	.0315115445	.0305189621
21	.0583184820	.0549223886	.0419588097	.0402000136
22	.0742864100	.0687845838	.0542785975	.0513385219
23	.0923479205	.0838668586	.0684752183	.0638042248
24	.1123900265	.0998749912	.0845103259	.0774141083
25	.1342663809	.1164978041	.1023092483	.0919473326
26	.1578095886	.1334324396	.1217686914	.1071624353
27	.1828430247	.1504065243	.1427650872	.1228149870
28	.2091913359	.1671951937	.1651629222	.1386739890
29	.2366890368	.1836317849	.1888224944	.1545356042
30	.2651868622	.1996118873	.2136066661	.1702322418
31	.2945557813	.2150912303	.2393863237	.1856435046
32	.3246887883	.2300784983	.2660443923	.2006879813
33	.3555007535	.2446245433	.2934783829	.2153312798
34	.3869267340	.2588096128	.3216015623	.2295759941
35	.4189191987	.2727301589	.3503429195	.2434554909
36	.4514446317	.2864865789	.3796461647	.2570254627
37	.4844799399	.3001729227	.4094680223	.2703551610
38	.5180090216	.3138692283	.4397760896	.2835191069
39	.5520197629	.3276367703	.4705465091	.2965899058
40	.5865016339	.3415161591	.5017616699	.3096325904
41	.6214439574	.3555279503	.5334081099	.3227007135
42	.6568348465	.3696752172	.5654747338	.3358342157
43	.6926607374	.3839474357	.5979514173	.3490589323
44	.7289064037	.3983250029	.6308280139	.3623874759
45	.7655553127	.4127837714	.6640937495	.3758211541
46	.8025901842	.4272990923	.6977369530	.3893525423
47	.8399936218	.4418490104	.7317450572	.4029683419
48	.8777487121	.4564164214	.7661047930	.4166521924
49	.9158395172	.4709901548	.8008025019	.4303871715
50	.9542514167	.4855650794	.8358244948	.4441577990
51	.9929712841	.5001414246	.8711574038	.4579514432
52	1.0319875099	.5147235643	.9067884812	.4717591096
53	1.0712898958	.5293185285	.9427058231	.4855756565
54	1.1108694613	.5439344841	.9788985031	.4993995348
55	1.1507181992	.5585793870	1.0153566198	.5132321784
56	1.1908288222	.5732599390	1.0520712671	.5270771849
57	1.2311945292	.5879809251	1.0890344461	.5409394202
58	1.2718088133	.6027449365	1.1262389377	.5548241621
59	1.3126653249	.6175524391	1.1636781566	.5687363725
60	1.3537577910	.6324021080	1.2013460047	.5826801534
61	1.3950799855	.6472913298	1.2392367384	.5966584128
62	1.4366257403	.6622167729	1.2773448589	.6106727357
63	1.4783889843	.6771749367	1.3156650321	.6247234346

DIRICHLET 1 (MEANS AND VARIANCES),   B= 9
P=1/17   P=1/18

N	MEAN	VARIANCE	MEAN	VARIANCE
9	.0000030600	.0000030600	.0000018294	.0000018294
10	.0000225001	.0000224996	.0000137206	.0000137204
11	.0000914297	.0000914213	.0000568302	.0000568270
12	.0002714801	.0002714064	.0001718881	.0001718585
13	.0006580092	.0006575763	.0004241068	.0004239269
14	.0013803537	.0013784483	.0009050996	.0009042804
15	.0025971075	.0025903625	.0017313853	.0017283876
16	.0044874526	.0044673154	.0030397947	.0030305544
17	.0072400994	.0071876804	.0049805793	.0049557731
18	.0110414768	.0109195644	.0077092351	.0076498035
19	.0160645925	.0158065432	.0113780489	.0112485970
20	.0224596054	.0219553136	.0161282157	.0158681502
21	.0303467471	.0294264707	.0220831568	.0215957441
22	.0398118596	.0382292130	.0293434275	.0284833285
23	.0509045282	.0483203303	.0379833881	.0365435662
24	.0636385749	.0596073943	.0480496383	.0457487823
25	.0779945471	.0719556627	.0595610823	.0560327932
26	.0939237560	.0851978873	.0725104041	.0672953411
27	.1113533920	.0991459936	.0868666768	.0794086480
28	.1301922522	.1136035002	.1025788070	.0922254571
29	.1503366451	.1283775755	.1195795093	.1055878465
30	.1716760977	.1432897645	.1377895255	.1193360951
31	.1940985539	.1581846432	.1571218277	.1333169404
32	.2174948387	.1729359261	.1774855869	.1473906799
33	.2417622373	.1874498374	.1987897330	.1614367238
34	.2668071227	.2016658191	.2209459791	.1753573685
35	.2925466360	.2155548685	.2438712368	.1890797334
36	.3189094815	.2291159560	.2674893897	.2025559483
37	.3458359496	.2423710679	.2917324370	.2157618020
38	.3732773091	.2553594527	.3165410542	.2286941494
39	.4011947276	.2681316222	.3418646399	.2413674228
40	.4295578828	.2807435973	.3676609404	.2538096093
41	.4583434162	.2932517895	.3938953503	.2660580415
42	.4875333638	.3057087984	.4205399893	.2781553098
43	.5171136743	.3181602884	.4475726530	.2901455497
44	.5470728948	.3306429957	.4749757217	.3020712913
45	.5774010784	.3431838182	.5027351011	.3139709901
46	.6080889380	.3557998647	.5308392506	.3258772929
47	.6391272488	.3684992807	.5592783395	.3378160287
48	.6705064819	.3812826392	.5880435547	.3498058695
49	.7022166369	.3941446733	.6171265677	.3618585639
50	.7342472334	.4070761395	.6465191590	.3739796219
51	.7665874189	.4200656285	.6762129852	.3861693191
52	.7992261471	.4331011756	.7061994677	.3984238864
53	.8321523900	.4461715700	.7364697807	.4107367593
54	.8653553506	.4592673071	.7670149096	.4230997825
55	.8988246507	.4723811686	.7978257563	.4355042847
56	.9255004785	.4855084534	.8288932675	.4479419637
57	.9665236865	.4986469098	.8602085669	.4604055510
58	1.0007358402	.5117964357	.8917630755	.4728892432
59	1.0351792187	.5249586241	.9235486115	.4853889147
60	1.0698467775	.5381362297	.9555574605	.4979021349
61	1.1047320829	.5513326256	.9877824175	.5104280302
62	1.1398292283	.5645513058	1.0202167987	.5229670327
63	1.1751327437	.5777954760	1.0528544290	.5355205606

DIRICHLET 1 (MEANS AND VARIANCES),   B=10

N	P=1/10 MEAN	VARIANCE	P=1/11 MEAN	VARIANCE
10	.0003628800	.0003627483	.0001399059	.0001398864
11	.0019958400	.0019918566	.0008394357	.0008387310
12	.0061871040	.0061488237	.0028108377	.0028029369
13	.0142702560	.0140666158	.0069444226	.0068961975
14	.0273158646	.0265697081	.0141375434	.0139376733
15	.0459502243	.0438388012	.0251376618	.0245057598
16	.0703098108	.0653663413	.0404391252	.0388038023
17	.1000944296	.0900755348	.0602422744	.0566131428
18	.1346726200	.1165359054	.0844643289	.0773301061
19	.1732015476	.1432027715	.1127841660	.1000638979
20	.2147610820	.1686862773	.1447070324	.1237739703
21	.2584901773	.1920056289	.1796401681	.1474280100
22	.3036972818	.2127458426	.2169706302	.1701550718
23	.3499206252	.2310667391	.2561349300	.1913631285
24	.3969320997	.2475763835	.2966706456	.2107968146
25	.4446957108	.2631249936	.3382440188	.2285277985
26	.4933004450	.2785850753	.3806530326	.2448881402
27	.5428876053	.2946688049	.4238103037	.2603686263
28	.5935875480	.3118100468	.4677129288	.2755070623
29	.6454739963	.3301167537	.5124069747	.2907876737
30	.6985425141	.3494002934	.5579534047	.3065664882
31	.7527145220	.3692727089	.6044006936	.3230314137
32	.8078610660	.3892807315	.6517674290	.3401995135
33	.8638357829	.4090349124	.7000359379	.3579472533
34	.9205061006	.4282992032	.7491557291	.3760633293
35	.9777748165	.4470242935	.7990538149	.3943099742
36	1.0355887014	.4653270729	.8496481598	.4124786147
37	1.0939348315	.4834318064	.9008606579	.4304291495
38	1.1528279628	.5015938263	.9526269419	.4481074743
39	1.2122932694	.5200251087	1.0049015718	.4655414025
40	1.2723498035	.5388415108	1.0576584120	.4828195365
41	1.3329993113	.5580447639	1.1108870435	.5000604107
42	1.3942226865	.5775401157	1.1645867487	.5173801795
43	1.4559835777	.5971790214	1.2187598614	.5348662805
44	1.5182366299	.6168102619	1.2734060930	.5525621112
45	1.5809370564	.6363234257	1.3285189260	.5704644923
46	1.6440485950	.6556739598	1.3840844778	.5885324302
47	1.7075479507	.6748860096	1.4400825891	.6067032730
48	1.7714250692	.6940354931	1.4964894215	.6249112588
49	1.8356796422	.7132198802	1.5532806397	.6431037106
50	1.9003154252	.7325252165	1.6104342944	.6612514340
51	1.9653344751	.7520011306	1.6679327529	.6793517536
52	2.0307331114	.7716506434	1.7257633702	.6974245938
53	2.0965005271	.7914357721	1.7839179523	.7155036234
54	2.1626199832	.8112947716	1.8423913490	.7336253993
55	2.2290717625	.8311640349	1.9011796757	.7518195080
56	2.2958367062	.8509975143	1.9602786647	.7701020117
57	2.3628992008	.8707783791	2.0196825291	.7884733312
58	2.4302487991	.8905204489	2.0793835266	.8069204141
59	2.4978800929	.9102597342	2.1393722024	.8254219936
60	2.5657910994	.9300398308	2.1996381274	.8439551474
61	2.6339809025	.9498967213	2.2601708509	.8625012978
62	2.7024474237	.9698480397	2.3209607824	.8810502026
63	2.7711860093	.9898896331	2.3819997805	.8996011973
64	2.8401891394	1.0099994710	2.4432813392	.9181617621

DIRICHLET 1 (MEANS AND VARIANCES),   B=10

N	P=1/12 MEAN	VARIANCE	P=1/13 MEAN	VARIANCE
10	.0000586071	.0000586037	.0000263226	.0000263220
11	.0003760625	.0003759211	.0001781841	.0001781523
12	.0013386035	.0013368116	.0006664769	.0006660327
13	.0034968591	.0034846311	.0018229599	.0018196367
14	.0074916529	.0074355280	.0040758284	.0040592160
15	.0139586732	.0137638287	.0079015746	.0078391397
16	.0234411496	.0228916621	.0137682397	.0135786753
17	.0363280912	.0350083610	.0220834593	.0215957801
18	.0528235992	.0500332666	.0331557990	.0320564920
19	.0729443026	.0676234313	.0471726817	.0449474198
20	.0965383405	.0872199287	.0641943577	.0600736922
21	.1233191477	.1081231037	.0841613125	.0770807440
22	.1529081588	.1295849141	.1069118003	.0954955652
23	.1848809783	.1509040962	.1322061478	.1147810250
24	.2188116522	.1715088037	.1597545925	.1343954216
25	.2543101130	.1910135816	.1892455747	.1538488487
26	.2910490964	.2092431434	.2203716888	.1727487435
27	.3287787232	.2262225940	.2528510148	.1908291386
28	.3673289957	.2421401054	.2864423192	.2079611978
29	.4066021788	.2572919821	.3209535120	.2241458049
30	.4465580976	.2720212015	.3562436229	.2394915618
31	.4871957792	.2866595426	.3922192780	.2541831505
32	.5285346900	.3014811928	.4288271282	.2684455523
33	.5705982604	.3166729075	.4660438874	.2825092849
34	.6134015515	.3323227165	.5038656140	.2965808847
35	.6569439698	.3484261186	.5422976474	.3108215585
36	.7012069888	.3649060709	.5813462655	.3253354648
37	.7461560599	.3816412846	.6210127082	.3401676257
38	.7917453831	.3984967042	.6612897841	.3553101929
39	.8379240158	.4153505878	.7021608966	.3707148294
40	.8846419009	.4321140875	.7436010380	.3863084498
41	.9318547175	.4487412168	.7855791232	.4020094990
42	.9795269006	.4652291360	.8280609840	.4177423207
43	1.0276326409	.4816104170	.8710123982	.4334478386
44	1.0761550815	.4979401084	.9144016568	.4490896184
45	1.1250842169	.5142809042	.9582013544	.4646552260
46	1.1744141407	.5306895274	1.0023892705	.4801535271
47	1.2241402843	.5472067035	1.0469483840	.4956090861
48	1.2742571648	.5638520291	1.0918661874	.5110550752
49	1.3247569606	.5806238850	1.1371335495	.5265260972
50	1.3756290109	.5975035454	1.1827433948	.5420520973
51	1.4268601385	.6144619539	1.2286894447	.5576541650
52	1.4784355536	.6314673706	1.2749652083	.5733425788
53	1.5303400303	.6484922154	1.3215633288	.5891170230
54	1.5825590534	.6655178704	1.3684753166	.6049685536
55	1.6350796992	.6825368105	1.4156916253	.6208826761
56	1.6878911115	.6995520792	1.4632019821	.6368428147
57	1.7409845468	.7165746589	1.5109958564	.6528335128
58	1.7943530536	.7336196310	1.5590629530	.6688428606
59	1.8479909178	.7507021280	1.6073936314	.6848638636
60	1.9018930291	.7678339573	1.6559791891	.7008946982
61	1.9560543126	.7850214992	1.7048119803	.7169380033
62	2.0104693298	.8022651158	1.7538853772	.7329995002
63	2.0651320983	.8195599507	1.8031936087	.7490863012
64	2.1200361250	.8368977245	1.8527315250	.7652052651

DIRICHLET 1 (MEANS AND VARIANCES),   B=10

N	$p=1/14$ MEAN	VARIANCE	$p=1/15$ MEAN	VARIANCE
10	.0000125453	.0000125452	.0000062929	.0000062929
11	.0000887135	.0000887057	.0000461479	.0000461458
12	.0003457011	.0003455816	.0001861298	.0001860952
13	.0009826381	.0009816725	.0005465961	.0005462974
14	.0022777851	.0022725968	.0013067470	.0013050394
15	.0045681323	.0045472644	.0026984370	.0026911555
16	.0082175529	.0081500248	.0049904420	.0049655374
17	.0135812472	.0133967969	.0084668877	.0083951995
18	.0209731700	.0205332962	.0134040288	.0132243608
19	.0306408600	.0297019977	.0200489480	.0196469877
20	.0427496740	.0409221962	.0286026162	.0277845208
21	.0573765429	.0540851000	.0392085531	.0376714092
22	.0745121825	.0689637539	.0519473408	.0492498415
23	.0940701065	.0852358278	.0668365563	.0623738751
24	.1159005746	.1025159274	.0838352841	.0768220984
25	.1398075775	.1203931105	.1028521651	.0923170809
26	.1655670262	.1384688401	.1237558653	.1085492146
27	.1929444639	.1563908088	.1463868541	.1252021889
28	.2217108626	.1738789337	.1705694475	.1419773332
29	.2516554198	.1907411807	.1961231944	.1586143850
30	.2825946877	.2068784662	.2228728527	.1749068660
31	.3143778142	.2222793837	.2506564131	.1907110393
32	.3468880865	.2370066787	.2793308603	.2059482572
33	.3800412984	.2511781210	.3087755895	.2206012607
34	.4137816900	.2649446888	.3388936063	.2347055634
35	.4480763178	.2784688479	.3696108013	.2483374164
36	.4829087184	.2919052765	.4008737086	.2615999799
37	.5182726381	.3053857667	.4326462148	.2746092780
38	.5541664540	.3190093094	.4649057000	.2874812969
39	.5905887138	.3328376414	.4976390493	.3003212796
40	.6275350202	.3468958703	.5308389098	.3132158997
41	.6649962922	.3611772593	.5645004734	.3262286130
42	.7029582727	.3756509042	.5986189613	.3393981295
43	.7414020364	.3902708793	.6331878901	.3527396440
44	.7803051866	.4049854764	.6681981035	.3662482365
45	.8196434124	.4197453669	.7036374848	.3799037237
46	.8593921063	.4345098492	.7394912132	.3936762017
47	.8995277992	.4492507222	.7757424013	.4075315656
48	.9400292469	.4639537144	.8123729440	.4214364140
49	.9808780855	.4786177264	.8493644257	.4353619064
50	1.0220590488	.4932523932	.8866989571	.4492863334
51	1.0635598014	.5078746046	.9243598516	.4631963418
52	1.1053704867	.5225046577	.9623320871	.4770869209
53	1.1474831041	.5371626429	1.0006025381	.4909603802
54	1.1898908357	.5518655310	1.0391599926	.5048246253
55	1.2325874214	.5666252478	1.0779949928	.5186910690
56	1.2755666579	.5814478336	1.1170995505	.5325724968
57	1.3188220636	.5963336142	1.1564667956	.5464811586
58	1.3623467181	.6112781774	1.1960906103	.5604272798
59	1.4061332619	.6262738682	1.2359652957	.5744181027
60	1.4501740158	.6413114909	1.2760853013	.5884574816
61	1.4944611780	.6563819288	1.3164450369	.6025459849
62	1.5389870464	.6714774551	1.3570387710	.6166813990
63	1.5837442261	.6865925933	1.3978606076	.6308594976
64	1.6287257899	.7017244779	1.4389045269	.6450749321

DIRICHLET 1 (MEANS AND VARIANCES),   B=10

N	$P=1/16$ MEAN	VARIANCE	$P=1/17$ MEAN	VARIANCE
10	.0000033004	.0000033004	.0000018000	.0000018000
11	.0000249591	.0000249585	.0000139765	.0000139763
12	.0001036653	.0001036545	.0000595372	.0000595337
13	.0003130610	.0003129630	.0001842095	.0001841755
14	.0007686528	.0007680620	.0004629152	.0004627009
15	.0016281131	.0016254624	.0010025849	.0010015797
16	.0030847992	.0030752832	.0019405394	.0019367737
17	.0053559091	.0053272234	.0034387236	.0034268988
18	.0086675236	.0085923976	.0056748270	.0056426234
19	.0132389092	.0130636405	.0088316793	.0087536807
20	.0192680971	.0188968415	.0130863283	.0129150774
21	.0269201516	.0261955052	.0185999861	.0182540416
22	.0363189139	.0350001613	.0255097018	.0248590572
23	.0475424587	.0452835816	.0339222638	.0327720161
24	.0606220992	.0569520838	.0439105283	.0419841433
25	.0755445066	.0698525602	.0555121191	.0524359517
26	.0922563549	.0837843157	.0687302755	.0640210955
27	.1106708269	.0985143654	.0835365093	.0765936355
28	.1306752982	.1137945771	.0998746704	.0899779523
29	.1521395402	.1293789576	.1176659959	.1039803423
30	.1749238388	.1450394924	.1368147199	.1184012426
31	.1988865151	.1605792254	.1572138455	.1330470480
32	.2238904396	.1758416638	.1787507264	.1477406031
33	.2498082584	.1907160502	.2013121631	.1623296478
34	.2765261800	.2051385002	.2247887855	.1766927478
35	.3039462938	.2190893959	.2490785706	.1907425059
36	.3319875037	.2325877261	.2740894184	.2044261061
37	.3605852464	.2456832506	.2997407766	.2177234598
38	.3896902295	.2584474412	.3259643698	.2306433844
39	.4192664553	.2709641246	.3527041360	.2432183508
40	.4492888089	.2833206475	.3799155088	.2554983759
41	.4797404698	.2956002219	.4075642048	.2675446253
42	.5106103789	.3078759090	.4356246823	.2794232363
43	.5418909409	.3202064977	.4640784309	.2911997820
44	.5735760938	.3326343255	.4929122369	.3029346899
45	.6056598199	.3451849177	.5221165442	.3146798100
46	.6381351223	.3578681776	.5516840019	.3264762117
47	.6709934474	.3706807611	.5816082625	.3383531835
48	.7042244972	.3836092192	.6118830600	.3503283180
49	.7378163558	.3966334868	.6425015754	.3624084996
50	.7717558385	.4097303299	.6734560709	.3745915652
51	.8060289727	.4228764309	.7047377590	.3868683934
52	.8406215268	.4360508773	.7363368603	.3992251762
53	.8755195165	.4492369183	.7682427989	.4116456549
54	.9107096342	.4624229460	.8004444835	.4241131369
55	.9461795700	.4756027439	.8329306255	.4366121604
56	.9819182078	.4887751116	.8656900539	.4491297250
57	1.0179156979	.5019430191	.8987119947	.4616560575
58	1.0541634211	.5151124663	.9319862928	.4741849281
59	1.0906538681	.5282912229	.9655035650	.4867135703
60	1.1273804602	.5414876062	.9992552808	.4992422839
61	1.1643373429	.5547094234	1.0332337765	.5117738153
62	1.2015191761	.5679631629	1.0674322117	.5243126134
63	1.2389209437	.5812534815	1.1018444831	.5368640565
64	1.2765377980	.5945829885	1.1364651104	.5494337266

DIRICHLET 1 (MEANS AND VARIANCES),   B=10
p=1/18   p=1/19

N	MEAN	VARIANCE	MEAN	VARIANCE
10	.0000010163	.0000010163	.0000005919	.0000005919
11	.0000080742	.0000080742	.0000047973	.0000047972
12	.0000351609	.0000351597	.0000212992	.0000212987
13	.0001111204	.0001111081	.0000685837	.0000685790
14	.0002850018	.0002849205	.0001791104	.0001790783
15	.0006295005	.0006291043	.0004025744	.0004024123
16	.0012416587	.0012401170	.0008075451	.0008068930
17	.0022406232	.0022356028	.0014811312	.0014789375
18	.0037628210	.0037486622	.0025266880	.0025203039
19	.0059552731	.0059198079	.0040598850	.0040434023
20	.0089679243	.0088875010	.0062036305	.0061651456
21	.0129458514	.0127782612	.0090824219	.0089999333
22	.0180220779	.0176973169	.0128166658	.0126524112
23	.0243115319	.0237206483	.0175174301	.0172106317
24	.0319064799	.0308890952	.0232819757	.0227401691
25	.0408735845	.0392049805	.0301902892	.0292796384
26	.0512525822	.0486314584	.0383027296	.0368379301
27	.0630564623	.0590945577	.0476588019	.0453933251
28	.0762729524	.0704876647	.0582769958	.0548945016
29	.0908670639	.0826780044	.0701555766	.0652633022
30	.1067844307	.0955145365	.0832741747	.0763990029
31	.1239551634	.1088366072	.0975960018	.0881837283
32	.1422979573	.1224826767	.1130705138	.1004885937
33	.1617242076	.1362984913	.1296363412	.1131801283
34	.1821419223	.1501441625	.1472243183	.1261265445
35	.2034592577	.1638997512	.1657604599	.1392034640
36	.2255875468	.1774691114	.1851687545	.1522987822
37	.2484437354	.1907819056	.2053736721	.1653164429
38	.2719521854	.2037938524	.2263023058	.1781789916
39	.2960458447	.2164853901	.2478861019	.1908288742
40	.3206668207	.2288590345	.2700621543	.2032285342
41	.3457664197	.2409357660	.2927740658	.2153594375
42	.3713047403	.2527508080	.3159724023	.2272202042
43	.3972499163	.2643491523	.3396147781	.2388240680
44	.4235771132	.2757811577	.3636656292	.2501958924
45	.4502673791	.2870984981	.3880957356	.2613689775
46	.4773064428	.2983506744	.4128815569	.2723818690
47	.5046835379	.3095822379	.4380044493	.2832753559
48	.5323903203	.3208308009	.4634498218	.2940898052
49	.5604199234	.3321258484	.4892062890	.3048629431
50	.5887661831	.3434883061	.5152648623	.3156281474
51	.6174230451	.3549307739	.5416182192	.3264132786
52	.6463841536	.3664583018	.5682600715	.3372400371
53	.6756426088	.3780695643	.5951846502	.3481238061
54	.7051908702	.3897582828	.6223863106	.3590739134
55	.7350207794	.4015147508	.6498592564	.3700942283
56	.7651236707	.4133273320	.6775973711	.3811840038
57	.7954905388	.4251838220	.7055941459	.3923388671
58	.8261122356	.4370725963	.7338426835	.4035518717
59	.8569796709	.4489834907	.7623357606	.4148145293
60	.8880839972	.4609083940	.7910659302	.4261177563
61	.9194167642	.4728415559	.8200256459	.4374526849
62	.9509700343	.4847796373	.8492073922	.4488113053
63	.9827364550	.4967215441	.8786038089	.4601869245
64	1.0147092900	.5086680972	.9082077991	.4715744393

TABLE E

This table gives the value of the Generalized Stirling Numbers defined in
Section 3 by

$$S_{n,r}^{(b)} = \frac{b^n}{b!} I_{1/b}^{(b)}(r,n)$$

for $b = 2(1)23,$

for $n = b(1)25,$

for $r = 1(1)[n/b].$

SOBEL, UPPULURI AND FRANKOWSKI

GENERALIZED STIRLING NUMBERS, B = 2

N → R	2	3	4	5	6	7	8	9	10	11	12
1	1	3	7	15	31	63	127	255	511	1023	2047
2			3	10	25	56	119	246	501	1012	2035
3					10	35	91	210	456	957	1969
4							35	126	336	792	1749
5									126	462	1254
6											462

N → R	13	14	15	16	17	18	19	20
1	4095	8191	16383	32767	65535	131071	262143	524287
2	4082	8177	16368	32751	65518	131053	262124	524267
3	4004	8086	16263	32631	65382	130900	261953	524077
4	3718	7722	15808	32071	64702	130084	260984	522937
5	3003	6721	14443	30251	62322	127024	257108	518092
6	1716	4719	11440	25883	56134	118456	245480	502588
7		1716	6435	17875	43758	99892	218348	463828
8				6435	24310	68068	167960	386308
9						24310	92378	260338
10								92378

N → R	21	22	23	24	25
1	1048575	2097151	4194303	8388607	16777215
2	1048554	2097129	4194280	8388583	16777190
3	1048344	2096898	4194027	8388307	16776890
4	1047014	2095358	4192256	8386283	16774590
5	1041029	2088043	4183401	8375657	16761940
6	1020680	2061709	4149752	8333153	16708810
7	966416	1987096	4048805	8198557	16531710
8	850136	1816552	3803648	7852453	16051010
9	646646	1496782	3313334	7116982	14969435
10	352716	999362	2496144	5809478	12926460
11		352716	1352078	3848222	9657700
12				1352078	5200300

GENERALIZED STIRLING NUMBERS, B = 3

N → R	3	4	5	6	7	8	9	10	11	12
1	1	6	25	90	301	966	3025	9330	28501	86526
2				15	105	490	1918	6825	22935	74316
3							280	2100	10395	42735
4										5775

N → R	13	14	15	16	17	18
1	261625	788970	2375101	7141686	21457825	64439010
2	235092	731731	2252341	6879678	20900922	63259533
3	158301	549549	1827826	5903898	18682014	58257810
4	45045	231231	981981	3741738	13307294	45172842
5			126126	1009008	5309304	23075052
6						2858856

GENERALIZED STIRLING NUMBERS, B = 3 (Cont'd)

N →	19	20	21	22	23
R					
1	193448101	580606446	1742343625	5228079450	15686335501
2	190957923	575363776	1731333808	5205011031	15638101281
3	179765973	550478241	1676305723	5083927299	15372843684
4	148417854	476330361	1502751363	4681265809	14445712787
5	89791416	325355316	1122632043	3740893299	12151860457
6	23279256	124710300	551496660	2181183147	8021782197
7			66512160	548725320	2979883335

N →	24	25
R		
1	47063200806	141197991025
2	46962537810	140988276150
3	46383762084	139730030100
4	44263259788	134908935700
5	38732272711	121702063725
6	28051272535	94587699525
7	13356888831	53515730125
8	1577585295	13146544125

GENERALIZED STIRLING NUMBERS, B = 4

N →	4	5	6	7	8	9	10	11	12
R									
1	1	10	65	350	1701	7770	34105	145750	611501
2					105	1260	9450	56980	302995
3									15400

N →	13	14	15	16	17	18
R						
1	2532530	10391745	42355950	171798901	694337290	2798806985
2	1487200	6914908	30950920	134779645	575156036	2417578670
3	200200	1611610	10335325	57962905	297797500	1439774336
4				2627625	35735700	300179880

N →	19	20	21	22
R				
1	11259666950	45232115901	181509070050	727778623825
2	10046531276	41388056231	169371383384	689568172832
3	6662393738	29844199346	130445781284	559533979466
4	2002016016	11633808186	61705547904	306902071476
5		488864376	6844101264	59152589496

N →	23	24	25
R			
1	2916342574750	11681056634501	46771289738810
2	2796362035104	11305163394129	45595968007260
3	2365296391535	9885290914059	40944327590760
4	1456171781064	6668268193587	29714641533060
5	405404363364	2416720442136	13124107356360
6		961976455444	1374252079200

SOBEL, UPPULURI AND FRANKOWSKI

GENERALIZED STIRLING NUMBERS, B = 5

N →	5	6	7	8	9	10	11	12	13
R									
1	1	15	140	1050	6951	42525	246730	1379400	7508501
2						945	17325	190575	1636635

N →	14	15	16	17	18
R					
1	40075035	210766920	1096190550	5652751651	28958095545
2	12122110	81431350	510880370	3049616570	17539336815
3		1401400	28028000	333533200	3073270200

N →	19	20	21	22
R				
1	147589284710	749206090500	3791262568401	19137821912055
2	98049492723	536181458345	2881837917245	15278338767076
3	24234675465	172096749825	1134040872965	7069307049805
4		2546168625	53469541125	666586946025

N →	23	24	25
R			
1	96416888184100	485000783495250	2436684974110751
2	80117864269828	416449389324276	2149359635463876
3	42240545297951	244205509154607	1375458924105651
4	6416039394765	52683671271231	388310385313851
5			5194672859376

GENERALIZED STIRLING NUMBERS, B = 6

N →	6	7	8	9	10	11	12	13
R								
1	1	21	266	2646	22827	179487	1323652	9321312
2							10395	270270

N →	14	15	16	17	18
R					
1	63436373	420693273	2734926558	17505749898	110687251039
2	4099095	47507460	466876410	4104160060	33309926650
3					190590400

N →	19	20	21	22
R				
1	693081601779	4306078895384	26585679462804	163305339345225
2	254752658160	1861763348445	13131569945130	90049230296025
3	5431826400	89625135600	1121672151600	11819314757250

N →	23	24	25
R			
1	998969857983405	6090236036084530	37026417000002430
2	603695815955540	3973576687375988	25764288866731800
3	110670237753075	950933767378595	7656731350017750
4		4509264634875	135277939046250

GENERALIZED STIRLING NUMBERS,   B = 7

N →	7	8	9	10	11	12	13	14
R								
1	1	28	462	5880	63987	627396	5715424	49329280
2								135135

N →	15	16	17	18	19
R					
1	408741333	3281882604	25708104786	197462483400	1492924634839
2	4729725	94594500	1422280860	17892864990	199124936010

N →	20	21	22	23
R				
1	11143554045652	82310957214948	602762379967440	4382641999117305
2	2026763158420	19282395272140	174073797222325	150741111934 9135
3		3621217600	1394168776000	30462587755600

N →	24	25
R		
1	31677463851804540	227832482998716310
2	12623010132252520	1284 9770508700600
3	4 97021168644000	6741279053509000

GENERALIZED STIRLING NUMBERS,   B = 8

N →	8	9	10	11	12	13	14	15
R								
1	1	36	750	11880	159027	1899612	20912320	216627840

N →	16	17	18	19
R				
1	2141764053	20415995028	189036065010	1709751003480
2	2027025	91891800	2343240900	44346982680

N →	20	21	22	23
R				
1	15170932662679	132511015347084	1142399079991620	9741955019900400
2	694740296250	9540421090200	11888539504 8420	1375295856374440

N →	24	25
R		
1	82318282158320505	690223721118368580
2	15006064187108995	156226380361251200
3	9161680528000	4580840264 00000

GENERALIZED STIRLING NUMBERS,   B = 9

N →	9	10	11	12	13	14	15	16
R								
1	1	45	1155	22275	359502	5135130	67128490	820784250

E 5

GENERALIZED STIRLING NUMBERS, B = 9 (Cont'd)

N →	17	18	19	20
R				
1	9528822303	106175395755	1144614626805	12011282644725
2		34459425	1964187225	62199262125

N →	21	22	23
R			
1	123272476465204	1241963303533920	12320068811796900
2	1446733012725	27610143335775	458380554006375

N →	24	25
R		
1	120622574326072500	11679214510929 73005
2	6859789072171035	9474520220252 5875

GENERALIZED STIRLING NUMBERS, B = 10

N →	10	11	12	13	14	15	16	17
R								
1	1	55	1705	39325	752752	12662650	193754990	2758334150

N →	18	19	20	21
R				
1	37112163803	477297033785	5917584964655	71187132291275
2			654729075	45831035250

N →	22	23	24
R			
1	8351437993 77954	9593401297313460	108254081784931500
2	1764494857125	49473074851200	1129764045234825

N →	25
R	
1	1203163392175387500
2	2229877374 8501250

GENERALIZED STIRLING NUMBERS, B = 11

N →	11	12	13	14	15	16	17
R							
1	1	66	2431	66066	1479478	28936908	512060978

N →	18	19	20	21
R				
1	8391004908	129413217791	1900842429486	26826851689001

GENERALIZED STIRLING NUMBERS,   B = 11   (Cont'd)

N →	22	23	24
R			
1	366282500870286	4864251308951100	63100165695775560
2	13749310575	1159525191825	53338158823950

N →	25
R	
1	8023559044384 62660
2	1774073543492250

GENERALIZED STIRLING NUMBERS,   B = 12

N →	12	13	14	15	16	17	18
R							
1	1	78	3367	106470	2757118	62022324	1256328866

N →	19	20	21	22
R				
1	23466951300	411016633391	6833042030178	108823356051137

N →	23	24	25
R			
1	1672162773483930	24930204590758260	36226262078 4874680
2		316234143225	316234143 22500

GENERALIZED STIRLING NUMBERS,   B = 13

N →	13	14	15	16	17	18	19
R							
1	1	91	4550	165620	4910178	125854638	2892439160

N →	20	21	22	23
R				
1	61068660380	1204909218331	22496861868481	401282560341390

N →	24	25
R	
1	6888836057922000	114485073343744260

GENERALIZED STIRLING NUMBERS,   B = 14

N →	14	15	16	17	18	19	20
R							
1	1	105	6020	249900	8408778	243577530	6302524580

GENERALIZED STIRLING NUMBERS,   B = 14 (Cont'd)

N →	21	22	23	24
R				
1	149304004500	3295165281331	68629175807115	1362091021641000

N →				25
R				
1				25958110360896000

GENERALIZED STIRLING NUMBERS,   B = 15

N →	15	16	17	18	19	20	21
R							
1	1	120	7820	367200	13916778	452329200	13087462580

N →	22	23	24	25
R				
1	345615943200	8479404429331	195820242247080	4299394655347200

GENERALIZES STIRLING NUMBERS,   B = 16

N →	16	17	18	19	20	21	22
R							
1	1	136	9996	527136	22350954	8099444464	26046574004

N →	23	24	25
R			
1	762361127264	206771182465555	526655161695960

GENERALIZES STIRLING NUMBERS,   B = 17

N →	17	18	19	20	21	22	23
R							
1	1	153	12597	741285	34952799	1404142047	49916988803

N →	24	25
R		
1	1610949936915	48063331393110

GENERALIZES STIRLING NUMBERS,   B = 18

N →	18	19	20	21	22	23	24
R							
1	1	171	15675	1023435	53374629	2364885369	92484925445

E 8

GENERALIZED STIRLING NUMBERS, B = 18 (Cont'd)

N →	25
R	
1	3275678594925

GENERALIZED STIRLING NUMBERS, B = 19

N →	19	20	21	22	23	24	25
R							
1	1	190	19285	1389850	79781779	3880739170	166218969675

GENERALIZED STIRLING NUMBERS, B = 20

N →	20	21	22	23	24	25
R						
1	1	210	23485	1859550	116972779	6220194750

GENERALIZED STIRLING NUMBERS, B = 21

N →	21	22	23	24	25
R					
1	1	231	28336	2454606	168519505

GENERALIZED STIRLING NUMBERS, B = 22

N →	22	23	24	25
R				
1	1	253	33902	3200450

GENERALIZED STIRLING NUMBERS, B = 23

N →	23	24	25
R			
1	1	276	40250

TABLE F

This table gives the (minimum) sample size n needed to satisfy a (δ^*, P^*)
requirement for the problem of finding the category with the smallest cell
probability in a multinomial distribution with k cells (the measure of
distance is $\delta = p_{[2]} - p_{[1]}$ and $b = k-1$ of them have common probability
in the LF configuration)

$$\text{for } k = 2(1)10, 15, 25,$$

$$\text{for } P^* = .75, .90, .975, .99,$$

$$\text{for } 4 \text{ values of } \delta^* \text{ that vary with } k.$$

Note: The top entry in each cell is based on the normal approximation given
in (4.24).

The lower entry in each cell is exact and is based on one of the last two
expressions in (4.5).

TABLE F

Minimum § Sample Size Needed for the Multinomial Selection Problem

δ^*	k = 2				
	$P^* = .75$	$P^* = .90$	$P^* = .95$	$P^* = .975$	$P^* = .99$
.10	45.05 [NA]	162.60	267.85	380.30	535.78
	44.88 [E]	162.96	268.85	**	**
.15	19.76	71.35	117.54	166.89	235.12
	20.25	72.34	118.70	168.65	238.16
.20	10.92	39.42	64.93	92.20	129.89
	10.83	40.37	66.39	94.31	132.54
.25	6.82	24.64	40.58	57.62	81.18
	6.78	24.97	42.22	60.01	84.18

δ^*	k = 3				
.10	66.89	157.94	231.60	307.41	409.77
	65.53	157.07	231.30	307.76	410.98
.15	29.04	67.60	98.73	130.76	174.01
	28.34	67.37	99.02	131.61	175.62
.20	15.81	36.22	52.67	69.59	92.45
	15.43	36.33	53.24	70.69	94.23
.25	9.69	21.80	31.54	41.57	55.13
	9.53	22.12	32.28	42.79	56.95

** Not computed

NA Top entries are values obtained by the normal approximation (4.24),

E Lower entries are exact values obtained by using (4.5).

§ Without taking curtailment into account.

TABLE **F** (continued)

	$k = 4$				
δ^*	$P^* = .75$	$P^* = .90$	$P^* = .95$	$P^* = .975$	$P^* = .99$
.05	277.34	577.89	812.78	1050.96	1369.15
	271.37	**	**	**	**
.10	67.23	137.06	191.46	246.59	320.27
	64.56	134.83	189.80	245.55	320.03
.15	28.60	56.84	78.79	101.03	130.79
	27.07	55.71	78.10	100.79	131.11
.20	15.13	29.19	40.10	51.18	66.04
	14.22	28.60	39.83	51.22	66.44

	$k = 5$				
.05	261.03	506.99	695.43	884.79	1136.14
	255.65	**	**	**	**
.10	62.85	117.68	159.49	201.49	257.29
	59.09	114.29	156.59	199.09	255.52
.15	26.03	46.94	62.86	78.88	100.23
	23.85	44.96	61.11	77.35	98.91
.20	13.15	22.66	29.94	37.33	47.24
	11.90	21.28	28.46	35.69	45.31

* * not computed

SOBEL, UPPULURI AND FRANKOWSKI

TABLE F (continued)

δ^*	$P^* = .75$	$P^* = .90$	$P^* = .95$	$P^* = .975$	$P^* = .99$
			k = 6		
.025	1010.82 **	1888.00 **	2551.84 **	3214.98 **	4091.20 **
.050	246.40 236.26	452.27 442.62	607.66 **	762.80 **	967.85 **
.075	106.20 99.66	191.19 185.00	255.18 249.56	319.07 314.05	403.57 399.27
.100	57.51 52.84	101.32 96.87	134.26 130.15	167.17 163.39	210.76 207.31
			k = 7		
.025	946.75 **	1708.50 **	2279.21 **	2846.64 **	3593.81 **
.050	229.02 217.19	404.57 393.20	535.69 525.19	666.00 **	837.67 **
.075	97.71 90.09	168.49 161.11	221.23 214.30	273.67 267.18	342.84 336.82
.100	52.21 46.75	87.58 82.09	113.92 108.54	140.16 134.83	174.86 169.47
			k = 8		
.025	886.81 **	1558.07 **	2057.05 **	2551.34 **	3200.43 **
.050	212.79 199.53	364.50 351.65	476.89 464.84	588.19 **	734.46 **
.075	89.82 81.22	149.38 140.84	193.43 185.14	237.09 229.02	294.59 286.67
.100	47.22 41.05	75.86 69.29	97.09 90.27	118.22 111.05	143.81 138.38

** not computed

TABLE F (continued)

δ*			k = 9		
	P* = .75	P* = .90	P* = .95	P* = .975	P* = .99
.025	832.32 **	1430.72 **	1872.72 **	2309.24 **	2881.20 **
.050	198.12 183.48	330.59 316.25	428.08 414.43	524.34 **	650.61 **
.075	82.59 73.10	133.05 123.34	170.14 160.44	206.85 197.08	255.17 245.12
.100	42.52 35.76	65.68 57.85	82.83 74.19	99.94 90.35	122.63 111.58

			k = 10		
.025	783.29 **	1321.76 **	1717.36 **	2107.06 **	2616.73 **
.050	184.78 168.96	301.33 285.69	386.64 371.56	470.70 456.16	580.82 **
.075	75.97 65.66	118.84 107.94	150.25 139.01	181.32 169.63	222.24 209.70
.100	38.05 30.92	56.62 47.12	70.48 59.17	84.36 71.10	102.83 86.79

* *   not computed

TABLE F (continued)

δ^*	k = 15				
	$P^* = .075$	$P^* = 90$	$P^* = .95$	$P^* = .975$	$P^* = .99$
.01	3946.15 **	6416.26 **	8205.33 **	9953.65 **	12224.63 **
.02	955.64 **	1525.01 **	1936.21 **	2337.83 **	2859.69 **
.03	409.60 **	640.12 **	806.22 **	968.48 **	1179.55 **
.04	220.77 **	337.01 **	420.72 **	502.61 **	609.40 **

	k = 20				
.01	3250.75 **	5109.71 **	6442.06 **	7737.16 **	9412.52 **
.02	775.77 **	1187.55 **	1481.66 **	1767.46 **	2112.38 **
.03	326.07 **	483.92 **	596.52 **	706.14 **	848.53 **
.04	170.80 **	244.33 **	297.09 **	348.80 **	416.47 **

	k = 25				
.01	1186.00 **	1818.51 **	2268.48 **	2704.24 **	3266.37 **
.02	650.14 **	962.39 **	1183.73 **	1398.16 **	1675.34 **
.03	266.56 **	377.82 **	456.96 **	534.08 **	634.61 **
.04	132.78 **	178.64 **	212.30 **	245.76 **	290.07 **

** not computed

INDEX